AF706112

Dynamics of Educational Change in Periods of Uncertainty and Crisis

Annette G. Walters
https://orcid.org/0009-0007-6068-8467
Washington County School District, USA

IGI Global
Scientific Publishing
Publishing Tomorrow's Research Today

Vice President of Editorial	Melissa Wagner
Director of Acquisitions	Mikaela Felty
Director of Book Development	Jocelynn Hessler
Production Manager	Mike Brehm
Cover Design	Jose Rosado

Published in the United States of America by
IGI Global Scientific Publishing
701 East Chocolate Avenue
Hershey, PA, 17033, USA
Tel: 717-533-8845 | Fax: 717-533-7115
Website: https://www.igi-global.com E-mail: cust@igi-global.com

Library of Congress Cataloging-in-Publication Data

Names: Walters, Annette, 1973- editor.
Title: Dynamics of educational change in periods of uncertainty and crisis / [edited by] Annette G. Walters.
Description: Hershey, PA : IGI Global Scientific Publishing, [2026] | Includes bibliographical references and index. | Summary: "This book aims to cultivate a sense of hope and possibility, inspiring educators to envision a future where education evolves beyond mere functionality into a transformative experience that enables individual growth"-- Provided by publisher.
Identifiers: LCCN 2025017778 (print) | LCCN 2025017779 (ebook) | ISBN 9798337313504 (hardcover) | ISBN 9798337313511 (paperback) | ISBN 9798337313528 (ebook)
Subjects: LCSH: Learning strategies. | Educational change. | Uncertainty. | Social change.
Classification: LCC LB1060 .D96 2025 (print) | LCC LB1060 (ebook) | DDC 370--dc23/eng/20250501
LC record available at https://lccn.loc.gov/2025017778
LC ebook record available at https://lccn.loc.gov/2025017779

British Cataloguing in Publication Data
A Cataloguing in Publication record for this book is available from the British Library.

Table of Contents

Detailed Table of Contents

This chapter per the author, shares the complexities of navigating leadership in an ever-evolving educational landscape demands a deep understanding of the needs of students, staff, and the broader learning community. This work include challenges that school communities face, during constant changes, which may affect the financial stability of the district, professional and personal well-being of educators, quality of the student learning experience, and the overall growth of the community. Leaders who confront these challenges must embody resilience and cultivate strong relationships, drawing on the support of the educational community and its stakeholders to ensure sustained success. This publication seeks to offer a thorough and insightful examination of the essential role of strong school leadership during periods of change, emphasizing the significance of collaborative partnerships and intentional educational strategies.

In times of crisis, educational systems often face unprecedented challenges that can lead to disorder and uncertainty. However, this turmoil presents a unique opportunity for transformation and innovation. By rethinking traditional education models, we can foster resilience, creativity, and adaptability among students and educators. Emphasizing critical thinking and problem-solving skills enables learners to navigate complexities and emerge with newfound insights. Collaborative approaches, using technology and community resources, can enhance engagement and inclusivity, bridging gaps exacerbated by crises. Moreover, integrating emotional intelligence and mental well-being into curricula can support students in coping with stress and ambiguity. Ultimately, embracing disorder as a catalyst for change can lead to a more robust, relevant, and responsive educational landscape, equipping future generations to thrive in an ever-evolving world.

Education is currently undergoing a significant transformation, characterized by various challenges including the integration of technology, resistance to change within educational institutions, declining enrollment, and the pressing need for community-focused strategies. This paper emphasizes the transformative potential of technology in educational practices while also acknowledging the risks it poses in widening existing disparities. Innovations such as adaptive learning platforms have the capacity to boost student engagement, yet they may inadvertently deepen inequalities among learners. Furthermore, the discussion extends to the role of service-learning and career technical education as a means to cultivate civic responsibility and tackle real-world challenges, despite facing obstacles related to institutional support and funding. The paper also addresses the implications of declining enrollment across educational institutions, submitting that by embracing community engagement and innovative teaching methodologies, institutions may enhance their relevance and support.

Periods of uncertainty and crisis have historically tested the boundaries of educational systems—but they have also revealed remarkable opportunities for transformation. This chapter explores the evolving nature of educational leadership in times marked by disruption, unpredictability, and rapid change. It delves into the mindset and approaches adopted by innovative leaders who have dared to rethink tradition, reframe uncertainty as possibility, and guide their institutions not only to survive but to thrive. Drawing on real-world examples, emotional insights, and forward-thinking strategies, the chapter underscores the essential qualities of leadership that resonate in turbulent times—adaptability, empathy, creativity, and vision. Ultimately, the chapter makes a compelling case for cultivating leadership that is not only reactive but regenerative—leadership that redefines what it means to guide educational communities through the unknown with clarity, care, and courage.

In times of educational upheaval, the prevailing systems are often stripped down to their bare essentials, exposing not just their strengths but also their deepest

Amid a sharp increase in threats of school violence, the intersection of social media and school safety has captured the nation's attention. Recent events highlight the challenge of identifying and responding to online threats. While some of these events may turn out to be hoaxes, they can cause significant disruptions, place added strain on law enforcement, and instigate fear in a community. The impact might be school closures, canceled events, increased security, or heightened anxiety. Schools and law enforcement agencies have adapted by implementing threat assessment strategies, increasing School Resource Officer's (SROs) presence, and incorporating technology to mitigate dangers. However, concerns regarding over-policing, privacy, and disproportionate disciplinary actions remain. Legal frameworks struggle to keep up and to strike a balance between First Amendment rights and criminal actions. This chapter explores the impact of social media threats, examines the role of SROs, and proposes strategies for effective prevention, response, and community engagement.

This chapter explores how the conflict in Ukraine has affected higher education, emphasising the real-life experiences of instructors and students dealing with unprecedented difficulties. Responses to an online survey show that distance learning is driven by need rather than preference. The participants mentioned significant challenges, such as frequent power outages, erratic Internet, increased anxiety, decreased motivation, and worries about personal safety. Faculty and students reported that increased workloads, poor communication, and irregular attendance also hampered effective teaching and learning during wartime. The chapter provides practical suggestions for overcoming these problems during and after conflict based on recent evidence. It suggests online mental health services, international partnerships, and infrastructure support to improve educational resilience. Although focused on Ukraine, the knowledge and lessons presented here can be applied to other armed conflict-affected contexts and offer helpful advice for maintaining education during emergencies.

Preface

INTRODUCTION

This edited volume, *Dynamics of Educational Change in Periods of Uncertainty and Crisis*, emerges at a moment when education systems across the globe are being tested, reshaped, and reimagined. The chapters collected here reflect a shared recognition: that uncertainty is no longer an occasional disruption but a defining condition of contemporary education. In this context, the task before us is not merely to respond to change, but to understand its dynamics and harness its possibilities.

The text that follows situates education within a landscape marked by complexity—where technological acceleration, societal transformation, and global crises converge to challenge long-standing assumptions. These forces have exposed tensions among educators, policymakers, and communities, often resulting in fragmentation and discord. Yet, within this disruption lies an opportunity: to cultivate resilience not as simple recovery, but as adaptive capacity—an ongoing commitment to innovation, reflection, and growth.

Contributors to this volume engage deeply with the idea that periods of instability can serve as catalysts for renewal. What we are witnessing is, in many respects, a renaissance in educational thought and practice. Emerging perspectives call for systems that are more inclusive, equitable, and responsive. Pedagogies are shifting toward learner-centered approaches, interdisciplinary inquiry, and meaningful integration of technology. At the same time, foundational questions—what education is for, whom it serves, and how it is enacted—are being revisited with urgency and care.

A central aim of this work is to move beyond narratives of crisis as deficit. Instead, it seeks to illuminate the possibilities embedded within disruption. The chapters explore how educators and leaders can navigate uncertainty with intentionality, transforming constraints into conditions for meaningful change. This requires not only innovation, but also a steadfast commitment to equity. The inequities laid bare by recent global events—particularly disparities in access to resources, technology, and support—underscore the necessity of systemic reform.

This volume also affirms that meaningful educational transformation must be collective. The voices represented here—scholars, practitioners, leaders, community members, and advocates—highlight the importance of collaboration across roles and contexts. Particular attention is given to perspectives that have historically been marginalized, recognizing that inclusive decision-making is essential to building systems where all learners can thrive.

As editor, I invite readers to engage with this collection not as a set of definitive answers, but as an ongoing conversation. The insights offered here are intended to provoke reflection, challenge assumptions, and inspire action. In times of uncertainty, education must remain a site of possibility—a space where individuals and communities can imagine and create more just and sustainable futures.

It is my hope that this volume contributes meaningfully to that endeavor.

ORGANIZATION OF THE BOOK

Presented here is an overview of the chapters that collectively shape *Dynamics of Educational Change in Periods of Uncertainty and Crisis*. Each contribution reflects a distinct yet interconnected perspective on how education systems respond, adapt, and evolve in the face of disruption.

Chapter 1 examines the intricate realities of leading within volatile educational environments, emphasizing that effective leadership during uncertainty requires not only strategic acumen but also a deep attentiveness to the well-being of students, educators, and communities. It highlights the multifaceted challenges that arise amid constant change—ranging from financial instability to the erosion of professional morale—and underscores the importance of resilience, relationship-building, and collaborative engagement as essential pillars for sustaining meaningful progress.

Chapter 2 reframes crisis not as a breakdown of systems, but as an opening for intellectual and structural renewal. It explores how disorder can be transformed into insight through the deliberate cultivation of critical thinking, creativity, and adaptability. By advocating for the integration of emotional well-being, technological tools, and community collaboration, the chapter positions crisis as a generative force capable of reshaping education into a more inclusive and responsive enterprise.

Chapter 3 engages with the paradoxical nature of contemporary education, where innovation and inequity often coexist. It interrogates the dual role of technology as both an enabler of engagement and a potential amplifier of disparity, while also addressing institutional resistance to change and declining enrollment trends. The discussion extends to community-centered practices such as service-learning and career-focused education, suggesting that renewed relevance and sustainability depend on meaningful connections between institutions and the communities they serve.

Chapter 4 turns to the evolving nature of educational leadership, presenting it as a dynamic and regenerative force in times of crisis. It explores how forward-thinking leaders embrace uncertainty as an opportunity to rethink established norms, drawing on qualities such as empathy, creativity, and vision. The chapter ultimately advances a conception of leadership that goes beyond survival, advocating for approaches that enable institutions to thrive and reimagine their purpose in uncertain contexts.

Chapter 5 centers the lived experiences of students navigating disrupted educational landscapes, offering a powerful reminder that resilience is often most visible at the learner level. By foregrounding student voices, this chapter challenges traditional hierarchies in educational design and calls for a future in which learner agency is foundational. It highlights the importance of building systems that are not only adaptable but also empathetic and inclusive in responding to student needs.

Chapter 6 addresses the critical role of empowering educators during times of crisis, recognizing that their capacity to adapt directly influences the continuity and quality of learning. It outlines practical strategies for fostering resilience, including the development of growth mindsets, collaborative practices, and innovative pedagogies. The chapter also draws attention to the emotional and psychological dimensions of teaching during disruption, reinforcing the need for leadership that prioritizes well-being alongside instructional effectiveness.

Chapter 7 explores the intersection of modern pedagogical approaches and emerging technologies in advancing equitable and high-quality education. It highlights how innovative teaching methods—paired with tools such as artificial intelligence, immersive technologies, and digital platforms—can enhance engagement and accessibility. In doing so, the chapter situates educational transformation within a broader global agenda, emphasizing the role of innovation in achieving inclusive and sustainable learning outcomes.

Chapter 8 offers a culturally grounded perspective on school leadership by examining how core values can shape supportive and resilient educational environments. It demonstrates how principles rooted in community, care, compassion, and collective responsibility can redefine school climate, fostering spaces where both students and educators feel valued and supported. The chapter reinforces the importance of culturally responsive leadership in navigating complex and uncertain contexts.

Chapter 9 focuses on the adaptive practices educators must adopt in response to rapid and ongoing change. It underscores the importance of flexibility, collaboration, and culturally responsive teaching, while also highlighting professional learning as a transformative process. Through reflection, storytelling, and shared problem-solving, educators are positioned as active agents of change capable of turning challenges into opportunities for growth and innovation.

Chapter 10 addresses the growing influence of social media on school safety and community dynamics, particularly in the context of perceived threats and misinformation. It explores the complex balance between ensuring security and safeguarding individual rights, while examining the evolving roles of schools, law enforcement, and policy frameworks. The chapter provides insight into strategies for prevention, response, and communication, emphasizing the need for thoughtful and coordinated approaches in managing digital-era challenges.

Chapter 11 concludes the volume by examining the profound impact of armed conflict on higher education, with a focus on the lived realities of those experiencing disruption firsthand. It highlights the resilience of students and educators navigating extreme conditions, while also identifying systemic barriers such as infrastructure limitations and psychological strain. The chapter offers practical recommendations for sustaining education in crisis contexts, emphasizing the importance of mental health support, international collaboration, and long-term recovery planning.

Together, these chapters provide a comprehensive exploration of how education systems confront uncertainty, revealing not only the depth of the challenges faced but also the breadth of possibilities for transformation.

CONCLUSION

As this volume draws to a close, it becomes evident that periods of uncertainty and crisis, while deeply disruptive, are also profoundly instructive. The collective insights presented across these chapters reaffirm that education is not a static institution, but a living, evolving system—one that is continuously shaped by the pressures, possibilities, and complexities of the world it inhabits.

What emerges most clearly is the necessity of intentional transformation. The challenges explored throughout this work—ranging from leadership dilemmas and technological shifts to inequities, global conflict, and the redefinition of learning spaces—are neither isolated nor temporary. They demand sustained reflection and deliberate action. Yet, rather than signaling decline, these conditions reveal an opportunity to reconstruct education in ways that are more just, adaptive, and human-centered.

A unifying thread across the contributions is the emphasis on resilience—not as endurance alone, but as the capacity to reimagine, to innovate, and to lead with purpose. Whether through empowering educators, centering student voices, leveraging technology, or grounding leadership in cultural and ethical values, the work presented here challenges conventional approaches and calls for renewed commitments. It underscores that meaningful change is not achieved through isolated efforts, but through collaboration that bridges disciplines, roles, and communities.

Equally important is the recognition that equity must remain at the forefront of all educational reform. The disparities illuminated during times of crisis cannot be treated as peripheral concerns; they are central to the future of education itself. Ensuring access, inclusion, and opportunity for all learners requires not only systemic change but also the courage to confront longstanding assumptions and practices.

In reflecting on the themes of this volume, it is clear that the future of education will be defined by those willing to engage uncertainty with clarity and conviction. The work ahead calls for leaders, educators, researchers, and communities to move beyond reactive responses and toward proactive, sustained transformation. It requires a shared vision—one that sees education not merely as a mechanism for knowledge transmission, but as a vital force for individual empowerment and societal progress.

As editor, I offer this volume as both a resource and an invitation: a resource for understanding the dynamics of change in turbulent times, and an invitation to participate in shaping what comes next. The questions raised here do not conclude with these pages; rather, they extend outward, urging continued inquiry, dialogue, and action.

In the face of uncertainty, education must remain a space of possibility. It is within this space that innovation takes root, resilience is cultivated, and futures are reimagined. It is my hope that this work contributes, in some measure, to that ongoing and essential endeavor.

Chapter 1
Leading During Volatility: Education Strategies and Practices in Uncertain Times

Annette G. Walters
http://orcid.org/0009-0007-6068-8467
Washington County School District, USA

ABSTRACT

This chapter per the author, shares the complexities of navigating leadership in an ever-evolving educational landscape demands a deep understanding of the needs of students, staff, and the broader learning community. This work include challenges that school communities face, during constant changes, which may affect the financial stability of the district, professional and personal well-being of educators, quality of the student learning experience, and the overall growth of the community. Leaders who confront these challenges must embody resilience and cultivate strong relationships, drawing on the support of the educational community and its stakeholders to ensure sustained success. This publication seeks to offer a thorough and insightful examination of the essential role of strong school leadership during periods of change, emphasizing the significance of collaborative partnerships and intentional educational strategies.

INTRODUCTION

In the realm of educational leadership, the criteria for success extend beyond mere skills and knowledge. As highlighted by Burton et al. (2020), effective leaders in educational settings must also possess a certain charisma and the ability to influence others. This combination of attributes is essential for fostering an environment conducive to learning and growth. Leaders who can engage and inspire their teams

DOI: 10.4018/979-8-3373-1350-4.ch001

are more likely to cultivate a positive atmosphere that encourages collaboration and innovation, ultimately benefiting the entire educational institution (Northouse, 2022).

Moreover, the role of charm and influence in educational leadership cannot be overstated. These qualities enable leaders to build strong relationships with faculty, students, and the broader community. By establishing trust and rapport, educational leaders can effectively communicate their vision and motivate others to work towards common goals (Aljbour et al., 2022; Fullan, 2016). This relational aspect of leadership is crucial, as it helps to create a sense of belonging and commitment among stakeholders, which is vital for the success of any educational initiative (Agosto & Roland, 2018; Leithwood & Louis, 2012).

While knowledge and expertise are fundamental to effective educational leadership, the importance of interpersonal skills such as charm and influence should not be overlooked (Northouse, 2022). Leaders who can navigate the complexities of human relationships and inspire those around them are better equipped to drive meaningful change within their institutions (Xhelilaj et al., 2021). As the educational landscape continues to evolve, the ability to connect with and motivate others will remain a key determinant of success for leaders in this field (Agosto & Roland, 2018; Northouse, 2022).

Leaders facing the complexities of evolving educational environments, such as dwindling enrollment, workforce reductions, budget constraints, and regulatory mandates, must exemplify resilience while fostering strong relationships within the educational community (Bagchi & Pattnaik, 2022). This resilience is not merely a personal attribute but an intentional endeavor that relies on the support and collaboration of all stakeholders (Aljbour et al., 2022; Fullan, 2016). To navigate these challenges effectively, leaders must prioritize the development of their staff and create a unified vision centered on achieving student success (Leithwood et al., 2020). This approach requires a deep understanding of staff capabilities and a commitment to the well-being of every individual within the educational ecosystem.

Further, to individual and team development, the importance of collaboration among members of the school community cannot be overstated (Bagchi & Pattnaik, 2022; Leithwood et al., 2020). This includes educators, administrators, school board members, parents, and local organizations. By fostering an environment of open communication and shared responsibility, leaders can cultivate a culture of collaboration that enhances problem-solving and innovation (Cable, 2013; Hargreaves & Fullan, 2012). Such partnerships are essential for addressing the multifaceted challenges that schools face today because they allow for the pooling of resources, ideas, and expertise (Epstein, 2018).

In periods of transformation within the educational sector, it is crucial to adopt deliberate strategies that simultaneously tackle various aspects of the learning environment (Leithwood et al., 2020). These strategies should encompass a comprehensive approach that includes mental health support, the enhancement of instructional quality, opportunities for professional development, and adherence to both state and national educational standards, as highlighted by Darling-Hammond et al. (2017).

Mental health support is particularly important, as it addresses the emotional and psychological well-being of both students and educators, fostering a more conducive learning atmosphere (Cable, 2013; Leithwood et al., 2020). By integrating mental health resources into the educational framework, schools can create an environment where individuals feel safe, valued, and ready to engage in the learning process (Kluch et al., 2022). Moreover, enhancing instructional quality is essential for ensuring that teaching methods are effective and responsive to the needs of diverse learners. This can be achieved through the implementation of evidence-based practices and innovative teaching strategies that cater to various learning styles (Bagchi & Pattnaik, 2022; Leithwood et al., 2020). Professional development opportunities for educators play a pivotal role in this enhancement, as they equip teachers with the latest pedagogical tools and techniques (Darling-Hammond et al., 2017). Continuous training and support not only improve instructional quality but also empower educators to adapt to the evolving educational landscape, ultimately benefiting student outcomes (Darling-Hammond et al., 2017; DeMatthews et al., 2021).

In addition to these elements, adherence to state and national educational standards is vital for maintaining a consistent and high-quality educational experience. These standards provide a framework for curriculum development and assessment, ensuring that all students receive a well-rounded education that prepares them for future challenges (DeMatthews et al., 2021). By aligning educational practices with these standards, institutions can foster accountability and transparency, which are essential for building trust among stakeholders, including parents, students, and the community (Bagchi & Pattnaik, 2022; Darling-Hammond et al., 2017).

Furthermore, maintaining fiscal responsibility is vital to ensure that resources are allocated in a manner that is both effective and sustainable. Educational institutions must navigate budget constraints while prioritizing investments that yield the greatest impact on student learning and well-being (Cable, 2013; Darling-Hammond et al., 2017). This requires strategic planning and collaboration among administrators, educators, and policymakers to identify areas where funding can be maximized. By adopting a holistic approach to resource allocation, schools can thrive amidst change, meeting the diverse needs of students and educators alike (DeMatthews et al., 2021; Kryshtanovych, 2022). Ultimately, a well-rounded strategy that integrates mental health support, instructional quality, professional development, adherence to standards, and fiscal responsibility will create a resilient educational environment capable of adapting to the challenges of transformation (Xhelilaj et al., 2021).

Maintaining a student-centered focus during periods of transition is critical because such times can create uncertainty and stress for both learners and educators

(Darling-Hammond et al., 2017; Kryshtanovych, 2022). Educational leaders must engage in deliberate planning and execution of strategies that prioritize student well-being and academic success (Fullan, 2016). This includes providing mental health supports, equipping teachers with training to adapt instructional practices, and ensuring alignment with academic standards to maintain educational quality (Kluch et al., 2022).

Ultimately, the success of educational strategies during times of change depends on collaborative efforts among all stakeholders (Epstein, 2018; Kryshtanovych, 2022). By fostering a culture of support and open communication among students, staff, and the broader community, educational institutions can navigate challenges more effectively (Epstein, 2018; Kellogg & Van Maanen, 2020). This intentional collaboration reinforces a commitment to nurturing an effective learning community where students remain the central focus of all educational initiatives (Agosto & Roland, 2018).

Implementing collaborative educational strategies is essential for achieving sustained success in educational settings (Aljbour et al., 2022; Paredes & Medina, 2016). These strategies typically encompass a range of initiatives, including ongoing professional development for educators, active engagement with the community, and inclusive decision-making processes that reflect a variety of viewpoints (Agosto & Roland, 2018; Hargreaves & Fullan, 2012). By ensuring that all stakeholders are actively involved, educational leaders can create strategies that are not only relevant but also specifically designed to address the unique needs of both students and the broader community (Rosa & Gifford, 2023). This approach fosters a sense of ownership and commitment among participants, ultimately leading to more effective and responsive educational practices (Ahammad et al., 2018; Aljbour et al., 2022).

One of the cornerstones of collaborative educational strategies is ongoing professional development for educators (Bagchi & Pattnaik, 2022). This continuous learning process equips teachers with the latest pedagogical techniques and tools, enabling them to adapt to the ever-evolving educational landscape (Agosto & Roland, 2018; Hargreaves & Fullan, 2012). Professional development should not be a one-time event but rather a sustained effort that encourages educators to share best practices, reflect on their teaching methods, and collaborate with peers (Aljbour et al., 2022; Bagchi & Pattnaik, 2022). By fostering a culture of lifelong learning, schools can ensure that educators remain motivated and well-prepared to meet the diverse needs of their students (Anlesinya et al., 2019; Rosa & Gifford, 2023).

Active engagement with the community is another critical component of successful collaborative strategies. Schools do not exist in isolation; they are part of a larger ecosystem that includes families, local businesses, and community organizations (Harris & Goodall, 2018; Sanders & Harvey, 2019). By forging strong partnerships with these entities, educational leaders can tap into a wealth of resources and expertise that can enhance the learning experience (Bagchi & Pattnaik, 2022; Cable, 2013).

Community engagement can take many forms, from inviting local professionals to speak in classrooms to organizing service-learning projects that connect students with real-world issues (Chen & Li, 2020; Henderson & Mapp, 2020). Such initiatives not only enrich the curriculum but also help students develop a sense of social responsibility and civic engagement (Chen & Li, 20020; Epstein, 2018). Inclusive decision-making processes are vital for ensuring that the voices of all stakeholders are heard and valued (Miller & Johnson, 2021). When educators, parents, students, and community members are involved in shaping educational policies and practices, the resulting strategies are more likely to resonate with those they affect (Henderson & Mapp, 2020; Miller & Johnson, 2021; Paredes & Medina, 2016).

This inclusivity fosters a collaborative environment where diverse perspectives are considered, leading to more innovative and effective solutions (Martin & Kemple, 2020). Moreover, when stakeholders feel that their input is valued, they are more likely to invest in the success of the educational initiatives, creating a shared sense of purpose and commitment (González-DeHass et al., 2017; Martin & Kemple, 2020).

The implementation of collaborative educational strategies is not merely a trend but a necessary approach for fostering sustained success in educational settings (Chen & Li, 20020; Martin & Kemple, 2020). By prioritizing ongoing professional development, engaging with the community, and embracing inclusive decision-making, educational leaders can create a dynamic and responsive learning environment (Chen & Li, 2020; Epstein, 2018; Martin & Kemple, 2020; Sanders & Harvey, 2019).

Professional Development

Professional development is fundamental to the success of educational systems, as it equips educators with the vital skills and knowledge required to adapt to the dynamic nature of teaching (Chen & Li, 2020; Darling-Hammond et al., 2017). In an era where teaching methods, curriculum frameworks, and the diverse needs of students are in constant flux, the importance of ongoing training for teachers cannot be overstated (Bagchi & Pattnaik, 2022; Cable, 2013). As highlighted by Darling-Hammond et al. (2017), educators must engage in continuous professional learning to maintain their effectiveness and relevance in the classroom. This commitment to professional development ensures that teachers are not only well-versed in current pedagogical strategies but also prepared to meet the challenges posed by an increasingly diverse student population.

By participating in continuous professional learning opportunities, educators can refine their instructional practices and explore innovative teaching approaches that resonate with their students. These opportunities may include workshops, collaborative learning communities, and online courses that focus on the latest research and trends in education. Such engagement allows teachers to share best practices, gain insights from their peers, and adapt their methodologies to better suit the unique learning styles of their students (Darling-Hammond et al., 2017).

Moreover, as educators become more adept at integrating technology and differentiated instruction into their teaching, they create a more inclusive and engaging learning environment that caters to the varied needs of all learners. This dedication to lifelong learning not only enhances the professional growth of teachers but also profoundly influences the quality of learning experiences for their students. When educators are equipped with the latest knowledge and skills, they are better positioned to inspire and motivate their students, fostering a culture of curiosity and critical thinking (Kryshtanovych, 2020). As a result, students benefit from enriched educational experiences that promote deeper understanding and retention of knowledge.

Ultimately, the commitment to professional development cultivates an environment where both educators and learners can thrive, leading to improved educational outcomes and a more robust educational system overall. In this way, investing in the professional growth of teachers is not merely an administrative responsibility; it is a crucial step toward ensuring that all students receive the high-quality education they deserve.

The significance of ongoing training, coaching, and mentorship in the educational sector is paramount, as these components not only enhance collaboration among educators but also encourage the sharing of effective practices (Hargreaves & Fullan, 2012; Medina-Garrido et al., 2021). In an era where educational demands are constantly evolving, the need for teachers to engage in continuous professional development has never been more critical. When educators participate in substantial training opportunities, they are not merely acquiring new skills; they are also cultivating a mindset that embraces change and innovation (Xhelilaj et al., 2021). This willingness to adapt is crucial, as it allows teachers to implement innovative teaching methods that can significantly improve student performance (Darling-Hammond et al., 2017; Medina-Garrido et al., 2021).

Moreover, the collaborative atmosphere fostered by ongoing training and mentorship creates a culture of mutual learning and support among educators. In this environment, teachers are encouraged to share their experiences, challenges, and successes, which leads to a rich exchange of knowledge and strategies (Medina-Garrido et al., 2021; Xhelilaj et al., 2021). Such collaboration not only enhances individual teaching practices but also builds a sense of community within schools (Medina-Garrido et al., 2021; Williams III et al., 2022). Educators who feel supported by their peers are more likely to take risks in their teaching, experiment with new approaches, and ultimately refine their craft (Rosa & Gifford, 2023).

As a result, this dynamic exchange of ideas and resources elevates the overall quality of education that students receive. When teachers are equipped with the latest pedagogical strategies and are inspired by their colleagues, they can create more engaging and effective learning experiences for their students (Xhelilaj et al., 2021). This, in turn, fosters an environment where both teachers and learners can thrive, leading to improved academic outcomes and a more vibrant school culture. In essence, the investment in ongoing training, coaching, and mentorship is not just beneficial for educators; it is a vital component in shaping the future of education and ensuring that all students have access to high-quality learning experiences (Medina-Garrido et al., 2021; Williams III et al., 2022).

A well-trained teaching staff is essential for fostering a school culture that emphasizes ongoing improvement and excellence. When educators engage in continuous professional development, they not only refine their instructional techniques but also expand their understanding of diverse learning styles and effective pedagogical strategies (Medina-Garrido et al., 2021). This enhancement of skills translates into more dynamic and engaging classroom experiences, where teachers are equipped to meet the varied needs of their students. As educators become more adept in their craft, they ignite a genuine enthusiasm for learning among their students, creating an infectious energy that permeates the classroom (Schildkamp & Visscher, 2016). This vibrant interaction between teachers and students cultivates an atmosphere that supports both academic achievement and social development, encouraging students to actively participate in their education and strive for success (Simons & Vanderlinde, 2020).

Moreover, a culture of ongoing improvement fosters collaboration among staff members, as teachers share best practices and innovative ideas (Sanders & Harvey, 2019). This collaborative spirit not only strengthens professional relationships but also leads to a more cohesive educational environment where everyone is invested in the collective success of the students (Sanders & Harvey, 2019; Simons & Vanderlinde, 2020). When teachers feel supported and empowered through professional development opportunities, they are more likely to take risks in their teaching, experiment with new approaches, and ultimately inspire their students to do the same (Medina-Garrido et al., 2021; Xhelilaj et al., 2021; Williams III et al., 2022).

Consequently, investing in professional development is not simply a bureaucratic requirement; it is a crucial approach to improving educational outcomes and nurturing a vibrant learning community. As highlighted by Darling-Hammond et al. (2017), such investment is foundational to creating an educational ecosystem where both teachers and students can thrive, leading to sustained improvements in student performance and overall school culture. In this way, the commitment to professional growth becomes a catalyst for transformative change, ensuring that schools remain dynamic and responsive to the needs of their communities (Medina-Garrido et al., 2021; Xhelilaj et al., 2021).

Community Engagement

The active participation of the community is crucial for the success of educational programs, particularly during times of significant change (Sanders & Harvey, 2019). As educational landscapes evolve, the need for a collaborative approach becomes increasingly evident. The progress of education is heavily dependent on the engagement and support of a diverse array of stakeholders, such as parents, local organizations, and community leaders (Epstein, 2018). These stakeholders bring unique perspectives and resources that can significantly enrich the educational experience. By promoting collaboration with these groups, schools can create an environment that not only enhances the educational journey for students but also ensures that educational objectives resonate with the values and expectations of the community (Schildkamp & Visscher, 2016; Weiss & Caspe, 2021).

Forming partnerships with local entities allows schools to better understand the needs and aspirations of the families they serve (Yazzie-Mintz & Nagaoka, 2021). This understanding is vital, as it fosters trust and improves communication between educators and community members. When schools actively seek input from parents and local organizations, they demonstrate a commitment to inclusivity and responsiveness, which can lead to stronger relationships. This, in turn, strengthens support for educational initiatives (Epstein, 2018; Yazzie-Mintz & Nagaoka, 2021). When community members feel a sense of ownership over educational efforts, they are more inclined to contribute resources, share their expertise, and propose innovative solutions.

Such engagement not only enhances the effectiveness of schools but also elevates the quality of education available to students (Weiss & Caspe, 2021). For instance, local businesses may offer internships or mentorship programs, while community organizations can provide additional resources or support services that address specific challenges faced by students and families (Yazzie-Mintz & Nagaoka, 2021). Once educational programs align with community values, they are more likely to receive enthusiastic backing, whether through volunteer efforts, financial contributions, or advocacy. Ultimately, the synergy created by active community participation not only enriches the educational landscape but also cultivates a sense of shared responsibility for the success of future generations (Schildkamp & Visscher, 2016). By harnessing the collective strengths of the community, schools can navigate change more effectively and create a more vibrant, supportive educational environment (Rosa & Gifford, 2023).

The relationship between professional development and community engagement plays a vital role in enhancing educational systems, creating a dynamic interplay that significantly benefits both educators and the communities they serve (Weiss & Caspe, 2021). As educators engage in ongoing professional development, they not only refine their teaching skills but also broaden their understanding of diverse pedagogical approaches and contemporary educational theories (Xhelilaj et al., 2021). This continuous learning process empowers them to integrate community insights and local knowledge into their teaching practices, making lessons more relevant and relatable to students. For instance, when teachers are equipped with the latest strategies and tools, they can better address the unique cultural, social, and economic contexts of their students, thereby fostering a more inclusive classroom environment (Medina-Garrido et al., 2021; Peters, 2010).

This reciprocal relationship between professional development and community engagement fosters a culture of continuous improvement within educational institutions (Medina-Garrido et al., 2021). As educators become adept at incorporating community feedback and insights, they are better positioned to adapt their curricula and teaching methods to meet the evolving needs of their students and the broader community (Leithwood et al., 2020). This adaptability is crucial in an era where societal changes occur rapidly, and educational initiatives must remain relevant to prepare students for the future.

Research by Hargreaves and Fullan (2012) underscores the importance of this synergy, highlighting that when schools actively engage with their communities, they can create educational programs that are not only effective but also reflective of the values and aspirations of the community they serve (Paredes & Medina, 2016; Martin & Kemple, 2020).

By adopting this collaborative framework, schools can cultivate a more inclusive and effective learning environment that benefits all stakeholders involved (Miller & Johnson, 2021). Students gain access to a richer educational experience that acknowledges and respects their backgrounds, while families feel more connected and invested in their children's education (Martin & Kemple, 2020). Additionally, the broader community reaps the rewards of a well-educated populace, which can lead to enhanced civic engagement and economic development (Kellogg & Van Maanen, 2020).

Ultimately, the integration of professional development and community engagement creates a robust educational ecosystem where continuous learning and collaboration thrive, ensuring that all participants, including students, families, educators, and community members, are empowered to contribute to and benefit from a vibrant educational landscape (Kellogg & Van Maanen, 2020).

Inclusive Decision-Making

Effective educational reform hinges on the principle of inclusive decision-making, which plays a crucial role in fostering a collaborative environment (Martin & Kemple, 2020). This leadership strategy actively involves a diverse array of stakeholders, including teachers, parents, students, and community members, allowing them to share their insights and experiences during the decision-making process (Chen & Li, 2020; Spillane, 2006). By embracing a variety of perspectives, educational leaders not only enhance the quality and relevance of their decisions but also cultivate stronger partnerships among all involved parties. When individuals feel that their opinions are acknowledged and valued, they are more inclined to support and champion necessary changes within the institution (Barge & Loges, 2018; Chen & Li, 2020). This sense of collective ownership promotes shared accountability and contributes to the development of a more cohesive educational community (González-DeHass et al., 2017; Leithwood & Louis, 2012).

The collaborative nature of decision-making often sparks innovative solutions, as the integration of diverse viewpoints can reveal insights that might otherwise remain hidden when relying on a singular perspective (Kellogg & Van Maanen, 2020; Spillane, 2006). For instance, when teachers share their classroom experiences alongside parents' insights into their children's needs, the resulting dialogue can lead to more effective teaching strategies and curriculum adjustments that resonate with students' realities (González-DeHass et al., 2017). This synergy not only enriches the decision-making process but also empowers stakeholders to take an active role in shaping the educational landscape (Henderson & Mapp, 2020). As a result, the institution becomes more adaptable and responsive to the evolving challenges it faces, ultimately leading to improved student outcomes and a more engaged community (Miller & Johnson, 2021).

Inclusive decision-making fosters a culture of trust and respect, which is essential for any successful reform initiative (Darling-Hammond et al., 2017; Schildkamp & Visscher, 2016). When stakeholders see that their contributions are valued, they are more likely to engage in open communication and collaboration, breaking down silos that often exist within educational settings (Schildkamp & Visscher, 2016). This collaborative spirit encourages a shared vision for the future, where everyone is invested in the success of the institution. As educational leaders prioritize inclusivity, they not only enhance the decision-making process but also lay the groundwork for a sustainable and dynamic educational environment that can thrive amidst change (Darling-Hammond et al., 2017; Paredes & Medina, 2016). Ultimately, the commitment to inclusive decision-making serves as a cornerstone for effective educational reform, ensuring that all voices are heard and that the collective wisdom of the community is harnessed for the benefit of all students (Martin & Kemple, 2020).

SHIFTING LANDSCAPES

The educational landscape is currently experiencing a profound transformation, influenced by a confluence of demographic shifts, technological advancements, policy reforms, and changing societal expectations (Barge & Loges, 2018; Chen & Li, 2020). As these factors reshape the way education is delivered and perceived, educational leaders are increasingly called upon to demonstrate both adaptability and strategic foresight (González-DeHass et al., 2017). This dual requirement is essential for effectively navigating the complexities of a rapidly evolving environment. Schools today face a myriad of challenges, including declining student enrollment, a shortage of qualified teachers, and the pressure of heightened accountability standards (Miller & Johnson, 2021).

Additionally, there is an urgent need to integrate digital technologies into teaching methodologies, which further complicates the educational landscape (Henderson & Mapp, 2020). In light of these evolving circumstances, it is imperative for leaders to remain agile, all while maintaining a clear vision focused on enhancing student achievement and ensuring the long-term sustainability of their institutions. Moreover, the role of educational institutions has expanded significantly beyond the confines of traditional academic instruction (Yazzie-Mintz & Nagaoka, 2021). Today, schools are tasked with addressing social-emotional learning, promoting equity initiatives, and preparing students for the workforce. This broadening of focus means that educational leaders must adeptly juggle multiple priorities, ensuring that their policies and practices not only align with educational standards but also meet the diverse needs of their communities (Henderson & Mapp, 2020; Miller & Johnson, 2021). Successfully navigating these complexities requires a steadfast commitment to strategic planning, collaborative decision-making, and data-informed practices that foster continuous improvement. By leveraging data effectively, leaders can make informed choices that enhance educational outcomes and address the unique challenges faced by their institutions (Peters, 2010; Walker & Hoover-Dempsey, 2017).

The swift evolution of the education sector underscores the essential role of resilience in leadership. In an era marked by rapid technological advancements, shifting societal expectations, and unforeseen global challenges, educational leaders must cultivate a robust capacity to navigate their institutions through periods of uncertainty (Walker & Hoover-Dempsey, 2017). This involves not only providing stability and support to students and staff but also fostering a sense of community and trust (Xhelilaj et al., 2021). Leaders are tasked with the dual responsibility of addressing immediate challenges, such as adapting to remote learning or managing resource constraints while simultaneously preparing for future developments that may impact the educational landscape (Williams III et al., 2022).

To effectively meet these demands, educational leaders must adopt a proactive stance, anticipate changes and equip their institutions with the tools necessary to adapt (Yazzie-Mintz & Nagaoka, 2021). This foresight requires a deep understanding of emerging trends, such as the integration of artificial intelligence in the classroom or the increasing importance of mental health resources for students. By staying informed and responsive, leaders can implement strategies that not only address current issues but also lay the groundwork for long-term success (Xhelilaj et al., 2021; Williams III et al., 2022).

Nurturing a culture that values adaptability and collective responsibility is crucial in this process. When educational leaders encourage collaboration among staff, students, and the wider community, they create an environment where innovation can flourish (Henderson & Mapp, 2020). This collaborative spirit empowers individuals to share ideas, experiment with new teaching methods, and support one another through challenges (Agosto & Roland, 2018; Chen & Li, 2020). As a result, institutions become more agile, capable of pivoting in response to new demands and opportunities (Epstein & Sheldon, 2019).

Ultimately, by placing a strong emphasis on the growth and learning of their students, these leaders can position their institutions not just to endure but to thrive amidst the continuous changes in the educational landscape (Agosto & Roland, 2018; Darling-Hammond et al., 2017). A resilient leadership approach fosters a sense of purpose and direction, inspiring both educators and learners to embrace change as an opportunity for growth. In doing so, educational leaders not only enhance the immediate educational experience but also prepare their institutions to meet the challenges of tomorrow with confidence and creativity (Epstein, 2018).

COLLABORATIVE PARTNERSHIPS

Collaborative partnerships are essential for enhancing the effectiveness of educational institutions and fostering student achievement. Schools that forge robust connections with families, community organizations, and local businesses are more adept at meeting the varied needs of their student populations. These alliances not only broaden the range of resources available but also create additional opportunities for students, ultimately forming a comprehensive support network that significantly improves educational outcomes (Epstein, 2018). For instance, partnerships with local businesses can provide students with internships and mentorship programs, while community organizations can offer tutoring and extracurricular activities that enrich the educational experience. Such collaborations enable schools to tap into a wealth of expertise and resources that would otherwise be inaccessible, thereby creating a more holistic approach to education (Henderson & Mapp, 2020).

The involvement of families is particularly crucial in driving academic success and boosting student engagement (Kellogg & Van Maanen, 2020). When parents and guardians take an active role in their children's education, students often exhibit increased motivation, better attendance rates, and enhanced academic performance (Miller & Johnson, 2021). Research has shown that students whose families are engaged in their schooling are more likely to develop positive attitudes toward learning and demonstrate resilience in the face of challenges (Leithwood et al., 2020). This underscores the importance of fostering a school culture that values and encourages family participation. Consequently, educational leaders must focus on developing strategies that promote meaningful family engagement and ensure consistent communication with community stakeholders to maximize these benefits (Epstein, 2018). By implementing initiatives such as workshops, family nights, and regular updates on student progress, schools can create an environment where families feel valued and empowered to contribute to their children's educational journey (Pera, 2019; Rosa & Gifford, 2023).

The impact of these collaborative efforts extends beyond individual student success; they also contribute to the overall health of the school community. When schools actively engage with families and local organizations, they cultivate a sense of belonging and shared responsibility among all stakeholders (Schildkamp et al., 2016). This collective investment in education fosters a positive school climate, which is essential for both student and teacher satisfaction (Sanders & Harvey, 2019). As a result, educational institutions that prioritize collaboration are not only better equipped to address the diverse needs of their students but also to create a thriving educational ecosystem that benefits everyone involved. In this way, the synergy created through collaborative partnerships becomes a powerful catalyst for sustained improvement in educational outcomes, ultimately leading to a more equitable and effective educational landscape (Simons & Vanderlinde, 2020).

Collaboration with community organizations and local agencies, alongside family partnerships, can significantly enhance the resources and support services available to students (Tian & Yu, 2021). By leveraging the strengths and expertise of various stakeholders, schools can create a network of support that extends beyond the classroom. These community-based collaborations often provide essential programs such as mentorship initiatives, which connect students with role models who can guide them through academic and personal challenges (Weiss & Caspe, 2021; Wong & Badr, 2022).

Additionally, opportunities for career exploration allow students to gain insights into various professions, helping them to make informed decisions about their futures. Access to mental health services is another critical component of these partnerships, as it addresses the emotional and psychological needs of students, fostering resilience and well-being. All of these elements contribute to a richer educational experience, ensuring that students are not only academically prepared but also equipped to navigate the complexities of life (Miller & Johnson, 2021; Paredes & Medina, 2016).

By fostering these connections, schools can develop a comprehensive support system that addresses both academic challenges and the social and emotional well-being of students. This holistic approach is vital for creating an environment where learners can thrive, as it recognizes that education is not solely about intellectual growth but also about nurturing the whole child (Paredes & Medina, 2016).

When students feel supported in all aspects of their lives, they are more likely to engage actively in their education, leading to improved academic performance and overall satisfaction with their school experience. Similarly, such collaborative efforts instill a collective sense of accountability for educational outcomes (Miller & Johnson, 2021). When educators, families, and community stakeholders unite in pursuit of shared objectives, they create a culture of collaboration that reinforces the importance of education within the community.

As schools evolve into more robust and resilient institutions through these partnerships, they become better equipped to cater to the diverse needs of all students (Kellogg & Van Maanen, 2020; Leithwood et al., 2020). This adaptability is crucial in today's rapidly changing world, where students face a myriad of challenges that can impact their learning (Henderson & Mapp, 2020). By working together, schools and community organizations can respond more effectively to these challenges, ensuring that every student has access to the resources they need to succeed (Paredes & Medina, 2016). Ultimately, the synergy created by these collaborative efforts not only enhances educational outcomes but also strengthens the fabric of the community, fostering a sense of belonging and shared purpose among all stakeholders involved in the educational journey (Kellogg & Van Maanen, 2020).

EDUCATION STRATEGIES

Effective educational strategies hinge on meticulous planning, the application of evidence-based practices, and ongoing assessment. These components are essential for creating a robust learning environment that not only meets curriculum standards but also addresses the diverse needs of students (Henderson & Mapp, 2020). Educational leaders play a crucial role in this process, as they are responsible for ensuring that instructional methods align with established benchmarks while remaining flexible enough to adapt to the unique learning profiles of their students (Kellogg & Van Maanen, 2020). By fostering a culture of collaboration and innovation among educators, leaders can create a dynamic atmosphere where teaching practices are continuously refined and improved (Epstein, 2018).

Research has shown that schools that adopt strategic instructional frameworks and utilize data-driven decision-making are more likely to see lasting enhancements in student achievement (Darling-Hammond et al., 2017). This evidence underscores the importance of integrating systematic approaches to instruction that are informed by empirical data. By analyzing student performance metrics and feedback, educators can identify areas of strength and weakness, allowing them to tailor their teaching strategies accordingly (Henderson & Mapp, 2020). Such a responsive approach enhances the learning experience while also promotes accountability among educators, as they are encouraged to reflect on their practices and make necessary adjustments to improve outcomes (Epstein, 2018).

A vital aspect of a successful educational strategy is the incorporation of professional development into instructional planning. When teachers engage in continuous professional learning, they acquire new teaching tools, refine their instructional techniques, and cultivate innovative methods to engage students (Darling-Hammond et al., 2017). This ongoing educational journey empowers educators to stay attuned to evolving educational trends and to implement effective strategies that foster student success (Hargreaves & Fullan, 2012).

Professional development should not be viewed as a one-time event but rather as an integral part of an educator's career. By prioritizing sustained learning opportunities, schools can ensure that their teachers are well-equipped to meet the challenges of a rapidly changing educational landscape, ultimately benefiting the students they serve. (Kryshtanovych et al., 2022)

The interplay of meticulous planning, evidence-based practices, and continuous assessment forms the backbone of effective educational strategies (Leithwood et al., 2020). By embracing these principles and investing in the professional growth of educators, schools can create an environment that not only enhances student achievement but also fosters a culture of lifelong learning. As educational leaders champion these initiatives, they pave the way for a more equitable and effective educational system that meets the needs of all learners (Kryshtanovych et al., 2020).

An important approach to enhancing educational effectiveness is the implementation of collaborative leadership frameworks. These distributed leadership models empower educators to engage actively in decision-making processes, allowing them to leverage their unique expertise and insights to drive school improvement initiatives (Miller & Johnson, 2021). By fostering an environment where teachers and staff share responsibilities in instructional leadership, schools can cultivate robust professional communities that not only promote innovation but also encourage collaborative problem-solving (Kryshtanovych, 2022; Spillane, 2006;). This shift from traditional hierarchical structures to more inclusive leadership practices enable educators to feel valued and invested in their roles, ultimately leading to a more dynamic and responsive educational environment (Leithwood et al., 2020; Miller & Johnson, 2021).

It is crucial for effective educational strategies to prioritize student-centered learning environments. Instructional methodologies should be thoughtfully crafted to foster critical thinking, creativity, and teamwork among students, ensuring that they are not merely passive recipients of information but active participants in their own learning journeys (Paredes & Medina, 2016). By emphasizing active learning and meaningful engagement, educators can develop classrooms that stimulate curiosity and facilitate a deeper comprehension of academic subjects (Darling-Hammond et al., 2017; Northouse, 2022). This approach not only enhances students' academic performance but also equips them with essential skills for the 21st century, such as collaboration and adaptability, which are vital in an increasingly complex and interconnected world. Ultimately, the integration of collaborative leadership and student-centered learning creates a synergistic effect that can significantly elevate the quality of education and better prepare students for future challenges (Paredes & Medina, 2016; Williams III et al, 2022).

SUMMARY

Educational leadership is essential in determining the effectiveness and longevity of schools within a landscape that is becoming increasingly intricate and fluid. Leaders in education are tasked with the challenge of harmonizing their expertise, strategic thinking, and interpersonal skills to navigate their institutions through transformative periods (Paredes & Medina, 2016; Williams III et al, 2022). The ability to connect with others through charisma, influence, and exceptional communication fosters strong relationships with various stakeholders, thereby galvanizing collective efforts toward common objectives. This multifaceted approach is vital for creating an environment where everyone is motivated to contribute to the school's mission and vision (Northouse, 2022).

In examining the attributes of successful educational leadership, several critical elements have surfaced. These include the promotion of professional development opportunities for staff, the encouragement of community involvement, the facilitation of inclusive decision-making processes, and the establishment of collaborative partnerships (Paredes & Medina, 2016; Wong & Badr, 2022). Each of these factors plays a significant role in nurturing a positive educational atmosphere, where educators, students, and community members can unite to pursue favorable outcomes. By investing in these areas, leaders can create a culture of continuous growth and shared responsibility, which is essential for fostering an environment conducive to learning (Xhelilaj et al, 2021).

As educational institutions confront a myriad of evolving challenges, it is imperative for leaders to remain flexible while steadfastly focusing on student achievement. By emphasizing collaboration, resilience, and a commitment to ongoing improvement, educational leaders can develop schools that not only adeptly respond to change but also offer enriching learning experiences for all students (Yazzie-Mintz & Nagaoka, 2021). Ultimately, the effectiveness of educational initiatives hinges on the collaborative efforts of leaders, educators, families, and communities, all working in concert to cultivate robust and equitable learning environments that benefit every learner (Wong & Badr, 2022).

REFERENCES

Agosto, V., & Roland, E. (2018). Intersectionality and educational leadership: A critical review. *Review of Research in Education, 42*(1), 255–285. DOI: 10.3102/0091732X18762433

Ahammad, M. F., Glaister, K. W., Sarala, R. M., & Glaister, A. J. (2018). Strategic talent management in emerging markets. *Thunderbird International Business Review, 60*(1), 5–8. DOI: 10.1002/tie.21927

Aljbour, A., French, E., & Ali, M. (2022). An evidence-based multilevel framework of talent management: A systematic review. *International Journal of Productivity and Performance Management, 71*(8), 3348–3376. DOI: 10.1108/IJPPM-02-2020-0065

Anlesinya, A., Dartey-Baah, K., & Amponsah-Tawiah, K. (2019). Strategic talent management scholarship: A review of current foci and future directions. *Industrial and Commercial Training, 51*(5), 299–314. DOI: 10.1108/ICT-11-2018-0095

Bagchi, M. J., & Pattnaik, A. (2022). Talent Analytics and its Affect on Employee Retention-Employee Perspective. *Journal of Positive School Psychology, 6*(3), 7311–7321.

Barge, J. K., & Loges, W. E. (2018). Community-based school programs and family engagement: Developing a model of educational success. *Journal of Family and Consumer Sciences, 110*(3), 18–28. DOI: 10.14307/JFCS.110.3.18

Cable, D. (2013). Traditional on-boarding of staff requires a shake-up. *Strategic HR Review, 12*(6). Advance online publication. DOI: 10.1108/SHR-06-2013-0068

Chen, X., & Li, L. (2020). Family involvement in school-based decision making: The role of school leadership and community engagement. *International Journal of Leadership in Education, 23*(3), 343–360. DOI: 10.1080/13603124.2019.1585597

Darling-Hammond, L., Hyler, M. E., & Gardner, M. (2017). *Effective teacher professional development*. Learning Policy Institute. DOI: 10.54300/122.311

DeMatthews, D., Carrola, P., Reyes, P., & Knight, D. (2021). School leadership burnout and job-related stress: Recommendations for district administrators and principals. *The Clearing House: A Journal of Educational Strategies, Issues and Ideas, 94*(4), 159–167. DOI: 10.1080/00098655.2021.1894083

Epstein, J. L. (2018). *School, family, and community partnerships: Preparing educators and improving schools* (3rd ed.). Routledge.

Epstein, J. L., & Sheldon, S. B. (2019). Present and accounted for: Improving student attendance with family and community involvement. *The Journal of Educational Research, 112*(1), 40–48. DOI: 10.1080/00220671.2018.1466029

Fullan, M. (2016). *The new meaning of educational change* (5th ed.). Teachers College Press.

González-DeHass, A. R., Willems, P. P., & Holbein, M. F. (2017). Examining the role of parental involvement in the academic success of Latino children. *Educational Psychology Review, 29*(2), 235–257. DOI: 10.1007/s10648-016-9364-0

Hargreaves, A., & Fullan, M. (2012). *Professional capital: Transforming teaching in every school.* Teachers College Press.

Harris, A., & Goodall, J. (2018). Parental involvement and academic achievement: A systematic review of research. *Journal of Educational Psychology, 110*(1), 17–32. DOI: 10.1037/edu0000209

Henderson, A. T., & Mapp, K. L. (2020). *A new wave of evidence: The impact of school, family, and community connections on student achievement revisited.* Educational Policy Institute.

Kellogg, D., & Van Maanen, K. (2020). Enhancing family-school collaboration: The role of community-based interventions in rural schools. *Rural Education Journal, 39*(1), 98–115. DOI: 10.1080/09120201.2020.174654

Kluch, Y., Taylor, E. A., Wright-Mair, R., & Boring, D. (2022). "I felt like... they left me to fend for myself": A mixed-method examination of sense of belonging among minoritized groups in sport academic programs. *Sport Management Education Journal, 17*(1), 1–16. DOI: 10.1123/smej.2021-0014

Kryshtanovych, M., Akimova, L., Akimov, O., Parkhomenko-Kutsevil, O., & Omarov, A. (2022). Features of creative burnout among educational workers in public administration system. *Creativity Studies, 15*(1), 116–129. DOI: 10.3846/cs.2022.15145

Leithwood, K., Harris, A., & Hopkins, D. (2020). Seven strong claims about successful school leadership revisited. *School Leadership & Management, 40*(1), 5–22. DOI: 10.1080/13632434.2019.1596077

Leithwood, K., & Louis, K. S. (Eds.). (2012). *Linking leadership to student learning*. Jossey-Bass.

Martin, M. A., & Kemple, K. M. (2020). Building capacity for school-family partnerships: A longitudinal study of an urban district's implementation of a parent involvement program. *Journal of Educational Administration, 58*(4), 457–475. DOI: 10.1108/JEA-10-2018-0197

Medina-Garrido, J. A., Biedma-Ferrer, J. M., & Rodríguez-Cornejo, M. V. (2021). I quit! Effects of work-family policies on the turnover intention. *International Journal of Environmental Research and Public Health, 18*(4), 1893. DOI: 10.3390/ijerph18041893

Miller, M. A., & Johnson, R. M. (2021). Fostering family-school partnerships: Effects of teacher communication practices on student achievement in early elementary grades. *School Community Journal, 31*(2), 75–95. DOI: 10.21747/0903-1068/31.2.75

Northouse, P. G. (2022). *Leadership: Theory and practice* (9th ed.). Sage Publications.

Paredes, C. S., & Medina, S. A. (2016). The role of family and community in supporting student success in schools with high levels of cultural diversity. *Education Policy Analysis Archives, 24*(79), 1–21. DOI: 10.14507/epaa.24.2659

Pera, A. (2019). Towards effective workforce management: Hiring algorithms, big data-driven accountability systems, and organizational performance. *Psychosociological Issues in Human Resource Management, 7*(2), 19–24. DOI: 10.22381/PIHRM7220193

Peters, A. (2010). Elements of successful mentoring of a female school leader. *Leadership and Policy in Schools, 9*(1), 108–129. DOI: 10.1080/15700760903026755

Rosa, M., & Gifford, E. (2023). Parental involvement in urban schools: A cross-cultural examination of engagement strategies. *Journal of Urban Education, 62*(4), 361–378. DOI: 10.1007/s11524-022-00442-2

Sanders, M. G., & Harvey, A. (2019). Community and family partnerships: Enhancing student achievement in high-poverty schools. *Educational Policy, 50*(6), 785–804. DOI: 10.1111/edpo.12297

Schildkamp, K., & Visscher, A. J. (2016). The role of family and community partnerships in school improvement: A framework for monitoring student progress. *Educational Management Administration & Leadership, 44*(2), 179–194. DOI: 10.1177/1741143214566754

Simons, L., & Vanderlinde, R. (2020). Teacher perspectives on parent-school partnerships: A comparative study across Europe. *Teaching and Teacher Education, 91*, 103049. DOI: 10.1016/j.tate.2020.103049

Spillane, J. P. (2006). *Distributed leadership*. Jossey-Bass.

Tian, L., & Yu, M. (2021). Investigating school-family partnerships in an urban setting: Parent perceptions and school effectiveness. *Journal of Educational Psychology*, *113*(3), 374–388. DOI: 10.1037/edu0000421

Walker, J. M. T., & Hoover-Dempsey, K. V. (2017). Parental involvement in education: A longitudinal analysis of involvement, academic achievement, and self-regulation. *Parenting, Science and Practice*, *17*(3), 228–242. DOI: 10.1080/15295192.2017.1301152

Weiss, H. B., & Caspe, M. (2021). Supporting family and community engagement in schools: A study of successful practices in K-12 education. *Journal of Family & Community Engagement*, *2*(2), 85–100. DOI: 10.32474/JFCE.2021.02.00003

Williams, J. A.III, Hill-Jackson, V., Caldwell, C., & Craig, C. J. (2022). Teacher recruitment and retention: Local strategies, global inspiration. *Journal of Teacher Education*, *73*(4), 333–337. DOI: 10.1177/00224871221118155

Wong, E., & Badr, H. (2022). The intersection of family involvement and school culture: Implications for educational success in immigrant communities. *International Journal of Intercultural Relations*, *83*, 32–45. DOI: 10.1016/j.ijintrel.2021.11.007

Xhelilaj, L. K., Petani, R., & Ntalla, M. (2021). Relationship between teacher's burnout, occupational stress, coping, gender and age. *Journal of Educational and Social Research*, *11*(4), 275.

Yazzie-Mintz, E., & Nagaoka, J. (2021). Engaging communities and families in school reform: Lessons from successful programs. *Educational Policy Research Journal*, *49*(2), 140–160. DOI: 10.1080/00357562.2021.1850152

Chapter 2
Transforming Disorder into Insight:
Rethinking Education in Times of Crisis

Binod Kumar
http://orcid.org/0009-0002-9721-1367
Radha Govind University, India

Oindrilla Ghosh
http://orcid.org/0000-0001-9609-0781
Christ University, Bangalore, India

ABSTRACT

In times of crisis, educational systems often face unprecedented challenges that can lead to disorder and uncertainty. However, this turmoil presents a unique opportunity for transformation and innovation. By rethinking traditional education models, we can foster resilience, creativity, and adaptability among students and educators. Emphasizing critical thinking and problem-solving skills enables learners to navigate complexities and emerge with newfound insights. Collaborative approaches, using technology and community resources, can enhance engagement and inclusivity, bridging gaps exacerbated by crises. Moreover, integrating emotional intelligence and mental well-being into curricula can support students in coping with stress and ambiguity. Ultimately, embracing disorder as a catalyst for change can lead to a more robust, relevant, and responsive educational landscape, equipping future generations to thrive in an ever-evolving world.

DOI: 10.4018/979-8-3373-1350-4.ch002

INTRODUCTION

In an era marked by unprecedented challenges be it climate change, global pandemics, social unrest, or technological upheaval education has become an essential lens through which to examine societal resilience and adaptability (Atiyeh et al., 2023). The phrase "Transforming Disorder into Insight" encapsulates the potential for educational systems to evolve amid crises, reframing adversity as an opportunity for growth and innovation (van der Vorst & Jelicic, 2019). Education is not merely about the accumulation of knowledge; it is fundamentally a dynamic process that shapes individuals and communities (Kai et al., 2018). The crises facing today's world call for a rethinking of traditional educational paradigms, emphasizing flexibility and creative problem-solving (Gamage et al., 2023).

Historically rooted in structured methodologies and curricula, education has often been slow to adapt to changing circumstances (Murray, 2023). However, the recent global crises have forced educators, policymakers, and communities to rethink what it means to teach and learn (Perkins, 2023). The rapid proliferation of online learning during the COVID-19 pandemic, for instance, highlighted both the potential and the pitfalls of digital education (ClassPoint, 2024). It compelled us to question the relevance of established pedagogical methods and the accessibility of educational resources (Ibrahim et al., 2023). In such an environment, inquiry-based learning, critical thinking, and emotional intelligence have emerged as crucial competencies, transcending the confines of rote memorization and standardized testing (Scribes, 2024).

Furthermore, the socio-political landscapes of our times demand that educational institutions not only impart knowledge but also foster civic engagement and social justice (Selwyn, 2022). The injustices laid bare by movements advocating for racial equality and climate activism reveal the vital role that education plays in empowering individuals to contribute meaningfully to society (Cardona et al., 2023). Therefore, rethinking education requires a broader understanding of its purpose: one that transcends mere knowledge acquisition to include the cultivation of empathetic, informed citizens capable of engaging in constructive dialogue and activism (Zootzky & Pfeiffer, 2024).

HISTORICAL EVOLUTION

The landscape of education has always been marked by both continuity and change, often driven by historical events and societal upheavals (D'Errico, 2007). Understanding this historical evolution is crucial for contextualizing today's educational crisis and rethinking its role in fostering resilience and insight (Cuban,

2001). At various points in history, education has adapted to meet the challenges posed by disorder, whether that disorder stems from war, economic instability, or social transformation (Esposito, Sangrà, and Maina, 2013).

One significant example is the impact of World War II on education (Dusek, 2006). In the wake of the war, many nations recognized the need for an educated populace capable of critical thinking and civic engagement (Perrotta and Evans, 2013). The devastation prompted a reevaluation of educational goals, leading to the establishment of reforms aimed at promoting democratic values, peace, and global cooperation (Mayes, 1995). Education systems in countries such as Germany and Japan underwent fundamental transformations, emphasizing not only academic knowledge but also moral education and social responsibility (Chiang et al., 2016). This shift highlighted the potential for education to act as a stabilizing force in society, helping individuals to engage thoughtfully in a post-war world (Januszewski and Persichitte, 2008).

The advent of the digital age has spawned a different set of challenges and possibilities (Latchem, 2014). The rapid integration of technology into education has transformed teaching and learning dynamics, offering new opportunities for engagement while also exacerbating disparities in access and resources (Friesen, 2009). The COVID-19 pandemic served as a significant stress test for educational systems worldwide, forcing a rapid shift to remote learning and highlighting the digital divide (Czerniewicz, 2010). In response to this crisis, educators discovered innovative ways to leverage technology for interactive and collaborative learning experiences, encouraging the development of digital literacy an essential skill in the 21st century (Oliver,2013).

Educational theorists like Paulo Freire argued that authentic education is a dialogical process that must adapt to the realities of learners' lives (Braden, 1995). As such, the ongoing crises faced by society necessitate a rethinking of educational practices to create spaces where students are empowered to explore real-world issues, reflect on their experiences, and actively participate in their learning (Arthur, 2009). This historical evolution of education reveals a complex interplay between disorder and insight, suggesting that embracing crisis as an opportunity for reform can lead to more meaningful and responsive educational experiences (Selwyn, 2014).

THEORETICAL FRAMEWORKS

Theoretical frameworks serve as essential tools for understanding and interpreting complex phenomena, particularly in the field of education (Hew and Cheung, 2014). During times of crisis, such as global pandemics, economic downturns, and sociopolitical upheavals, the educational landscape often undergoes significant transformation (Sergio et al., 2011).

One of the most relevant theoretical frameworks during crises is Constructivism, which posits that learners construct knowledge through experiences and interactions with their environment (Christensen et al., 2008). In a time of crisis, when traditional educational structures may falter, Constructivism encourages educators to create learning environments where students actively engage with real-world problems (Thumlert, Castell, and Jenson, 2015). For instance, project-based learning can be implemented to allow students to explore and solve pressing issues stemming from the crisis, fostering critical thinking and collaboration (Ladyman, 2007). This experiential, student-centered approach not only enhances learning outcomes but also equips students with the skills necessary for adapting to unforeseen challenges (Heidegger, 1996).

Another crucial framework is Social Constructivism, which highlights the importance of social interactions in the learning process (Haraway, 1988). Crises often disrupt social dynamics, but they also provide opportunities for communities to come together and share knowledge (Dépelteau, 2008). Educators can leverage technology to facilitate collaborative learning through virtual classrooms, enabling students to engage with peers and mentors across geographical boundaries (Heidegger, 1959). This collective intelligence creates a rich learning environment, allowing insights to emerge from diverse perspectives and experiences (Whitehead, 1929). Furthermore, Social Constructivism underscores the role of cultural contexts in shaping understanding, reminding educators to be mindful of equity and inclusion in their approaches, particularly during turbulent times (Surry and Baker, 2016).

Transformative Learning Theory is also pertinent during crises (Wu 2009). Developed by Jack Mezirow, this framework emphasizes the importance of critical reflection in fostering personal and social transformation (Marx, 1964). In the face of crisis, learners are often confronted with assumptions and beliefs that may no longer hold true (Allenby and Sarewitz, 2011). Educators can create spaces for critical dialogue, encouraging students to reflect on their values, beliefs, and the information presented to them (Marx, 1975). This process can lead to profound insights, empowering learners to reframe their understanding of the world and their place within it (Bayne, 2015). By focusing on transformative learning, educators can help students to not only cope with disorder but also to emerge as agents of change in their communities (Heidegger, 1996).

Lastly, systems thinking offers a holistic perspective on education during crises (Zhao et al., 2002). This framework encourages individuals to see the interconnections between various elements within a system, recognizing that changes in one area can have ripple effects across others (Mayes, 1995). In the educational context, this approach can help educators identify how various factors such as policy, technology, and social dynamics interact and contribute to the learning environment during a crisis (Latour, 2004). By applying systems thinking, educators can develop more integrated and adaptive strategies that consider the complex realities faced by students, ultimately transforming disorder into a deeper understanding and more effective educational practices (Heidegger, 1971).

GOVERNANCE

Governance plays a crucial role in shaping educational systems, especially during crises that demand rapid adaptation and innovation (Friesen, 2009). The structures, policies, and processes that define governance can significantly influence how educational institutions respond to disorder and transform challenges into opportunities for growth (Ihde, 1990). In times of crisis, effective governance becomes essential for ensuring not only the continuity of education but also the enhancement of its quality and relevance to societal needs (Weiss, 2008).

At the core of effective educational governance is the principle of flexibility (Verbeek 2008). Traditional governance models, often characterized by rigid hierarchies and bureaucratic procedures, may hinder rapid responses to crises (An and Li, 2014). In contrast, adaptive governance frameworks that emphasize collaboration, stakeholder engagement, and decentralized decision-making are more suited to navigating uncertainty (Hegel, 1967). For instance, during the COVID-19 pandemic, many educational institutions shifted to remote learning (Heidegger, 1977). Those governed by flexible structures were able to quickly implement strategies, such as training educators in digital tools and providing resources for students, ultimately facilitating smoother transitions (Gehlen, 1988).

Moreover, participatory governance is critical in times of crisis (Kant, 1990). Involving all stakeholders students, parents, educators, and community members in decision-making processes fosters a sense of ownership and accountability (Freire, 2005). This inclusiveness enables diverse perspectives to inform educational policies and practices, thereby enhancing their relevance and effectiveness (Donnelly, 2007). For example, in addressing the social inequities exacerbated by crises, community input can help identify specific needs and create tailored interventions that support vulnerable populations, ensuring that no one is left behind (Willet, 2018).

Finally, the role of technology in governance cannot be overstated (Reich, Garrison, and Neubert, 2016). In times of crisis, digital tools can facilitate communication, collaboration, and resource sharing among educational stakeholders (Chan & Tsi, 2023). Governance frameworks that embrace technology can enhance transparency and responsiveness (Halagatti et al., 2023). Virtual platforms allow for real-time feedback and facilitate ongoing dialogue between educators and community members, ensuring that policies are continually refined to meet the evolving needs of the educational landscape (Dewey, 1916).

ACCOUNTABILITY

Accountability is a crucial pillar in the realm of education, especially when the paradigms shift during times of crisis (van der Vorst & Jelicic, 2019). In the context of transforming disorder into insight, accountability serves multiple roles, acting as both a guiding principle and a measure of effectiveness (Gamage et al., 2023). As educational institutions face unprecedented challenges be they pandemics, social upheavals, or economic downturns the need for accountability becomes even more pronounced (Selwyn, 2022).

At its core, accountability in education entails responsibility not just for educators and administrators but for students, parents, and the broader community (Zootzky & Pfeiffer, 2024). When a crisis disrupts traditional educational approaches, stakeholders must reassess their roles and ensure that they remain committed to the educational mission (Cardona et al., 2023). This commitment fosters a culture of responsibility, leading to a collaborative environment where all parties are engaged in the learning process (Pedaste et al., 2015).

In rethinking education during crises, accountability can manifest through transparent communication (Smith, 2022). Institutions must openly share information about how they are addressing the disruptions and what measures are being taken to maintain educational standards (Mason, 2008). This transparency breeds trust, allowing students and their families to feel assured that their educational needs are being prioritized (Heidegger, 1996). Regular updates about curricular changes, technological adaptations, and mental health support initiatives serve as a foundation for a collective response to adversity (Halagatti et al., 2023).

Moreover, accountability in educational institutions encourages reflection and adaptation (Heidegger, 1971). Educators must examine the efficacy of their teaching strategies, especially when conventional methods falter under the pressures of crisis (Thomson, 2004). This reflective practice involves assessing what works and what does not not only through student performance metrics but also through qualitative feedback from students and families (Pearce & Chiavaroli, 2023). By embracing a mindset of continuous improvement, educators can pivot their approaches, harnessing disorder to foster resilience and creativity (Moqbel & Al-Kadi, 2023).

CHALLENGES

The rapid evolution of artificial intelligence (AI) in education presents a transformative opportunity to glean insights from disorderly situations, such as those presented by crises like the global pandemic (Wrathall, 2011). However, leveraging AI in educational contexts does not come without its challenges (Murgatroyd, 2023). One primary concern is the accessibility and equity of technology (Huang and Liu, 2016). In many regions, especially in low-income areas, students lack access to necessary devices and high-speed internet (Deng, 2018). This digital divide can exacerbate educational inequities, where students who are already disadvantaged face even greater challenges in accessing quality education (ClassPoint, 2024).

Another significant challenge lies in data privacy and ethics (Koh & Doroudi, 2023). AI systems typically rely on vast amounts of data to function effectively, raising concerns about how this data is collected, stored, and utilized (Heidegger, 2000). In educational settings, this is particularly critical, as students are a vulnerable population (Roll et al., 2011). Misuse of personal data or lack of consent can lead to breaches of trust and legal complications (Trajtenberg, 2018). Furthermore, there is the risk of algorithmic bias, where AI systems may inadvertently perpetuate existing societal biases (Pezze, 2006). If the data used to train these systems reflects historical prejudices or unequal treatment, the outputs produced may reinforce rather than alleviate disparities (Cardona et al., 2023).

The complexity of integrating AI into educational curricula poses another hurdle (Mitcham, 2006). Teachers and educational institutions may lack the training or resources necessary to effectively implement AI technologies in the classroom (Oliver and Conole, 2003). This gap in knowledge can hinder the potential benefits of AI, as educators may be unsure how to interpret the insights generated by these systems or apply them in a meaningful way (Eynon, 2015). Continuous professional development is essential, yet often lacking, leading to a disconnect between AI advancements and their practical application in education (Feenberg, 2002).

Additionally, the speed at which AI technology evolves can leave educational stakeholders scrambling to keep up (Clegg and Steel, 2002). Policymakers and administrators must work to ensure that regulations are in place to protect students while also fostering innovation (Lindgren et al., 2016). Balancing the need for swift adaptation with the need for careful oversight can be a daunting task, especially during crises when time is of the essence (Harrer, 2023).

APPLICATIONS OF AI

Despite the challenges posed by integrating artificial intelligence (AI) into education during crises, the applications of this technology offer significant potential to transform disorder into invaluable insights (Ihde, 2003). One prominent application is the use of AI-driven analytics to assess students' learning patterns and individual needs (Selwyn, 2010). During crises, such as the COVID-19 pandemic, educators can leverage AI to monitor students' progress in real time, identifying those who may be falling behind (Selwyn, 2022). These insights can drive timely interventions, allowing educators to tailor their approaches to meet the unique challenges faced by each student (Gamage et al., 2023).

AI's capacity for personalization is particularly noteworthy (Lee & Soylu, 2023). Adaptive learning platforms powered by AI can deliver customized educational content, ensuring that students receive instruction tailored to their learning pace and style (Kai et al., 2018). In times of crisis, when traditional teaching methods may falter, these personalized systems help maintain engagement and motivation, ultimately promoting better learning outcomes (Chen & Perez, 2023).

Furthermore, AI can facilitate improved communication and collaboration among students, teachers, and parents (Dogan et al., 2023). For instance, AI chatbots can serve as immediate support tools, providing answers to frequently asked questions and aiding students in navigating their educational resources (Oliver, 2011). This enhances the learning experience by ensuring that help is readily available, reducing the anxiety that often accompanies educational disruptions during crises (Cuban, 2001).

In addition to direct academic support, AI can assist in analyzing broader educational trends during crises (Beckera and Birdib, 2018). By aggregating data on student performance and engagement, AI can help identify systemic challenges faced by schools and districts (Latour, 2005). Educators and administrators can utilize these insights to implement evidence-based strategies, ultimately enhancing the resilience and adaptability of educational institutions in uncertain times (Oliver, 2015).

Moreover, AI can play a crucial role in scaling up professional development for teachers (Lewin and Lundie, 2016). Virtual training programs powered by AI can provide personalized learning experiences for educators, enhancing their digital literacy and equipping them with the skills necessary to effectively integrate technology into their classrooms (Ouyang et al., 2022). During a crisis, this translates to a more prepared teaching workforce able to respond to the evolving needs of their students (Salinas-Navarro et al., 2024).

Another innovative application of AI is the use of virtual reality (VR) and augmented reality (AR) to create immersive learning experiences (Kirkwood and Price, 2014). These technologies can enhance student engagement and understanding in subjects that might be difficult to teach through traditional methods, particularly in remote learning scenarios (Greco & Cinganotto, 2023). For example, students studying biology could explore a virtual ecosystem, experiencing concepts first-hand that would otherwise be abstract in a digital form (Zimmerman, Forlizzi, and Evenson, 2007).

AI can also enhance assessment and feedback mechanisms (Minocha and Reeves, 2010). Traditional methods of evaluation often lag in providing timely and constructive feedback (Ibrahim et al., 2023). AI-driven platforms can analyze student submissions, offering insights and suggestions almost instantaneously (Agalianos, Whitty, and Noss, 2006). This immediate feedback loop not only helps students understand their strengths and weaknesses better but also enables educators to adjust their teaching strategies on the fly (Nikolopoulou, 2024a).

THE USE OF AUGMENTED REALITY (AR)

Augmented Reality (AR) is transforming educational experiences by seamlessly overlaying digital information onto the physical world (Hooda et al., 2022). This immersive technology enhances learning by providing interactive experiences that engage students on multiple sensory levels (Reveley, 2013). In classrooms, AR bridges the gap between theoretical knowledge and practical application (Martínez-Comesaña et al., 2023). For instance, by using AR applications, biology students can visualize the intricacies of cellular structures in 3D, facilitating a deeper understanding of complex concepts (Peters et al., 2020).

AR also fosters collaborative learning. Students can work together on projects that require them to share and manipulate digital objects, promoting teamwork and communication skills (Orchard et al., 2020). Additionally, AR can cater to various learning styles visual learners can benefit from 3D models, while kinesthetic learners can interact physically with the augmented content. This adaptability is vital in a diverse classroom setting where each student has unique preferences and needs (Jackson, 2020a).

Moreover, AR can promote equity in education. Students in remote or under-resourced areas can access high-quality educational content through AR devices or apps, leveling the educational playing field (Rudolph et al., 2024). For example, museum exhibits can come alive with AR, enabling students in different locations to experience world-class art and history without ever leaving their schools (Perkins, 2023). This accessibility can inspire a passion for learning and exploration that might otherwise be stifled (Varaki et al., 2021).

Despite these challenges, the benefits of AR in education are substantial (Hung et al., 2020). By creating immersive and interactive learning experiences, AR not only makes education more engaging but also enhances students' retention and understanding of material (Kato et al., 2020). Through AR, education can evolve from traditional rote learning to a dynamic system that encourages exploration and discovery (Papastephanou et al., 2022). As schools continue to adapt to technological advancements, AR stands out as a powerful tool for enriching educational practices (Nikolopoulou, 2024b).

THE USE OF VIRTUAL REALITY (VR)

Virtual Reality (VR) is reshaping education by providing immersive experiences that transport students beyond the confines of their classrooms (Buchanan et al., 2021). By wearing VR headsets, learners can explore historical events, distant planets, or intricate cellular processes in an engaging and participatory manner (Biesta et al., 2021). This technology offers a sense of presence that traditional educational tools cannot replicate, leading to deeper understanding and retention of complex information (Roth et al., 2020).

One of the most significant applications of VR in education is in experiential learning (Sardoc et al., 2020). Students can engage in virtual field trips to historical sites, laboratories, or even outer space, providing contextual understanding that enhances theoretical learning (Waghid et al., 2020). For instance, a history class can virtually walk through ancient Rome, experiencing the architecture and culture firsthand (Roll et al., 2011). This experiential approach promotes empathy and a personal connection to the subject matter, which enhances cognitive engagement (Jandric et al., 2021).

Additionally, VR is particularly advantageous in training for careers that require high-stakes decision-making and fine motor skills, such as in medicine or engineering (Tesar, 2021). Medical students can practice surgical procedures in a risk-free virtual environment, allowing them to refine their skills without endangering real patients (Roy, 2020). Similarly, engineering students can simulate the building of structures and systems, testing their designs in a virtual space before implementing them in the real world (Gloria Ladson-Billings, 2021).

Yet, as with any technology, the successful implementation of VR in education requires thoughtful consideration (Harari, 2014). The cost of VR equipment and the development of content can be prohibitive for many educational institutions, especially underfunded ones (Perkins, 2023). Additionally, there is a learning curve associated with using VR technology, both for educators and students, requiring significant investment in training and infrastructure.

THE USE OF MIXED REALITY (MR)

Mixed Reality (MR) combines the strengths of Augmented Reality (AR) and Virtual Reality (VR) to create interactive environments where physical and digital worlds converge (Garroutte & Westcott, 2013). In education, MR offers learners a unique way to engage with content, allowing them to manipulate digital objects in real-world settings (Marker, 2018). This capability provides a more nuanced approach to learning, as students can interact with educational material at a level that is both creative and analytical (Barad, 2019).

For instance, in subjects like chemistry, MR can enable students to visualize and manipulate atoms and molecules, allowing them to conduct experiments virtually (Braidotti, 2019). This hands-on approach can deepen students' understanding by allowing them to "see" the interactions occurring at a molecular level, making abstract concepts tangible (Rosiek et al., 2020). In history education, MR can bring historical figures or events to life, creating immersive narratives that enhance student engagement and comprehension (Deleuze, 1995).

MR also promotes collaborative learning. For group projects, students can work together within the same MR environment, facilitating teamwork and communication (Jackson & Mazzei, 2012). This shared experience encourages discussion and debate, essential skills in developing critical thinking (Crawford et al., 2023). The ability to collaborate in a mixed-reality space prepares students for the increasingly interdisciplinary nature of modern workplaces (Hooda et al., 2022).

Despite its potential, the implementation of MR in education faces several hurdles (Siegel, 1983). The cost of MR technology and content creation can be high, limiting accessibility for many institutions (Perkins, 2023). There's also the challenge of ensuring that educators are prepared to integrate MR into their teaching effectively (Atiyeh et al., 2023). Professional development must keep pace with technological advancements, ensuring that teachers can leverage MR tools to enrich their curricula (van der Vorst & Jelicic, 2019).

Concerns about screen time and its effects on students' health and well-being also arise with the use of MR (Gamage et al., 2023). Schools must be mindful of maintaining a balance between technology use and traditional educational methods, ensuring that students' overall learning experiences remain well-rounded (Salinas-Navarro et al., 2024).

CONCLUSION

In times of crisis, such as the recent global pandemic, the traditional education system has faced unprecedented challenges (Murray, 2023). These disruptions have compelled educators, students, and policymakers to rethink educational practices and seek new methodologies that can adapt to rapidly changing circumstances (Ibrahim et al., 2023). The integration of AR, VR, and MR technologies provides significant opportunities for transforming disorder into insight, offering solutions that not only address immediate challenges but also prepare learners for a complex future (Scribes, 2024).

The crisis has demonstrated the limitations of conventional educational approaches (Cardona et al., 2023). As educators sought to maintain continuity of learning, many turned to digital platforms and immersive technologies (Chan & Tsi, 2023). This newfound reliance on technology has highlighted the potential of AR, VR, and MR to create engaging and interactive learning experiences that transcend the constraints of physical classrooms (Halagatti et al., 2023). The ability to simulate real-world scenarios and provide immediate feedback allows students to learn at their own pace, fostering resilience and adaptability critical skills in a rapidly changing world (Zootzky & Pfeiffer, 2024).

However, embracing these technologies requires careful consideration. Educators must prioritize inclusivity and accessibility, ensuring that all students, regardless of background, can benefit from tech-enhanced learning experiences (Smith, 2022). Additionally, teacher training and professional development will be essential in helping educators navigate this new landscape.

SUGGESTIONS

In times of crisis, such as during the COVID-19 pandemic or other global upheavals, traditional educational systems often reveal significant vulnerabilities (Moqbel & Al-Kadi, 2023). However, these periods of turbulence also present unique opportunities for transformation and innovation in education. To genuinely rethink education in these challenging times, several suggestions can be implemented that focus on flexibility, inclusivity, and the integration of technology (Murgatroyd, 2023).

First and foremost, educational institutions should embrace a flexible curriculum that prioritizes critical thinking and problem-solving skills over rote memorization (Koh & Doroudi, 2023). By providing students with open-ended projects and real-world problem-solving tasks, educators can cultivate resilience and adaptability in learners. Such an approach encourages students to navigate uncertainty and develop the capacity to think critically about complex issues skills that are essential in times of crisis (Harrer, 2023).

Second, fostering an inclusive and supportive learning environment is crucial. Crises often exacerbate existing inequalities, and educational institutions must prioritize equity in their programming (Lee & Soylu, 2023). This can be achieved by employing differentiated instruction and personalized learning plans that cater to diverse student needs. Schools should also invest in training educators to recognize and support students who are experiencing trauma or mental health challenges (Dogan et al., 2023). Creating flexible support systems, such as mentorship programs and peer support groups, can also enhance students' emotional well-being during turbulent times (Ouyang et al., 2022).

Incorporating technology into the educational framework is another vital suggestion. Hybrid and fully online learning models offer unprecedented flexibility, making it essential that educators are equipped to leverage tools effectively (Salinas-Navarro et al., 2024). Professional development for teachers in digital literacy and pedagogical strategies for online education should be prioritized to ensure high engagement and learning outcomes. Additionally, schools can explore partnerships with tech companies to access resources that enhance teaching and learning experiences, such as virtual simulations and interactive platforms (Greco & Cinganotto, 2023).

Moreover, fostering partnerships with community organizations can enhance the relevance and applicability of education during crises. By connecting classroom learning with real-world community challenges, students can engage in service-learning projects that address current societal issues (Martínez-Comesaña et al., 2023). This not only provides valuable insights and learning opportunities for students but also strengthens community ties and enhances the societal impact of education.

Finally, consistent assessment and reflection are essential components of this transformative process. Educators should encourage both self-assessment among students and collective evaluation of educational practices at the institutional level (Hooda et al., 2022). Periodic feedback mechanisms can ensure that educational strategies remain relevant and responsive to the evolving crises and needs of society.

FUTURE RESEARCH DIRECTIONS

As we move further into an era punctuated by crises, including climate change, political instability, and global pandemics, there are numerous avenues for future research to explore the implications and innovations in education (Nikolopoulou, 2024b). These research directions can help educational institutions and policymakers better understand how to adapt and thrive in challenging circumstances.

One critical area for research is the longitudinal impact of crisis-induced educational changes on student outcomes (Jackson, 2020a). By studying students over extended periods, researchers can gain insights into the long-term effects of hybrid learning models, flexible curricula, and social-emotional learning programs on both academic success and emotional resilience. This research can guide future pedagogical approaches and help institutions better prepare for unforeseen challenges.

Additionally, investigating the effectiveness of community partnerships in education during crises is essential (Rudolph et al., 2024). Future research can explore case studies highlighting successful models that have demonstrated improved student engagement, community involvement, and academic achievement. Understanding the best practices derived from these collaborations can inform how educational institutions can leverage local resources and networks in times of uncertainty.

Furthermore, an analysis of the role of technology in enhancing education during crises warrants deeper exploration (Varaki et al., 2021). While technology presents myriad opportunities, it also poses challenges, such as access disparities and diminished face-to-face social experiences. Research can focus on evaluating various educational technologies' efficacy and exploring innovative solutions to bridge digital divides, ensuring every student has equitable access to learning resources.

Another relevant research direction is the examination of teachers' experiences and professional development during crises (Kato, 2020). Understanding how educators adapt their teaching methodologies and cope with changing roles can provide essential insights into the support they need to facilitate effective learning environments. Research can also explore the relationship between teacher well-being and student outcomes, emphasizing the importance of mental health and resilience among educators (Ghosh & Kumar, 2025).

Finally, investigating the impacts of student-led initiatives during crises is a potential area for fruitful research. By examining how students respond to crises through activism, creativity, and leadership, scholars can uncover valuable lessons about fostering agency and empowering youth. Such research can not only highlight the potential for student engagement but also inform education policies that prioritize student voice and leadership in shaping their learning experiences (Salinas-Navarro et al., 2024).

REFERENCES

Agalianos, A., Whitty, G., & Noss, R. (2006). The Social Shaping of Logo. *Social Studies of Science, 36*(2), 241–267. DOI: 10.1177/0306312706053809

Allenby, B., & Sarewitz, D. (2011). *The Techno-Human Condition*. MIT Press. DOI: 10.7551/mitpress/8714.001.0001

An, T., & Li, Y. (2014). On Theoretical Prospect of Educational Technology From the Perspective of Philosophy of Technology. *Educational Research, 4*, 37–42.

Arthur, W. B. (2009). *The Nature of Technology: What it is and How it Evolves*. Penguin.

Atiyeh, B., Emsieh, S., Hakim, C., & Chalhoub, R. (2023). A narrative review of artificial intelligence (AI) for objective assessment of aesthetic endpoints in plastic surgery. *Aesthetic Plastic Surgery, 47*(6), 2862–2873. DOI: 10.1007/s00266-023-03328-9

Barad, K. (2019). After the end of the world: Entangled nuclear colonialisms, matters of force, and the material force of justice. *Theory & Event, 22*(3), 524–550.

Bayne, S. (2015). What's the Matter with 'Technology-Enhanced Learning'? *Learning, Media and Technology, 40*(1), 5–20. DOI: 10.1080/17439884.2014.915851

Beckera, R., & Birdib, A. (2018). Flipping the Classroom: Old Ideas, New Technologies. *International Review of Economics Education, 29*, 1–5. DOI: 10.1016/j.iree.2018.06.001

Biesta, G., Heugh, K., Cervinkova, H., Rasinski, L., Osborne, S., Forde, D., Wrench, A., Carter, J., Anders Safstrom, C., Soong, H., O'Keeffe, S., Paige, K., Rigney, I., O'Toole, L., Hattam, R., Peters, M. A., & Tesar, M. (2021). Philosophy of education in a new key: Publicness, Social Justice, and Education; A South-North Conversation. *Educational Philosophy and Theory*. Advance online publication. DOI: 10.1080/00131857.2021.1929172

Braden, A. (1995). Review of Instructional Technology: The Definition and Domains of the Field. *Educational Technology Research and Development, 43*(1), 81–83. DOI: 10.1007/BF02300483

Braidotti, R. (2019). *Posthuman knowledge*. Polity Press.

Buchanan, R. A., Forster, D. J., Douglas, S., Nakar, S., Boon, H. J., Heath, T., Heyward, P., D'Olimpio, L., Ailwood, J., Eacott, S., Smith, S., Peters, M., & Tesar, M. (2021). Philosophy of education in a new key: Exploring new ways of teaching and doing ethics in education in the 21st century. *Educational Philosophy and Theory*. Advance online publication. DOI: 10.1080/00131857.2021.1880387

Cardona, M. A., Rodríguez, R. J., & Ishmael, K. (2023). *Artificial intelligence and the future of teaching and learning*. Lumina Foundation.

Chan, C. K. Y., & Tsi, L. H. Y. (2023). The AI revolution in education: Will AI replace or assist teachers in higher education? arXiv:2305.01185 [cs.CY]. https://doi.org//arXiv.2305.01185DOI: 10.48550

Chen, J. J., & Perez, C. (2023). Enhancing assessment and personalized learning through artificial intelligence. *Childhood Education*, *99*(6), 72–79. DOI: 10.1080/00094056.2023.2282903

Chiang, F. K., Zhu, G. X., Wang, Q., Cui, Z. F., Cai, S., & Yu, S. Q. (2016). Research and Trends in Mobile Learning from 1976 to 2013: A Content Analysis of Patents in Selected Databases. *British Journal of Educational Technology*, *47*(6), 1006–1019. DOI: 10.1111/bjet.12311

Christensen, C., Horn, M., & Johnson, C. (2008). *Disrupting Class: How disruptive innovation will change the way the world learns*. McGraw-Hill.

Clegg, S., & Steel, J. (2002). "Flexibility as Myth? New Technologies and Post-Fordism in Higher Education." *Proceedings of the Networked Learning Conference*.

Crawford, J., Cowling, M., & Allen, K.-A. (2023). Leadership is needed for ethical ChatGPT: Character, assessment, and learning using artificial intelligence (AI). *Journal of University Teaching & Learning Practice*, *20*(3), 457–476. DOI: 10.53761/1.20.3.02

Cuban, L. (2001). *Oversold and Underused: Computers in the Classroom*. Harvard University Press. DOI: 10.4159/9780674030107

Czerniewicz, L. (2010). Educational Technology–Mapping the Terrain with Bernstein as Cartographer. *Journal of Computer Assisted Learning*, *26*(6), 523–534. DOI: 10.1111/j.1365-2729.2010.00359.x

D'Errico, F. (2007). The Origin of Humanity and Modern Cultures: Archaeology's View. *Diogenes*, *54*(2), 122–133. DOI: 10.1177/0392192107077652

Deleuze, G. (1995). *Negotiations*. Columbia University Press.

Deng, L. (2018). Artificial intelligence in the rising wave of deep learning: The historical path and future outlook. *IEEE Signal Processing Magazine*, *35*(1), 180–177. DOI: 10.1109/MSP.2017.2762725

Dépelteau, F. (2008). Relational Thinking: A Critique of co-Deterministic Theories of Structure and Agency. *Sociological Theory*, *26*(1), 51–73. DOI: 10.1111/j.1467-9558.2008.00318.x

Dewey, J. (1916). Democracy and Education: An Introduction to the Philosophy of Education. New York, *NY: the Free Press.*

Dogan, M. E., Goru Dogan, T., & Bozkurt, A. (2023). The use of artificial intelligence (AI) in online learning and distance education processes: A systematic review of empirical studies. *Applied Sciences (Basel, Switzerland)*, *13*(5), 56–88. DOI: 10.3390/app13053056

Donnelly, K. (2007). Australia's Adoption of Outcomes Based Education: A Critique. *Issues in Educational Research*, *17*(2), 183–206.

Dusek, V. (2006). *Philosophy of Technology: An Introduction*. Blackwell Publishing.

Esposito, A., Sangrà, A., & Maina, M. (2013). "Chronotopes in Learner-Generated Contexts. A Reflection About the Interconnectedness of Temporal and Spatial Dimensions to Provide a Framework for the Exploration of Hybrid Learning Ecologies of Doctoral e-Researchers." *LC Research Paper Series,* 6: 15–28.

Eynon, R. (2015). The Quantified Self for Learning: Critical Questions for Education. *Learning, Media and Technology*, *40*(4), 407–411. DOI: 10.1080/17439884.2015.1100797

Feenberg, A. (2002). *Transforming Technology: A Critical Theory Revisited*. Oxford University Press. DOI: 10.1093/oso/9780195146158.001.0001

Freire, P. (2005). *Pedagogy of the Oppressed (30th Anniversary Edition)*. Continuum.

Friesen, N. (2009). *Re-thinking e-Learning Research: Foundations, Methods, and Practices*. Peter Lang.

Gamage, K. A. A., Dehideniya, S. C. P., Xu, Z., & Tang, X. (2023). ChatGPT and higher education assessments: More opportunities than concerns? *Journal of Applied Learning and Teaching*, *6*(2), 654–679. DOI: 10.37074/jalt.2023.6.2.32

Garroutte, E., & Westcott, K. (2013). The story is a living being: Companionship with stories in Anishinaabe Studies. In Doerfler, J., Sinclair, N. J., & Stark, H. K. (Eds.), *Centering Anishinaabeg studies: Understanding the world through stories* (pp. 61–80).

Gehlen, A. (1988). *Man: His Nature and Place in the World*. Columbia University Press.

Ghosh, O & Kumar, B. (2025). The Impact of Remote Experimentation on Student Engagement. *Revolutionizing Education With Remote Experimentation and Learning Analytics,* 523-544, DOI: DOI: 10.4018/979-8-3693-8593-7.ch031

Greco, S., & Cinganotto, L. (2023). Re-thinking education in the age of AI. *Journal of E-Learning and Knowledge Society,* I-IV Pages. DOI: 10.20368/1971-8829/1135873

Halagatti, M., Gadag, S., Mahantshetti, S., Hiremath, C. V., Tharkude, D., & Banakar, V. (2023). Artificial intelligence: The new tool of disruption in educational performance assessment. In Tyagi, P., Grima, S., Sood, K., Balamurugan, B., Özen, E., & Eleftherios, T. (Eds.), *Smart analytics, artificial intelligence and sustainable performance management in a global digitalised economy (110A)* (pp. 261–287). Emerald Publishing Limited., DOI: 10.1108/S1569-37592023000110A014

Harari, Y. N. (2014). *Sapiens: A brief history of humankind*. Random House.

Haraway, D. (1988). Situated Knowledges: The Science Question in Feminism and the Privilege of Partial Perspective. *Feminist Studies*, *14*(3), 575–599. DOI: 10.2307/3178066

Harrer, S. (2023). Attention is not all you need: The complicated case of ethically using large language models in healthcare and medicine. *EBioMedicine*, *90*, 104512. DOI: 10.1016/j.ebiom.2023.104512

Hegel, G. (1967). *The Phenomenology of the Mind* (Baillie, J. B., Trans.). Harper Torchbooks.

Heidegger, M. (1959). *An Introduction to Metaphysics* (Manheim, R., Trans.). Yale University Press.

Heidegger, M. (1971). *Poetry, Language, Thought* (Hofstader, A., Trans.). Harper and Row.

Heidegger, M. (1977). *The Question Concerning Technology, and Other Essays*. Harper Torchbooks.

Heidegger, M. (1996). *Being and Time* (Stambaugh, J., Trans.). State University of New York Press.

Heidegger, M. (2000). *Elucidations of Holderlin's Poetry* (Hoeller, K., Trans.). Humanity Books.

Hew, K., & Cheung, W. (2014). Students' and Instructors' Use of Massive Open Online Courses (MOOCs): Motivations and Challenges. *Educational Research Review*, *12*, 45–58. DOI: 10.1016/j.edurev.2014.05.001

Hooda, M., Rana, C., Dahiya, O., Rizwan, A., & Hossain, M. S. (2022). Artificial intelligence for assessment and feedback to enhance student success in higher education. *Mathematical Problems in Engineering*, *2022*, 5215722. Advance online publication. DOI: 10.1155/2022/5215722

Huang, M., & Liu, T. F. (2016). Education Facing to Thinking of Heidegger and Its Theoretical Predicaments. *Educational Research*, *12*, 12–21.

Hung, R., Zhengmei, P., Kato, M., Nishihira, T., Okabe, M., Di, X., Kwak, D.-J., Hwang, K., Tschong, Y., Chien, C.-H., Peters, M. A., & Tesar, M. (2020). Philosophy of education in a new key: East Asia. *Educational Philosophy and Theory*. Advance online publication. DOI: 10.1080/00131857.2020.1772028

Ibrahim, H., Asim, R., Zaffar, F., Rahwan, T., & Zaki, Y. (2023). Rethinking homework in the age of artificial intelligence. *IEEE Intelligent Systems*, *38*(2), 24–27. DOI: 10.1109/MIS.2023.3255599

Ihde, D. (1990). *Technology and the Lifeworld*. Indiana University Press. DOI: 10.2979/3108.0

Ihde, D. (2003). A Phenomenology of Technics. In Kaplan, D. (Ed.), *Readings in the Philosophy of Technology* (pp. 137–160). Rowman & Littlefield Publisher.

Jackson, A. Y., & Mazzei, L. A. (2012). *Thinking with theory in qualitative research*. Routledge Press.

Jackson, L., Alston, K., Bialystok, L., Blum, L., Burbules, N. C., Chinnery, A., Hansen, D. T., Hytten, K., Mayo, C., Norris, T., Stitzlein, S. M., Thompson, W. C., Waks, L., Peters, M. A., & Tesar, M. (2020a). Philosophy of education in a new key: Snapshot 2020 from the United States and Canada. *Educational Philosophy and Theory*. Advance online publication. DOI: 10.1080/00131857.2020.1821189

Jackson, Z. I. (2020b). *Becoming Human: Matter and meaning in an anti-black world*. Duke University Press.

Jandric, P., Jaldemark, J., Hurley, Z., Bartram, B., Matthews, A., Jopling, M., Manero, J., MacKenzie, M., Irwin, J., Rothmüller, N., Green, B., Ralston, S. J., Pyyhtinen, O., Hayes, S., Wright, J., Peters, M. A., & Tesar, M. (2021, December 06). Philosophy of education in a new key: Who remembers Greta Thunberg? Education and environment after the coronavirus. *Educational Philosophy and Theory*, *53*(14), 1421–1441. Advance online publication. DOI: 10.1080/00131857.2020.1811678

Januszewski, A., & Persichitte, K. (2008). A History of the AECT's Definitions of Educational Technology. In Januszewski, A., & Molenda, M. (Eds.), *Educational Technology: A Definition with Commentary* (pp. 259–282). Routledge.

Kai, S., Almeda, M. V., Baker, R. S., Heffernan, C., & Heffernan, N. (2018). Decision tree modeling of wheel spinning and productive persistence in skill builders. *Journal of Educational Data Mining*, *10*(1), 789–822. DOI: 10.5281/zenodo.3344810

Kant, I. (1990). *Kant on Education*. Heath and Co.

Kato, M., Saito, N., Matsushita, R., Ueno, M., Izawa, S., Maruyama, Y., Sugita, H., Ono, F., Muroi, R., Miyazaki, Y., Yamana, J., Peters, M. A., & Tesar, M. (2020). Philosophy of education in a new key: Voices from Japan. *Educational Philosophy and Theory*. Advance online publication. DOI: 10.1080/00131857.2020.1802819

Kirkwood, A., & Price, L. (2014). Technology-Enhanced Learning and Teaching in Higher Education: What is 'Enhanced' and How Do We Know? A Critical Literature Review. *Learning, Media and Technology*, *39*(1), 6–36. DOI: 10.1080/17439884.2013.770404

Koh, E., & Doroudi, S. (2023). Learning, teaching, and assessment with generative artificial intelligence: Towards a plateau of productivity. *Learning: Research and Practice*, *9*(2), 109–116. DOI: 10.1080/23735082.2023.2264086

Ladson-Billings, G. (2021). I'm here for the hard re-set: Post pandemic pedagogy to preserve our culture. *Equity & Excellence in Education*, *54*(1), 68–78. DOI: 10.1080/10665684.2020.1863883

Ladyman, J. (2007). Ontological, Epistemological, and Methodological Positions. In Kuipers, T. A. F. (Ed.), *General Philosophy of Science* (pp. 303–376). North Holland. DOI: 10.1016/B978-044451548-3/50008-2

Latchem, C. (2014). BJET Editorial: Opening up the Educational Technology Research Agenda. *British Journal of Educational Technology*, *41*(1), 3–11. DOI: 10.1111/bjet.12122

Latour, B. (2004). Why Has Critique Run Out of Steam? From Matters of Fact to Matters of Concern. *Critical Inquiry*, *30*(2), 225–248. DOI: 10.1086/421123

Latour, B. (2005). *Reassembling the Social*. Oxford University Press. DOI: 10.1093/oso/9780199256044.001.0001

Lee, J., & Soylu, M. Y. (2023). *ChatGPT and assessment in higher education*. Center for 21st Century Universities. https://www.c21u.gatech.edu/sites/default/files/publication/2023/03/C21U%20ChatGPT%20White%20Paper_ Final.pdf

Lewin, D., & Lundie, D. (2016). Philosophies of digital pedagogy. *Studies in Philosophy and Education*, *3*(35), 235–240. DOI: 10.1007/s11217-016-9514-7

Lindgren, R., Tscholl, M., Wang, S., & Johnson, E. (2016). Enhancing Learning and Engagement Through Embodied Interaction Within a Mixed Reality Simulation. *Computers & Education*, *95*, 174–187. DOI: 10.1016/j.compedu.2016.01.001

Marker, M. (2018). There is no place of nature; there is only the nature of place: Animate landscapes as methodology for inquiry in the Coast Salish territory. *International Journal of Qualitative Studies in Education : QSE*, *31*(6), 453–464. DOI: 10.1080/09518398.2018.1430391

Martínez-Comesaña, M., Rigueira-Díaz, X., Larrañaga-Janeiro, A., Martínez-Torres, J., Ocarranza-Prado, I., & Kreibel, D. (2023). Impact of artificial intelligence on assessment methods in primary and secondary education: Systematic literature review. *Revista de Psicodidáctica* (English Ed.), 28(2), 93–103. DOI: 10.1016/j.psicoe.2023.06.002

Marx, K. (1964). *Economic and Philosophic Manuscripts of 1844*. International Publishers.

Marx, K. (1975). *Economic and Philosophical Manuscripts of 1844* (Livingstone, R., & Benton, G., Trans.). Penguin.

Mason, M. (2008). Complexity Theory and the Philosophy of Education. *Educational Philosophy and Theory*, *40*(1), 4–18. DOI: 10.1111/j.1469-5812.2007.00412.x

Mayes, T. (1995). Learning Technology and Groundhog day. In Strang, W., Simpson, V. B., & Slater, D. (Eds.), *Hypermedia at Work: Practice and Theory in Higher Education* (pp. 21–37). University of Kent Press.

Minocha, S., & Reeves, A. (2010). Design of Learning Spaces in 3D Virtual Worlds: An Empirical Investigation of Second Life. *Learning, Media and Technology*, *35*(2), 111–137. DOI: 10.1080/17439884.2010.494419

Mitcham, C. (2006). From Phenomenology to Pragmatism: Using Technology as an Instrument. In Selinger, E. (Ed.), *Postphenomenology: A Critical Companion to Ihde* (pp. 21–33). State University of New York Press.

Moqbel, M. S. S., & Al-Kadi, A. M. T. (2023). Foreign language learning assessment in the age of ChatGPT: A theoretical account. *Journal of English Studies in Arabia Felix*, *2*(1), 76–88. DOI: 10.56540/jesaf.v2i1.62

Murgatroyd, S. (2023). Rethinking teaching in the age of artificial intelligence. *Revista Paraguaya De Educación A Distancia* (REPED), 4(2), 4–10. DOI: 10.56152/reped2023-vol4num2-art1

Murray, M. D. (2023). Generative AI art: Copyright infringement and fair use. *SMU Science and Technology Law Review*, *26*(2), 259. DOI: 10.25172/smustlr.26.2.4

Nikolopoulou, K. (2024a). Generative artificial intelligence in higher education: Exploring ways of harnessing pedagogical practices with the assistance of ChatGPT. *International Journal of Changes in Education*, *1*(2), 103–111. DOI: 10.47852/bonviewIJCE42022489

Nikolopoulou, K. (2024b). Generative artificial intelligence in higher education: Exploring ways of harnessing pedagogical practices with the assistance of ChatGPT. *International Journal of Changes in Education*, *1*(2), 103–111. DOI: 10.47852/bonviewIJCE42022489

Oliver, M. (2011). Technological Determinism in Educational Technology Research: Some Alternative Ways of Thinking about the Relationship Between Learning and Technology. *Journal of Computer Assisted Learning*, *27*(5), 373–384. DOI: 10.1111/j.1365-2729.2011.00406.x

Oliver, M. (2013). Learning Technology: Theorising the Tools We Study. *British Journal of Educational Technology*, *44*(1), 31–43. DOI: 10.1111/j.1467-8535.2011.01283.x

Oliver, M. (2015). From Openness to Permeability: Reframing Open Education in Terms of Positive Liberty in the Enactment of Academic Practices. *Learning, Media and Technology*, *40*(3), 365–384. DOI: 10.1080/17439884.2015.1029940

Oliver, M., & Conole, G. (2003). Evidence-Based Practice and e-Learning in Higher Education: Can We and Should We? *Research Papers in Education*, *18*(4), 385–397. DOI: 10.1080/0267152032000176873

Orchard, J., Gaydon, P., Williams, K., Bennett, P., D'Olimpio, L., C¸elik, R., Shah, Q., Neusiedl, C., Suissa, J., Peters, M. A., & Tesar, M. (2020). Philosophy of education in a new key: A 'covid collective' of the philosophy of education society of Great Britain (PESGB). *Educational Philosophy and Theory*. Advance online publication. DOI: 10.1080/00131857.2020.1838274

Ouyang, F., Zheng, L., & Jiao, P. (2022). Artificial intelligence in online higher education: A systematic review of empirical research from 2011 to 2020. *Education and Information Technologies*, *27*(6), 7893–7925. DOI: 10.1007/s10639-022-10925-9

Papastephanou, M., Zembylas, M., Bostad, I., Benhur Oral, S., Drousioti, K., Kouppanou, A., Strand, T., Wain, K., & Peters, M. A. (2022, July 03). (Open Reviewer) & Tesar, M. (Open Reviewer) (2020). Philosophy of education in a new key: Education for justice now. *Educational Philosophy and Theory*, *54*(8), 1083–1098. Advance online publication. DOI: 10.1080/00131857.2020.1793539

Pearce, J., & Chiavaroli, N. (2023). Rethinking assessment in response to generative artificial intelligence. *Medical Education*, *57*(10), 889–891. DOI: 10.1111/medu.15092

Pedaste, M., Mäeots, M., Siiman, L., Jong, T., Riesen, S., Kamp, E., Manoli, C., Zacharia, Z., & Tsourlidaki, E. (2015). Phases of Inquiry-Based Learning: Definitions and the Inquiry Cycle. *Educational Research Review*, *14*(2), 47–61. DOI: 10.1016/j.edurev.2015.02.003

Perkins, M. (2023). Academic integrity considerations of AI large language models in the post-pandemic era: ChatGPT and beyond. *Journal of University Teaching & Learning Practice*, *20*(2), 345–367. DOI: 10.53761/1.20.02.07

Perrotta, C., & Evans, M. A. (2013). Orchestration, Power, and Educational Technology: A Response to Dillenbourg. *Computers & Education*, *69*, 520–522. DOI: 10.1016/j.compedu.2013.04.007

Peters, M. A., Arndt, S., Tesar, M., Jackson, L., Hung, R., Mika, C., Ozolins, J. T., Teschers, C., Orchard, J., Buchanan, R., Madjar, A., Novak, R., Besley, T., & Sturm, S. (2020). Philosophy of education in a new key. A collective project of the PESA executive. *Educational Philosophy and Theory*. Advance online publication. DOI: 10.1080/00131857.2020.1759194

Pezze, B. D. (2006). Heidegger on Gelassenheit. *Minerva*, *10*, •••. Retrieved March 15, 2020, from https://www.minerva.mic.ul.ie/vol10/Heidegger.html

Reich, K., Garrison, J., & Neubert, S. (2016). Complexity and Reductionism in Educational Philosophy—John Dewey's Critical Approach in 'Democracy and Education' Reconsidered. *Educational Philosophy and Theory*, *48*(10), 997–1012. DOI: 10.1080/00131857.2016.1150802

Reveley, J. (2013). Understanding Social Media Use as Alienation: A Review and Critique. *E-Learning and Digital Media*, *1*(10), 83–94. DOI: 10.2304/elea.2013.10.1.83

Roll, I., Aleven, V., McLaren, B. M., & Koedinger, K. R. (2011). Improving students' help-seeking skills using metacognitive feedback in an intelligent tutoring system. *Learning and Instruction*, *21*(2), 267–280. DOI: 10.1016/j.learninstruc.2010.07.004

Rosiek, J. L., Snyder, J., & Pratt, S. L. (2020). The new materialisms and Indigenous theories of non-human agency: Making the case for respectful anti-colonial engagement. *Qualitative Inquiry*, *26*(3–4), 331–346. DOI: 10.1177/1077800419830135

Roth, K., Mollvik, L., Alshoufani, R., Adami, R., Dineen, K., Majlesi, F., Peters, M. A., & Tesar, M. (2020). Philosophy of education in a new key: Constraints and possibilities in present times with regard to dignity. *Educational Philosophy and Theory*. Advance online publication. DOI: 10.1080/00131857.2020.1851189

Roy, A. (2020). The pandemic is a portal. Financial Times. Available at: https://www.ft.com/content/10d8f5e8-74eb-11ea-95fe-fcd274e920ca

Rudolph, J., Ismail, M. F. B. M., & Popenici, S. (2024). Higher education's generative artificial intelligence paradox: The meaning of chatbot mania. *Journal of University Teaching & Learning Practice*, *21*(6), 1–35. DOI: 10.53761/54fs5e77

Salinas-Navarro, D. E., Vilalta-Perdomo, E., Michel-Villarreal, R., & Montesinos, L. (2024). Using generative artificial intelligence tools to explain and enhance experiential learning for authentic assessment. *Education Sciences*, *14*(1), 887–900. DOI: 10.3390/educsci14010083

Sardoc, M., Coady, C. A. J., Bufacchi, V., Moghaddam, F. M., Cassam, Q., Silva, D., Miscevic, N., Andrejc, G., Kodelja, Z., Vezjak, B., Peters, M. A., & Tesar, M. (2020). Philosophy of education in a new key: On radicalization and violent extremism. *Educational Philosophy and Theory*. Advance online publication. DOI: 10.1080/00131857.2020.1861937

Scribes, S. (2024). A discourse on generative AI. *Medium*. https://medium.com/@sundryscribes/a-discourse-on-generative-ai-4ca4a8911e1a

Selwyn, N. (2010). Looking Beyond Learning: Notes Towards the Critical Study of Educational Technology. *Journal of Computer Assisted Learning*, *26*(1), 65–73. DOI: 10.1111/j.1365-2729.2009.00338.x

Selwyn, N. (2014). Education and 'the Digital'. *British Journal of Sociology of Education*, *35*(1), 155–164. DOI: 10.1080/01425692.2013.856668

Selwyn, N. (2022). The future of AI and education: Some cautionary notes. *European Journal of Education*, *57*(4), 620–631. DOI: 10.1111/ejed.12532

Sergio, M., Gabriel, D., Sancristobal, E., Rosario, G., Manuel, C., & Juan, P. (2011). New Technology Trends in Education: Seven Years of Forecasts and Convergence. *Computers & Education*, *57*(3), 1893–1906. DOI: 10.1016/j.compedu.2011.04.003

Siegel, H. (1983). On the obligations of the professional philosopher of education. *Journal of Thought*, *18*(2), 31–37.

Smith, J. (2022). International education trends: Shaping the future of higher education. *International Journal of Open Publication and Exploration*, *10*(1), 1–5.

Surry, D., & Baker, F.III. (2016). The Co-Dependent Relationship of Technology and Communities. *British Journal of Educational Technology*, *47*(1), 13–28. DOI: 10.1111/bjet.12349

Tesar, M. (2021). Philosophy as a method: Tracing the histories of intersections of 'philosophy', 'methodology' and 'education. *Qualitative Inquiry*, *27*(5), 544–553. DOI: 10.1177/1077800420934144

Thomson, I. (2004). Heidegger's Perfectionist Philosophy of Education in Being and Time. *Continental Philosophy Review*, *7*(4), 439–467. DOI: 10.1007/s11007-005-6886-8

Thumlert, K., Castell, S., & Jenson, J. (2015). Short Cuts and Extended Techniques: Rethinking Relations Between Technology and Educational Theory. *Educational Philosophy and Theory*, *47*(8), 786–803. DOI: 10.1080/00131857.2014.901163

Trajtenberg, M. (2018). *AI as the next GPT: A political-economy perspective*. National Bureau of Economic Research., DOI: 10.3386/w24245

van der Vorst, T., & Jelicic, N. (2019). Artificial intelligence in education: Can AI bring the full potential of personalized learning to education? https://www.econstor.eu/handle/10419/205222

Varaki, B. S., Sadeqzadeh Qamsari, A., Sefidkhosh, M., Mahdi Sajjadi, S., Mohammadi Chaboki, R., Javidi Kalatehjafarabadi, T., Saffarheidari, H., Mohammadamini, M., Karimzadeh, O., Barkhordari, M., Zarghami-Hamrah, S., Peters, M. A., & Tesar, M. (2021). Philosophy of education in a new key: Reflection on higher education in Iran. *Educational Philosophy and Theory*. Advance online publication. DOI: 10.1080/00131857.2021.1905517

Verbeek, P. (2008). Cyborg Intentionality: Rethinking the Phenomenology of Human–Technology Relations. *Phenomenology and the Cognitive Sciences*, *7*(3), 387–395. DOI: 10.1007/s11097-008-9099-x

Waghid, Y., Davids, N., Mathebula, T., Terblanche, J., Higgs, P., Shawa, L., Herbert Manthalu, C., Waghid, Z., Ngwenya, C., Divala, J., Waghid, F., Peters, M. A., & Tesar, M. (2020). Philosophy of education in a new key: Cultivating a living philosophy of education to overcome coloniality and violence in African Universities. *Educational Philosophy and Theory*. Advance online publication. DOI: 10.1080/00131857.2020.1793714

Webster, S. (2009). *Educating for meaningful lives through existential spirituality*. Sense Publishers. DOI: 10.1163/9789460910043

Weiss, D. (2008). Human—Technology—World. *Techné (Blacksburg, Va.)*, *12*(2), 110–119. DOI: 10.5840/techne20081228

Whitehead, A. N. (1929). *Process and Reality: An Essay in Cosmology*. Macmillan.

Willet, R. (2018). Learning Through Making in Public Libraries: Theories, Practices, and Tensions. *Learning, Media and Technology*, *43*(3), 250–262. DOI: 10.1080/17439884.2017.1369107

Wrathall, A. (2011). We Still Need Education in Thinking: Heidegger's Education in a Technological Age. *Journal of Educational Studies*, *7*(1), 3–14.

Wu, G. S. (2009). *Lectures on the Philosophy of Technology*. China Renmin University Press.

Zhao, Y., Pugh, K., Sheldon, S., & Byers, J. (2002). Conditions for Classroom Technology Innovations. *Teachers College Record*, *104*(3), 482–515. DOI: 10.1111/1467-9620.00170

Zimmerman, J., Forlizzi, J., & Evenson, S. (2007). "Research Through Design as a Method for Interaction Design Research in HCI." *Proceedings of the SIGCHI Conference on Human Factors in Computing Systems*. DOI: 10.1145/1240624.1240704

Zootzky, G., & Pfeiffer, A. (2024). Educational transformation through AI: Preparing for a new era of learning. *INTED2024 Proceedings,* 1202–1207. DOI: 10.21125/inted.2024.0382

KEY TERMS AND DEFINITIONS

Adaptive Learning: A personalized approach to education that adjusts the curriculum and resources based on students' individual learning needs, preferences, and performance. It emphasizes flexibility and responsiveness, particularly useful in crisis situations where traditional methods may falter.

Crisis Literacy: The skills and knowledge required to navigate and assess information and situations during a crisis. In education, this involves teaching students to critically analyze sources of information, understand the dynamics of crises, and respond effectively to misinformation and uncertainty.

Holistic Education: An approach that focuses on the development of the whole person, including emotional, social, physical, and intellectual growth. During times of crisis, holistic education emphasizes mental health, well-being, and community connections, advocating for a more comprehensive understanding of what it means to educate effectively.

Resilience: The capacity to recover quickly from difficulties and adapt effectively to change or adversity. In the educational context, resilience refers to both students' and educational institutions' ability to cope with and flourish despite disruptions such as pandemics or social upheaval.

Transformative Pedagogy: An educational approach that seeks to stimulate critical thinking and empower learners to challenge existing assumptions and systems. This method encourages students to engage deeply with content, fostering a sense of agency and promoting societal change, especially in challenging times.

Chapter 3
The Paradox of Education:
Challenges, Changes, and Hope for the Future

Annette G. Walters
http://orcid.org/0009-0007-6068-8467
Washington County School District, USA

ABSTRACT

Education is currently undergoing a significant transformation, characterized by various challenges including the integration of technology, resistance to change within educational institutions, declining enrollment, and the pressing need for community-focused strategies. This paper emphasizes the transformative potential of technology in educational practices while also acknowledging the risks it poses in widening existing disparities. Innovations such as adaptive learning platforms have the capacity to boost student engagement, yet they may inadvertently deepen inequalities among learners. Furthermore, the discussion extends to the role of service-learning and career technical education as a means to cultivate civic responsibility and tackle real-world challenges, despite facing obstacles related to institutional support and funding. The paper also addresses the implications of declining enrollment across educational institutions, submitting that by embracing community engagement and innovative teaching methodologies, institutions may enhance their relevance and support.

DOI: 10.4018/979-8-3373-1350-4.ch003

INTRODUCTION

Education is currently navigating a pivotal moment, facing a range of enduring challenges alongside evolving commitments that demand attention. The integration of technology into educational frameworks, the reluctance of some to embrace change, fluctuating enrollment figures, and the dynamic nature of community relationships are all factors that are significantly altering the educational landscape (Selwyn, 2016; West et al., 2025). While these issues may appear overwhelming, they also present unique opportunities for innovation and revitalization within the sector (Derreth et al., 2024).

By critically analyzing the influence of technological progress, the importance of community involvement, and the potential of service-learning initiatives, insights can be gained into how educational institutions might adapt to better serve the needs of future learners and communities (Donelan & Kear, 2024; Selwyn, 2016). The impact of technology on education cannot be overstated, as it has the potential to enhance learning experiences and broaden access to knowledge (Donelan & Kear, 2024).

However, the successful integration of these technological tools often encounters resistance from educators who may be hesitant to alter traditional teaching methods. This resistance can hinder the adoption of innovative practices that could otherwise enrich the educational experience. Furthermore, fluctuating enrollment rates pose a significant challenge, as institutions must find ways to attract and retain students in an increasingly competitive environment. Addressing these issues requires a concerted effort to foster a culture of adaptability and openness to change among educators, ensuring that they are equipped to leverage technology effectively (Selwyn, 2016).

In addition to technological advancements, the role of community partnerships is crucial in shaping the future of education. Engaging with local communities can provide valuable resources, support, and real-world contexts for learning, thereby enhancing the educational experience for students (West et al., 2025). Service-learning initiatives, which connect academic coursework with community service, exemplify how education can be enriched through active participation and collaboration. By prioritizing community engagement and fostering a culture of involvement, educational institutions can not only address long-standing challenges but also cultivate a sense of purpose and connection among students (Cattaneo, 2021).

This paper delves into these themes, highlighting the necessity of embracing change and collaboration to create a more responsive and effective educational system for generations to come (Christensen et al., 2011). These initiatives foster a sense of civic responsibility among students while also allowing them to apply theoretical knowledge in practical settings, bridging the gap between classroom learning and real-world application. By actively involving community members in the educational process, institutions can create a more relevant and impactful learning environment that resonates with students' lives and aspirations. This symbiotic relationship not only benefits students but also strengthens community ties, creating a network of support that can enhance educational outcomes (West et al., 2025).

The evolving nature of community relationships necessitates that educational institutions remain responsive to the needs and expectations of their surrounding environments. As demographics shift and societal challenges emerge, schools and universities must adapt their curricula and outreach efforts to reflect these changes. This responsiveness can manifest in various ways, such as developing programs that address local workforce needs or incorporating diverse perspectives into the curriculum. Through prioritizing community engagement, educational institutions can cultivate a sense of belonging and relevance, which is essential for student retention and success (Rios & Neas, 2024).

While education is indeed at a crossroads, the challenges it faces can serve as catalysts for transformative change. Embracing technological advancements, fostering community partnerships, and implementing service-learning initiatives, educational institutions can not only navigate current obstacles but also position themselves as leaders in innovation (West et al., 2025). This proactive approach will not only enhance the educational experience for students but also ensure that institutions remain vital contributors to the communities they serve. As we look to the future, it is imperative that educators, administrators, and policymakers work collaboratively to harness these opportunities, ultimately creating a more inclusive and effective educational landscape for all learners.

THE IMPACT OF TECHNOLOGY ON EDUCATION

Technology holds the promise of transforming the educational landscape, yet its integration has faced numerous obstacles. As noted by Christensen, Horn, and Johnson (2011), innovations such as adaptive learning platforms and online courses have significantly altered the ways in which students interact with educational content (Selwyn, 2016). These disruptive technologies have the potential to enhance engagement and personalize learning experiences. The challenge lies in incorporating these advancements into established educational frameworks, where resistance from educators can hinder progress. Many teachers may feel apprehensive about adopting unfamiliar methods, fearing that they may disrupt traditional teaching practices or diminish the quality of education.

However, the evolution of experiential learning, as emphasized by Kolb (1984), necessitates the incorporation of digital tools to remain pertinent in a world increasingly dominated by technology. This shift is essential not only for enhancing student engagement but also for ensuring that educational practices evolve alongside technological advancements. By integrating digital resources into experiential learning, educators can create more dynamic and interactive learning environments that cater to diverse learning styles. However, this transition requires a commitment to professional development and support for educators to effectively utilize these new tools in their teaching.

Despite the potential benefits of technology in education, its widespread adoption can inadvertently deepen existing inequalities. Selwyn (2016) highlights the issue of the digital divide, which reveals significant disparities in access to technology across different regions and socioeconomic backgrounds. This uneven access complicates efforts to establish equitable learning environments, as students from underprivileged backgrounds may lack the necessary resources to fully benefit from technological advancements. While technology can facilitate personalized learning experiences, it is crucial to address these disparities to prevent the reinforcement of existing inequalities and ensure that all students have the opportunity to thrive in a technology-enhanced educational landscape (Donelan & Kear, 2024; Selwyn, 2016).

Without addressing the digital divide, the promise of technology in education risks becoming a privilege reserved for those who already have advantages, thereby perpetuating systemic inequalities. Schools in affluent areas may have the latest devices and high-speed internet, while those in less affluent regions struggle with outdated equipment and limited connectivity. This disparity not only affects students' ability to engage with digital learning tools but also impacts their overall educational outcomes, creating a cycle of disadvantage that is difficult to break (Donelan & Kear, 2024).

Furthermore, the integration of technology into education must be approached with a critical lens, considering the implications of data privacy and security. As educational institutions increasingly rely on digital platforms, the collection and management of student data become paramount concerns. Educators and administrators must navigate the complexities of ensuring that student information is protected while still leveraging data to enhance learning experiences. This balancing act requires clear policies and guidelines, as well as ongoing training for educators to understand the ethical dimensions of technology use in the classroom (Donelan & Kear, 2024).

While technology has the potential to revolutionize education by fostering personalized and engaging learning experiences, its successful integration is fraught with challenges. Addressing resistance from educators, ensuring equitable access, and safeguarding student data are critical steps that must be taken to harness the full benefits of technological advancements (Donelan & Kear, 2024). By prioritizing these areas, educational institutions can create a more inclusive and effective learning environment that prepares students for a future where technology plays an integral role in their lives. Ultimately, the goal should be to leverage technology not just as a tool for enhancement, but as a means to create a more equitable and responsive educational system that meets the diverse needs of all learners (Derreth et al., 2024).

COMMUNITY ENGAGEMENT AND SERVICE-LEARNING

A significant aspect of modern education is the increasing emphasis on community engagement through service-learning initiatives (Smith, 2022). This educational strategy prioritizes experiential learning, encouraging students to work alongside local communities to tackle pressing societal issues. By participating in service-learning, students not only enhance their academic knowledge but also cultivate a profound sense of civic duty, effectively bridging their studies with impactful community involvement (Smith, 2022; Vogelgesang & Astin, 2000). This dual benefit of academic enrichment and social responsibility is particularly vital in today's interconnected world, where students are not only learners but also potential agents of change. Engaging with real-world challenges allows students to apply theoretical concepts in practical settings, thereby deepening their understanding and fostering a sense of empathy towards diverse populations.

Service learning and career technical education (CTE) share several key attributes that enhance the educational experience for students (Douglas & Slate, 2025). Both methodologies prioritize real-world learning, enabling students to apply theoretical knowledge to practical situations, thereby fostering the acquisition of essential skills through hands-on projects and tasks (Gottfried & Bozick, 2019). This experiential approach encourages active participation, allowing students to engage directly with their learning environment rather than relying solely on traditional lectures or textbooks (Brown & Souto-Otero, 2020; Bringle et al., 2009). Furthermore, both frameworks are instrumental in cultivating vital competencies such as problem-solving, teamwork, communication, leadership, and critical thinking, which are crucial for success in any career.

The landscape of education is undergoing profound transformations driven by technological advancements. Innovations such as digital tools, artificial intelligence, and virtual reality are reshaping both teaching methodologies and workforce training programs. These technologies enable Career and Technical Education (CTE) programs to replicate real-world work environments, providing students with the chance to hone intricate skills in safe, controlled settings (Douglas & Slate, 2025). For instance, immersive technologies like virtual reality significantly enhance training in fields such as manufacturing and technical disciplines by allowing students to engage with simulated equipment and scenarios, thereby bridging the gap between theoretical knowledge and practical application. This hands-on approach not only boosts student engagement but also prepares them for the complexities of modern workplaces, where adaptability and technical proficiency are paramount (Brown & Souto-Otero, 2020).

However, despite advancements, challenges persist that threaten the effectiveness of CTE programs. Many educational institutions struggle with insufficient funding, inadequate infrastructure, and a shortage of qualified instructors, which hampers the delivery of high-quality CTE programs. These systemic issues can lead to disparities in educational quality, particularly in underserved communities where resources are already limited. Additionally, certain demographic groups continue to be underrepresented in technical fields, underscoring the necessity for research that explores CTE pathways in higher education (Smith, 2022; West et al., 2025). This research is crucial for advocating for broader access and equitable participation among diverse populations, ensuring that all students have the opportunity to benefit from the advancements in CTE (Wright & Lee, 2018).

Nonetheless, there is considerable optimism regarding the future of education, as CTE embodies a model that aligns academic learning with the demands of the workforce. By fostering experiential learning, CTE strengthens the ties between educational institutions, industries, and communities, creating a collaborative ecosystem that benefits all stakeholders. This model not only prepares students for immediate employment but also instills a lifelong learning mindset, essential in a rapidly evolving job market. By merging academic rigor with the development of technical skills, CTE equips students with essential competencies such as critical thinking, problem-solving, collaboration, and adaptability skills that are increasingly vital in today's job market (Smith, 2022).

Ultimately, the challenge of education lies in harmonizing traditional academic values with the need for practical workforce readiness. CTE offers a viable solution to this dilemma by establishing an educational framework that prioritizes both theoretical understanding and practical application. This dual focus not only enhances student employability but also contributes to a more skilled workforce, capable of meeting the challenges of an ever-changing economic landscape (Wright & Lee, 2018). As educational institutions continue to adapt to these changes, the integration of technology and innovative teaching practices will be key to unlocking the full potential of CTE, ensuring that it remains a cornerstone of modern education (Douglas & Slate, 2025).

By collaborating with local organizations, businesses, and community groups, students gain insights into community needs and professional settings, thereby bridging the gap between education and the workforce (Kezar & Rhoads, 2001; Smith, 2022). Ultimately, these approaches not only prepare students for future careers by equipping them with practical experience and workplace skills but also promote personal growth, instilling a sense of responsibility, confidence, and civic awareness through meaningful engagement in real-world projects. Service learning offers significant advantages by fostering a reciprocal relationship between students and the community. This educational approach not only enhances the learning experience for students but also addresses real-world needs within the community (Wright & Lee, 2018).

Through active participation in service projects, students gain practical skills and insights that complement their academic knowledge, while simultaneously contributing to the welfare of the community. This collaboration encourages civic responsibility and social awareness, as students engage with diverse populations and understand the challenges faced by their communities. Ultimately, service learning cultivates a sense of belonging and mutual support, creating a more cohesive and resilient community while empowering students to become informed and active citizens (Smith, 2022).

The path to successful community engagement is fraught with obstacles. Mann and Bowen (2021) highlight that inadequate funding and insufficient institutional backing can severely limit the sustainability and effectiveness of these programs over time. Without adequate resources, educational institutions may struggle to maintain meaningful partnerships with community organizations, which are essential for the success of service-learning initiatives. The lack of institutional support can lead to a diminished commitment from faculty and students alike, undermining the potential impact of these programs.

Additionally, Cattaneo (2021) points out that while service-learning can promote critical thinking and foster active citizenship, its success largely depends on the strength of the partnerships formed between educational institutions and community organizations. Effective collaboration requires ongoing communication, mutual respect, and a shared vision, which can be challenging to establish and maintain (Mann & Bowen, 2021).

Bringle and Hatcher (2009) stress the importance of ensuring that these collaborations are advantageous for both students and the community, as the challenge remains in nurturing these relationships amid evolving priorities and constrained resources. It is crucial for educational institutions to adopt a holistic approach that not only focuses on the academic benefits for students but also addresses the needs and aspirations of the communities they serve (Smith, 2022).

This requires a commitment to long-term engagement rather than one-off projects, fostering a culture of reciprocity where both parties can learn from and support each other. By prioritizing these collaborative efforts, educational institutions can create a more sustainable model for service-learning that not only enriches the student experience but also contributes meaningfully to community development (Bringle & Hatcher, 2009; Cattaneo, 2021). Ultimately, the success of service-learning initiatives hinges on the ability to navigate these complexities, ensuring that both students and communities thrive in the process.

STAGNANCY AND RESISTANCE TO CHANGE

One of the most significant obstacles to reform in education is the resistance to change from educators. As highlighted by Hartley (2011), many faculty members cling to traditional teaching methods, demonstrating a hesitance to adopt innovative pedagogical strategies or new technologies. This resistance can be attributed to several factors, including insufficient opportunities for professional development, anxiety about their proficiency with unfamiliar tools, and a strong emotional connection to conventional teaching practices. Educators often find comfort in established routines, which can create a significant barrier to the adoption of new approaches that may initially seem daunting or untested. This reluctance not only stifles individual growth but also limits the potential for transformative educational experiences for students.

To effectively tackle this issue, Kezar and Rhoads (2001) advocate for the establishment of a culture of innovation within higher education institutions. Such a culture would encourage faculty members to explore and implement new teaching techniques without the fear of failure. By fostering an environment that values experimentation and supports risk-taking, institutions can empower educators to step outside their comfort zones. This can be achieved through targeted professional development programs that provide educators with the necessary resources and training to navigate new technologies and pedagogical strategies. Furthermore, creating collaborative spaces where faculty can share their experiences and learn from one another can significantly enhance their confidence and willingness to innovate.

Additionally, as Furco (2001) points out, service-learning offers a compelling alternative teaching model that can effectively connect theoretical knowledge with practical application. This approach enables educators to incorporate meaningful, real-world experiences into their curricula, thereby enriching the learning process. By engaging students in community-based projects, educators can foster a deeper understanding of course material while simultaneously addressing societal needs. This not only enhances student engagement but also cultivates a sense of civic responsibility among learners. Ultimately, by embracing innovative teaching methods like service-learning and fostering a supportive culture of change, educational institutions can dismantle the barriers that hinder the transition to more engaging, student-centered learning experiences.

DECLINING ENROLLMENT AND THE HOPE FOR THE FUTURE

In recent years, numerous colleges and universities have experienced a notable decrease in enrollment, influenced by a confluence of factors including demographic shifts, escalating tuition fees, and growing skepticism regarding the value of a college degree. This decline is particularly pronounced among certain demographics, such as first-generation college students and those from lower-income backgrounds, who may find the financial burden of higher education increasingly daunting.

As highlighted by Vogelgesang and Astin (2000), this trend has prompted educational institutions to reevaluate their strategies for attracting and retaining students. In light of these challenges, many colleges and universities are actively seeking innovative ways to enhance accessibility and relevance in education. This includes initiatives aimed at cultivating a sense of belonging and encouraging community involvement among their student bodies, which are essential for fostering a supportive and engaging educational environment.

The recent drop in enrollment should not be perceived merely as a challenge; instead, it offers a significant opportunity for the higher education sector to reassess and redefine its core mission and objectives. This moment of reflection is crucial, as it allows institutions to consider how they can better serve their students and the communities in which they operate. According to Saltmarsh and Hartley (2011), the future trajectory of higher education is closely tied to its ability to embrace a more expansive democratic role.

This role transcends traditional educational boundaries, emphasizing the importance of engaging students not only in their academic pursuits but also in fostering meaningful connections with the surrounding communities. Such engagement extends well beyond the confines of traditional classroom environments, encouraging students to apply their knowledge and skills in real-world contexts.

By adopting this broader perspective, educational institutions have the potential to reshape the learning experience, making it not only intellectually stimulating but also socially relevant and impactful. This transformation can lead to a more holistic approach to education, where the benefits are felt both within the academic sphere and in the wider society. For instance, partnerships with local organizations can provide students with hands-on experiences that enhance their learning while simultaneously addressing community needs. This symbiotic relationship not only enriches the educational experience but also empowers students to become active, informed citizens who contribute positively to society.

Ultimately, this shift in focus can create a more inclusive and dynamic educational environment, where diverse voices and perspectives are valued. As institutions strive to redefine their missions, they can cultivate a culture of collaboration and innovation that inspires students to think critically and act compassionately. In doing so, they not only enhance the academic experience but also foster a sense of belonging and purpose among students, enriching the lives of both students and community members alike. By viewing the current enrollment challenges as a catalyst for change, the higher education sector can emerge stronger, more relevant, and better equipped to meet the needs of future generations.

One effective approach to achieving this transformation is through the implementation of community-oriented initiatives, particularly service-learning programs. These programs allow students to apply their academic knowledge in real-world contexts, thereby enhancing their learning while simultaneously addressing community needs. By involving students in such initiatives, institutions can significantly enhance the relevance of education in the contemporary landscape.

This revitalizes the purpose and impact of higher education in the 21st century while also fostering a generation of graduates who are not only well-prepared for their careers but also committed to contributing positively to society. In this way, the current enrollment challenges can serve as a catalyst for meaningful change, ultimately leading to a more engaged, informed, and socially responsible student body.

A PATH FORWARD

The landscape of higher education is currently navigating a host of significant challenges, including the integration of technology, outdated teaching methodologies, dwindling enrollment figures, and financial constraints. These issues have created a complex environment where traditional models of education are being questioned and re-evaluated. For instance, the rapid advancement of technology has outpaced many institutions' ability to adapt, leaving educators and students alike struggling to keep up with new tools and platforms that could enhance learning experiences.

Additionally, the reliance on conventional teaching methods often fails to engage a diverse student body, leading to a decline in enrollment as prospective students seek more relevant and dynamic educational opportunities. Financial constraints further exacerbate these challenges, as institutions grapple with budget cuts and the rising costs of education, making it increasingly difficult to provide quality resources and support for students.

Yet, amidst these obstacles, there is a glimmer of optimism for the future. By prioritizing community involvement, promoting service-learning initiatives, and adopting technological advancements that facilitate personalized education, institutions can adapt to the demands of an ever-evolving world. Community engagement not only enriches the educational experience but also fosters a sense of belonging and purpose among students, encouraging them to apply their learning in real-world contexts (Brown & Souto-Otero, 2020).

Service-learning initiatives bridge the gap between academic theory and practical application, allowing students to develop critical skills while contributing positively to their communities (Butin, 2010; Malkus, 2020). Furthermore, the integration of technology can transform the educational landscape by offering tailored learning experiences that cater to individual student needs, thereby enhancing engagement and retention (Derreth et al., 2024; Hoachlander & Hamilton, 2020).

This necessary transformation hinges not only on the backing of educational institutions but also on the dedication of educators to embrace change and foster innovation in their teaching practices. Educators play a pivotal role in this evolution, as their willingness to adopt new methodologies and technologies can significantly impact student outcomes. Professional development opportunities that focus on innovative teaching strategies and the effective use of technology can empower educators to create more engaging and relevant learning environments (Rios & Neas, 2024). By cultivating a culture of continuous improvement and collaboration, institutions can ensure that their faculty are equipped to meet the challenges of modern education head-on.

Moving forward, it is essential to unite efforts in tackling these challenges while seizing the opportunities. Collaboration among educators, administrators, and community stakeholders can lead to the development of comprehensive strategies that address the multifaceted issues facing higher education today. By working collaboratively, these groups can create a more resilient and responsive educational system that not only meets the needs of students but also anticipates future demands (Hartley, 2011).

The true potential of education lies in its ability to maintain strong connections with the communities it serves, ensuring that students are equipped for a productive life behind the educational walls with the intentional support of educators, policymakers, and the broader community (Christensen et al., 2011). It requires a collective commitment to reimagining the educational experience in ways that are both innovative and inclusive.

Educators must embrace professional development opportunities that equip them with the skills to effectively integrate new technologies into their teaching practices. This shift not only enhances their pedagogical approaches but also empowers them to create more interactive and engaging learning environments (Aspelund & Reed, 2019). Policymakers play a crucial role in this transformation by advocating for funding and resources that support innovative educational programs and initiatives. By prioritizing investments in education, they can help alleviate some of the financial pressures that institutions face, allowing them to focus on enhancing the quality of education.

Moreover, the involvement of the community is essential in shaping a responsive educational framework. Local businesses, nonprofits, and civic organizations can collaborate with educational institutions to create internship opportunities, mentorship programs, and real-world projects that enrich the curriculum (Doubeni et al., 2022). Such partnerships not only provide students with valuable experiences but also ensure that the education they receive is relevant to the needs of the local economy. This symbiotic relationship fosters a culture of continuous improvement and adaptation, where feedback from the community informs curriculum development and teaching strategies (Fuller & Unwin, 2017).

In conclusion, while the challenges facing higher education are formidable, they also present an opportunity for meaningful change (Hoachlander & Hamilton, 2020). By embracing a holistic approach that prioritizes community engagement, innovative teaching methodologies, and the strategic use of technology, educational institutions can not only survive but thrive in this dynamic landscape (Fuller & Unwin, 2017). The path forward requires a shared vision and collaborative effort among all stakeholders, ensuring that higher education remains a vital and transformative force in society. As we navigate this complex terrain, the potential for a more inclusive, relevant, and effective educational experience is within reach, promising a brighter future for students (Castello et al., 2017; Rios & Neas, 2024).

REFERENCES

Aspelund, K., & Reed, P. (2019). Career and technical education and workforce readiness. *Journal of Career and Technical Education, 34*(1), 45–60.

Bringle, R. G., Clayton, P. H., & Price, M. F. (2009). Partnerships in service learning and civic engagement. Partnerships: *A Journal of Service-Learning and Civic Engagement, 1*(1), 1–20.

Bringle, R. G., & Hatcher, J. A. (2009). Innovative practices in service-learning and civic engagement. *New Directions for Higher Education, 2009*(147), 37–46. DOI: 10.1002/he.356

Brown, P., & Souto-Otero, M. (2020). The misalignment between education and employment. *Oxford Review of Education, 46*(4), 425–444. DOI: 10.1080/03054985.2019.1697397

Cattaneo, L. (2021). Community engaged learning and student civic development. *Frontiers in Education, 6*, 626984. Advance online publication. DOI: 10.3389/feduc.2021.626984

Christensen, C. M., Horn, M. B., & Johnson, C. W. (2011). *Disrupting class: How disruptive innovation will change the way the world learns*. McGraw-Hill.

Derreth, R. T., Nelson, K. E., Nguy n, C. H., Adams, A. W., & Levin, M. B. (2024). Community partner experiences in a service-learning development program. *Journal of Higher Education Outreach & Engagement, 28*(2), 45–63.

Donelan, H., & Kear, K. (2024). *Online group projects in higher education: Persistent challenges and implications for practice*. Technology, Knowledge and Learning.

Doubeni, C. A., Adams, S. A., Alcaraz, K. I., Artiga, S., Bailey, S., Carney, J., Daniels, R., DeVoe, J., Dulin, M., Fuchs, S., George, S., Gennarelli, R., Geller, S., Goodman, M., Hamm, M., Harvey, H., Jatlaoui, T., Jones, S., Khan, M., & Zgierska, A. (2022). Community engagement education in academic health centers, colleges, and universities. *Journal of Clinical and Translational Science, 6*(1), e192. DOI: 10.1017/cts.2022.424

Douglas, E. S., & Slate, J. R. (2025). A post-pandemic analysis of career and technical education performance by student gender. *American Journal of STEM Education, 18*, 15–50. DOI: 10.32674/jk1c1x22

Fuller, A., & Unwin, L. (2017). Apprenticeship and the future of vocational education. *Journal of Vocational Education and Training*, *69*(1), 1–5. DOI: 10.1080/13636820.2016.1274362

Furco, A. (2001). Service-learning: A balanced approach to experiential education. *New Directions for Higher Education*, *2001*(114), 25–39.

Gottfried, M. A., & Bozick, R. (2019). Supporting the transition to college and careers through career technical education. *Educational Evaluation and Policy Analysis*, *41*(2), 191–213. DOI: 10.3102/0162373719832237

Hartley, M. (2011). Reclaiming democratic purposes in higher education. *Educational Theory*, *61*(4), 391–406.

Hoachlander, G., & Hamilton, S. (2020). Career pathways and workforce development in modern education. *Education Policy Analysis Archives*, *28*(75).

Malkus, N. (2020). *The evolution of career and technical education in U.S. high schools*. American Enterprise Institute.

Mann, J., & Bowen, G. (2021). Emphasizing the impact of service-learning and community engagement partnerships amid challenges facing higher education. *The International Journal of Research on Service-Learning and Community Engagement*, *9*(1). Advance online publication. DOI: 10.37333/001c.31379

Rios, M., & Neas, S. (2024). Community-engaged learning in graduate education: A review and call to action. *The International Journal of Research on Service-Learning and Community Engagement*, *12*(1), 7. Advance online publication. DOI: 10.37333/001c.127560

Saltmarsh, J., & Hartley, M. (2011). *To serve a larger purpose: Engagement for democracy and the transformation of higher education*. Temple University Press.

Selwyn, N. (2016). *Education and technology: Key issues and debates*. Bloomsbury Academic.

Smith, S. (2022). Promoting civic engagement through service learning. *Frontiers of Educational Research*, *4*(1), 27–33. DOI: 10.26549/fer.v4i1.3798

Vogelgesang, L. J., & Astin, A. W. (2000). Comparing the effects of community service and service-learning. *Michigan Journal of Community Service Learning*, *7*(1), 25–34. https://quod.lib.umich.edu/m/mjcsl

West, A., Chute, D. E., Daniels, J., & Bower, K. M. (2025). Community-engaged research in early home visiting: A scoping review of peer-reviewed literature. *Prevention Science : The Official Journal of the Society for Prevention Research, 26*(5), 703–715. DOI: 10.1007/s11121-025-01812-z

Wright, S., & Lee, M. (2018). Experiential learning and technical education in secondary schools. *The Journal of Educational Research, 111*(5), 561–570. DOI: 10.1080/00220671.2018.1451907

Chapter 4
Innovative Educational Leadership:
Steering Institutions through Times of Uncertainty

Mustafa Kayyali
http://orcid.org/0000-0003-3300-262X
Maaref University of Applied Sciences, Syria

ABSTRACT

Periods of uncertainty and crisis have historically tested the boundaries of educational systems—but they have also revealed remarkable opportunities for transformation. This chapter explores the evolving nature of educational leadership in times marked by disruption, unpredictability, and rapid change. It delves into the mindset and approaches adopted by innovative leaders who have dared to rethink tradition, reframe uncertainty as possibility, and guide their institutions not only to survive but to thrive. Drawing on real-world examples, emotional insights, and forward-thinking strategies, the chapter underscores the essential qualities of leadership that resonate in turbulent times—adaptability, empathy, creativity, and vision. Ultimately, the chapter makes a compelling case for cultivating leadership that is not only reactive but regenerative—leadership that redefines what it means to guide educational communities through the unknown with clarity, care, and courage.

INTRODUCTION

In the quiet before a crisis, leadership often feels routine—meetings follow agendas, decisions are predictable, and plans are mapped out with a sense of controlled progression. But when uncertainty erupts, whether through a global pandemic, political

DOI: 10.4018/979-8-3373-1350-4.ch004

upheaval, economic disruption, war, or ecological disaster, that calm predictability evaporates (D'Auria & De Smet, 2020). Schools, colleges, and universities—long seen as bastions of stability—suddenly become fragile, reactive organisms, and the true test of leadership emerges not in how well one manages the expected, but in how one adapts to the unthinkable. Crises, by their nature, dismantle the conventional. They accelerate the collapse of outdated systems and bring unresolved vulnerabilities to the surface. The 21st century, particularly in the past two decades, has offered educational institutions an unrelenting sequence of such events, from global financial crises to public health emergencies to deep sociopolitical divides. These moments have not only destabilized the infrastructure of education but have also fundamentally shaken the confidence of the communities within them. In such conditions, traditional leadership paradigms that rely on control, linear planning, and risk aversion quickly become inadequate. What emerges in their place is a new demand—one for innovation, empathy, agility, and a willingness to lead from uncertainty rather than despite it.

This chapter begins with a simple but profound premise: that times of uncertainty, while deeply disruptive, can also be powerful incubators for transformation. When institutions are forced out of their comfort zones, when the familiar vanishes and long-standing hierarchies falter, the possibility for reinvention takes root. However, this potential is not realized automatically. It must be stewarded—patiently, courageously, and creatively—by leaders who do not merely manage crises but rise within them. Innovative educational leadership in such times does not come neatly packaged in frameworks or toolkits. It is not the application of pre-existing models, but the courage to design new ones. It is found in school principals who hold space for grief and fear while maintaining a sense of direction. It emerges in university presidents who are willing to pivot entire curricula to remote models within weeks, without losing sight of equity and inclusion. It resonates in department chairs who consult their teams not out of obligation, but out of genuine respect for shared wisdom. These acts of leadership are often quiet, sometimes messy, and rarely glamorous—but they are transformative. They reflect a deeper reorientation: from authority to authenticity, from control to co-creation, from certainty to curiosity. The introduction of innovation into educational leadership during times of uncertainty is not about adopting technology for its own sake or embracing buzzwords. It is about rethinking the very role of education in society, and the responsibility of those who guide it. It means asking uncomfortable questions: What do learners truly need—not only academically, but emotionally and socially? What structures no longer serve our evolving missions? How do we lead with clarity when the future is anything but clear?

In exploring these questions, this chapter will trace the emerging patterns and mindsets that define innovative educational leadership in crisis settings. It will examine how leaders cultivate trust in uncertain environments, how they communicate with transparency while acknowledging their own limits, and how they turn systemic shocks into catalysts for long-overdue change. Importantly, it will highlight the human dimension of leadership—the emotional labor, the ethical dilemmas, the need for empathy, and the strength required to remain hopeful amid chaos. We are living in an era where uncertainty is not the exception, but increasingly the norm. The question is no longer *if* institutions will face disruption, but *how* they will respond when they do. The leaders who will shape the future of education are those who embrace uncertainty not as a threat, but as a canvas. They are the ones who see crisis as both rupture and opportunity—who understand that innovation, in its truest form, is not a strategy but a stance. A way of being. A way of leading.

THE CRISIS AS CATALYST: REIMAGINING THE ROLE OF EDUCATIONAL LEADERSHIP

There is something uniquely revealing about a crisis. It lays bare the assumptions we've grown accustomed to, dismantles the structures we thought were unshakable, and exposes the cracks we never bothered to notice when things were running smoothly. In education, crises have always been moments of reckoning—but they are also moments of reimagining (Sayed et al, 2021). It is within these periods of disorientation and disruption that the role of educational leadership undergoes its most meaningful transformations. At the onset of any major crisis—be it a global pandemic, a natural disaster, political upheaval, or economic collapse—there is often an instinct to pause, regroup, and try to return to "normal." But this longing for normalcy is, in many ways, a mirage. What we call "normal" is often just the accumulation of habits, routines, and systems that have gone unquestioned for too long. Crises, uncomfortable as they are, confront us with a rare and powerful opportunity: to stop, look around, and ask whether the systems we are leading are truly serving the people within them.

Educational leaders, more than anyone else, stand at the intersection of this questioning (Agosto & Roland, 2018). They are responsible not only for managing logistics and ensuring continuity but also for carrying the emotional weight of their communities—students, teachers, parents, staff. They are called upon to provide direction when no clear map exists, to speak with clarity even when they, too, feel uncertain, and to offer reassurance while holding space for the difficult truths that must be acknowledged. This is not leadership as we typically define it in calm times. It is something far deeper, more intimate, and profoundly human. One of the most remarkable features of leadership during crisis is its demand for immediacy (Dalcher, 2020). Decisions that would normally take months of deliberation are made in hours. Long-term strategies are rewritten overnight. Communication becomes raw, direct, and deeply personal. Leaders are no longer insulated by bureaucratic layers—they are exposed, visible, and utterly accountable. This exposure, while daunting, also becomes a crucible in which a new kind of leadership is forged—one that is grounded not in positional power, but in relational trust.

What many educational leaders have discovered, sometimes painfully, is that traditional notions of leadership are ill-equipped for crisis contexts. The default posture of control and command falters in the face of chaos. Plans unravel. Timelines dissolve. Metrics lose their meaning. And in that space of unknowing, a different kind of wisdom is required—one that values adaptability over rigidity, listening over directives, presence over performance. In many ways, the crisis becomes a mirror. It reflects not only the conditions of our institutions but the character of those who lead them. A crisis does not create values; it reveals them. It does not bestow empathy or humility or vision—it shows whether those qualities were already there, dormant perhaps, but ready to rise (Kayyali, 2024 A). And when they do rise, when a leader meets the moment not with bravado but with vulnerability, not with answers but with questions, something extraordinary happens. The community begins to trust—not because the leader is infallible, but because they are real (Frei & Morriss, 2020). Reimagining educational leadership in the aftermath—or even in the midst—of a crisis is not about returning to the pre-crisis status quo with slight modifications. It is about daring to ask what leadership should look like if it were truly designed for complexity, for diversity, for rapid change. It is about stepping away from the myth of the all-knowing, always-prepared leader and toward a model that embraces the leader as learner, as co-navigator, as someone who is just as embedded in the journey as those they lead. This reimagining often begins with a shift in how leaders perceive their role. Instead of seeing themselves as the primary source of direction, they begin to see themselves as facilitators of collective intelligence. They become bridge-builders across silos, amplifiers of quiet voices, and stewards of culture and belonging. They focus not just on what is urgent, but on what is essential. They know that resilience is not built through speed alone, but through depth—through creating spaces where people feel safe to speak, to grieve, to disagree, and ultimately, to heal.

One of the most significant realizations during crisis leadership is that the emotional landscape of a school or university is just as important as its physical infrastructure (Bataille & Cordova, 2023). A building can be repaired, a schedule rearranged—but the psychological toll of uncertainty, fear, and loss must be acknowledged and tended to. Leaders who do this—who check in with their staff not out of protocol but out of care, who ask their students not only what they need to learn but how they are feeling—these are the leaders who transform a crisis from a breaking point into a turning point. This kind of leadership does not arise from textbooks or training manuals. It is shaped in the moment, in the tension between fear and responsibility, in the small, repeated acts of showing up. It is not about having all the answers, but about asking the right questions with humility. It is about knowing when to take the lead and when to step back and let others rise. It is about honoring lived experience, not just professional expertise. We saw this kind of leadership during the COVID-19 pandemic, when school principals spent sleepless nights reorganizing classrooms for safety, when university deans held virtual office hours just to listen, when teachers became tech experts overnight to ensure that learning didn't stop (Kayyali, 2020). We saw it in moments of political unrest, when educators created spaces for dialogue and reflection instead of silence. We see it whenever a leader chooses to meet a crisis not with defensiveness, but with openness. And we feel it—in the relief of being heard, the comfort of shared struggle, and the possibility of imagining something better together.

But reimagining leadership also demands that we confront uncomfortable truths. Crises often highlight pre-existing inequities—disparities in access to technology, in mental health support, in financial resources (Radfar et al, 2021). Leaders who wish to be transformative must resist the urge to patch these holes temporarily and instead commit to long-term structural change. They must ask: Who was left behind in this crisis? Whose needs were not met? What systemic barriers must be dismantled—not just for now, but for good? In this way, the crisis becomes not only a challenge but a call to justice. It becomes a moment to interrogate the values that have shaped our institutions and to reimagine them in ways that center equity, inclusion, and compassion. Leaders who rise to this challenge are not just managing crisis—they are reshaping the very meaning of education. They understand that leadership is not about preserving systems, but about transforming them in service of those they were meant to serve.

This is not easy work. It demands emotional labor, ethical clarity, and a tolerance for ambiguity. It requires leaders to be both visionary and grounded, to hold both urgency and patience. But it is precisely this tension that gives crisis-era leadership its creative edge. In the push and pull of responding to immediate needs while imagining long-term transformation, a new kind of leader is born—one who is not afraid to experiment, to fail, to adapt. If we are to move forward as educators, as institutions, and as societies, we must hold onto the lessons that crisis has taught us. We must resist the temptation to "return to normal" and instead ask: What have we learned about ourselves, our communities, and our responsibilities? How can we lead differently, more humanely, more courageously? Educational leadership will never be the same again—and perhaps that is a gift. Crises have shaken us, yes, but they have also awakened us (Rice, 2023). They have stripped away the superficial and left us with the raw material of what truly matters: relationships, purpose, trust, and the will to grow. In the hands of courageous leaders, these raw materials become the foundation of something entirely new. The crisis, then, is not the end of the story.

It is the turning of the page. And those who lead with humility, imagination, and heart will be the ones who write the next chapter—not just for their institutions, but for the future of education itself.

ADAPTIVE LEADERSHIP IN UNCHARTED WATERS: NAVIGATING AMBIGUITY AND COMPLEXITY

There is a particular kind of silence that follows a disruptive event. Not the silence of stillness, but the kind that hums with uncertainty—when people look to one another unsure of what's next, when plans no longer feel relevant, when questions outweigh answers. This is the space where adaptive leadership becomes not only necessary but transformative. In these uncharted waters, where old maps no longer apply, leadership is no longer about steering a ship along a well-marked course. It's about reading the tides in real time, adjusting sails mid-storm, and helping others remain calm even when the horizon is unclear.

In education, the storms have come in many forms—some sudden and seismic, others slow-building and systemic. The COVID-19 pandemic, for instance, was not just a health crisis; it was an earthquake that shook the very foundations of teaching, learning, and institutional management (Kayyali, 2022). It forced leaders to make decisions with limited information, to balance competing pressures, and to simultaneously care for their communities' logistical, emotional, and ethical needs. But the truth is, even outside of such globally disruptive events, educational leaders are increasingly navigating a world defined by complexity—by shifting policies, evolving technologies, changing student demographics, and the growing demand for inclusivity, sustainability, and relevance. Adaptive leadership, in its truest form, is about relinquishing the illusion of control. It's about understanding that not all problems can be solved with technical expertise or precedent. Some challenges are what Ronald Heifetz calls "adaptive challenges"—issues that cannot be resolved simply by applying known solutions (Johnston & Berger, 2021). Instead, they require new learning, experimentation, and, most importantly, a change in values, beliefs, or behavior. In a crisis or in times of profound transition, these kinds of challenges dominate the landscape (Kayyali, 2024 B). The role of the leader, then, is not to provide all the answers, but to create the conditions in which people can begin to explore them together.

This shift—from being the one with answers to being the one who fosters inquiry—is neither natural nor easy. Many educational leaders have been conditioned to equate authority with certainty (Bush, 2020). They feel pressure to present confidence, to protect their teams from doubt, and to deliver quick fixes. But adaptive leadership requires the opposite: an openness to vulnerability, a willingness to say

"I don't know," and the humility to invite others into the process. It is less about heroism and more about hosting conversations that matter. It is about navigating ambiguity not by pretending it doesn't exist, but by guiding others through it with courage and clarity. This kind of leadership is also inherently relational. It relies on trust—not the performative trust of press releases and policy statements, but the earned trust that comes from presence, honesty, and consistency. In uncertain times, people do not expect their leaders to be flawless. But they do expect them to be real. They want to know that their leader is listening, thinking, evolving. They want to be seen not as passive recipients of decisions but as active participants in shaping the way forward.

Adaptive leaders understand that the strength of an institution lies not in rigid systems, but in resilient people (Dunn, 2020). And resilient people are not born in isolation—they are nurtured in cultures that value learning, feedback, and psychological safety. One of the quiet powers of adaptive leadership is the ability to hold space for discomfort, to allow tensions to surface without rushing to resolve them prematurely. There is wisdom in letting complexity breathe, in resisting the impulse to oversimplify, in trusting that clarity will emerge—not instantly, but eventually—through honest dialogue and collective reflection. That doesn't mean adaptive leaders are passive. On the contrary, they are deeply engaged—but in a way that centers on inquiry rather than instruction. They ask powerful questions: What are we not seeing? Whose voices are missing? What assumptions are we making? They stay close to the edges of their communities—where change is first felt and where innovation often begins. They are attuned not just to the loudest voices but to the subtle signals, the quiet discomforts, the emerging patterns. In doing so, they become sense-makers in a time of sensory overload. Consider, for example, a university dean in a region undergoing political upheaval. Policies are changing rapidly. Funding is uncertain. Students are scared. Faculty are divided. The old models of planning no longer hold. Rather than issuing top-down edicts, the dean begins convening weekly listening sessions—not to dictate, but to listen. These sessions become spaces of co-creation. Faculty propose ideas. Students express concerns. A sense of shared responsibility emerges. Slowly, new frameworks take shape—not perfect, not permanent, but responsive and inclusive. This is adaptive leadership in action: leadership that is fluid, distributed, and deeply human.

Similarly, imagine a school principal navigating a surge in student mental health issues. The curriculum is packed. Teachers are exhausted. Parents are anxious. Rather than pushing through with business as usual, the principal pauses. They reduce non-essential meetings. They bring in mental health professionals for the staff. They give students time to process, to grieve, to connect. They adjust expectations without lowering standards. And most of all, they acknowledge what everyone is feeling instead of pretending everything is fine. That acknowledgment alone becomes an act of leadership. It tells the community: we are in this together, and we will find our way forward—not because we know exactly how, but because we are committed to walking through the unknown side by side. Adaptive leadership also requires stamina. It is not a sprint but an ongoing, iterative process (Sott & Bender, 2025). It involves making decisions with incomplete information, absorbing the emotional weight of others, and being constantly prepared to pivot. This can be exhausting. That's why adaptive leaders need their own forms of support—spaces where they can reflect, be challenged, and reconnect with their purpose. Leadership in ambiguity is not sustainable unless it is grounded in a deep sense of self-awareness and care. To lead others with compassion, one must also lead oneself with compassion. And perhaps that's the most radical element of adaptive leadership: it is deeply personal. It's not about applying a model but embodying a mindset. It asks leaders to draw not only from their intellect, but from their values, their lived experience, and their inner clarity. It demands integrity, the ability to hold complexity without collapsing into chaos or denial. It is both pragmatic and poetic: pragmatic in its responsiveness, poetic in its capacity to imagine new possibilities even amidst uncertainty. Of course, adaptive leadership is not without its risks (London, 2023). It can be misunderstood as indecisiveness. It can be slowed by resistance, by the discomfort that comes with challenging norms. It may not always deliver immediate results. But its power lies in its ability to transform institutions not just structurally, but culturally. It shifts the narrative from control to collaboration, from perfection to progress, from authority to authenticity.

In many ways, the rise of adaptive leadership signals a broader shift in how we understand education itself. No longer just a transmission of knowledge, education becomes a dynamic, living ecosystem—one that evolves in response to its context, its people, its moment (Bandyopadhyay et al, 2021). And leaders in such ecosystems must be gardeners, not gatekeepers. They must cultivate rather than command, nurture rather than dictate. So, when we speak of navigating uncharted waters, we are not just describing the external conditions of crisis—we are also speaking of the internal journey leaders must undertake. The courage to step into ambiguity. The humility to listen. The vision to reimagine. The discipline to stay the course when the path is unclear. These are not just leadership strategies—they are human capacities. And in this moment of global uncertainty, they are the very capacities that our educational institutions need most. We may not know what the future holds. But we do know that it will demand more of us—more openness, more agility, more humanity. Adaptive leadership is not a response to a single crisis. It is a response to the age we are living in. It is a compass, not a map. A conversation, not a command. A promise, not of certainty, but of care. And perhaps, in the end, that is the most enduring gift of adaptive leadership: it reminds us that even in the most uncertain waters, we are never alone. That leadership, at its best, is not a position but a practice. A practice of presence. A practice of listening. A practice of becoming.

BUILDING INSTITUTIONAL RESILIENCE THROUGH INNOVATION AND COLLABORATION

Resilience is a word often spoken in times of crisis, but seldom understood in its deepest form (Buheji, 2018). In educational institutions, it has long been confused with simple endurance or the ability to "bounce back" after disruption. Yet true resilience is far more nuanced. It is not about returning to what was, but about evolving into what can be—stronger, wiser, and more aligned with the needs of an ever-changing world. It is not a destination but a process, cultivated through deliberate choices, inclusive culture, and the collective will to adapt. At the heart of this evolution lies two interdependent forces: innovation and collaboration. Innovation is not only about technology or new programs. Nor is it reserved for moments of grand reinvention. Often, it begins quietly—in a conversation between departments, in a shift in mindset, in the courage to question why things have always been done a certain way. In education, where systems tend to resist change and tradition is often mistaken for excellence, innovation can feel risky, even unwelcome. But when done with purpose and care, it becomes the lifeblood of resilience.

Institutions that embody resilience are those that do not merely react to change—they anticipate it, shape it, and use it as a chance to reimagine their purpose (Cahill & Leccardi, 2020). They create room for experimentation, for curiosity, for ideas that do not fit neatly within existing structures. They understand that failure is not something to be avoided at all costs, but a natural part of discovery. They reward learning not only in students but in their leaders, faculty, and staff. Most importantly, they create ecosystems where innovation is not the privilege of a select few, but the shared responsibility of the whole. What does this look like in practice? Sometimes it means rethinking the design of physical spaces to better support collaboration, creativity, or emotional well-being. Sometimes it means integrating student voice into curriculum development or policy decisions. Sometimes it's about small, agile changes: piloting a flexible schedule, integrating mindfulness into classrooms, reworking grading systems to focus on growth rather than competition. And sometimes it means asking more difficult, systemic questions—about equity, access, and the structural barriers that prevent some members of the community from thriving. But innovation without collaboration rarely leads to resilience. If innovation is the spark, collaboration is the oxygen that keeps it alive. Educational institutions are complex systems, composed of people with different roles, identities, experiences, and ideas. Collaboration ensures that these differences do not become points of division but sources of insight. When people from across disciplines, departments, and hierarchies come together—not just to share updates but to co-create solutions—something powerful happens. Innovation becomes more grounded, more inclusive, more sustainable.

Resilient institutions are not those that centralize power and protect silos. They are those that decentralize creativity and encourage dialogue. They invest in structures that make collaboration not only possible but meaningful. They cultivate a culture where feedback is not feared but welcomed, where listening is just as valued as speaking, and where leadership is distributed, not hoarded. Too often, collaboration is treated as a soft skill, something secondary to strategy. But in reality, it is one of the most vital strategic tools available—especially in uncertain times. During the pandemic, schools that already had strong collaborative cultures were able to pivot more quickly, share burdens more evenly, and maintain a sense of collective purpose even when apart (Darling-Hammond et al, 2020). Universities that had fostered cross-functional teams and interdepartmental trust were better equipped to manage disruptions, address student needs holistically, and make courageous decisions together. What made them resilient was not just their capacity to adapt, but their commitment to adapt together (Kayyali, 2024 C). Collaboration also invites vulnerability, and this is where many institutions falter. To collaborate authentically, individuals—and by extension, institutions—must let go of perfectionism, ego, and fear of conflict. They must be willing to engage in honest conversations, to own mistakes, and to approach disagreements as opportunities for growth rather than threats to authority. It is in these vulnerable spaces that true resilience is forged—not through glossy strategic plans, but through messy, honest, deeply human processes.

Moreover, building resilience through innovation and collaboration means expanding who gets to participate in shaping the institution's future. Students must not be treated as passive consumers but as co-authors of the educational experience. Faculty must not be confined to their disciplines but encouraged to work across boundaries (Simula & Scott, 2021). Staff, often the quiet backbone of schools and universities, must be recognized not only for their operational roles but for the insights they bring as day-to-day observers of the institutional pulse. Resilience also demands that institutions look beyond their own walls. Collaboration should extend into the broader community—parents, local organizations, policymakers, industry partners (Alkaher & Gan, 2020). In a world where education is increasingly interconnected with economic, social, and environmental systems, no institution can thrive in isolation. Those that cultivate meaningful partnerships are better able to innovate, to stay relevant, and to weather external shocks with the support of a broader ecosystem. These partnerships also enrich learning, exposing students to real-world challenges, diverse perspectives, and opportunities for civic engagement. It is essential, too, to recognize that innovation and collaboration are not linear processes. They are iterative, evolving, and sometimes uncomfortable. There will be missteps, resistance, and fatigue. People will disagree. Resources will be stretched. But these are not signs of failure—they are signs that something real is happening. That the institution is alive, not stagnant. That it is willing to be in the difficult work of change, rather than hiding behind the façade of tradition.

Leaders play a critical role here—not as directors of change, but as facilitators of a culture where change is possible (Bagga et al, 2023). They must model curiosity, openness, and empathy. They must create space for experimentation without fear of punishment. They must recognize and elevate the often invisible work of those who build bridges, who listen well, who ask the questions no one else dares to ask. They must champion equity, ensuring that innovation is not only for the privileged but benefits all members of the institution. Resilience is not something you can mandate through policy. It grows slowly, through relationships, through trust, through a shared sense of mission. It is sustained not by charismatic leadership alone, but by distributed leadership that empowers others to act. It is built in ordinary moments—the hallway conversations, the team debriefs after a failed project, the decision to try again when something didn't work the first time. It is in these moments that institutions shape their ability to withstand future crises—not by denying their fragility, but by embracing their humanity. One of the quiet truths of resilient institutions is that they tend to carry a deep sense of purpose (Mitchell, 2022). They are not just trying to survive disruption—they are anchored by a why that gives meaning to the work. That purpose may evolve, but it remains a compass. It reminds everyone involved that the work they are doing matters. That education is not just about outcomes but about people. That innovation is not just about being modern, but about being responsive. And that collaboration is not just a method—it is a mindset. In the end, institutional resilience is not a fixed trait—it is a living capacity. It must be cultivated, renewed, and protected. It requires time, trust, and an ongoing commitment to the kind of leadership that prioritizes relationships over hierarchy, creativity over compliance, and collective well-being over individual performance. It asks that we honor not only the efficiency of systems but the dignity of those who move within them. And so, as we look to the future of education in a world increasingly shaped by volatility, complexity, and transformation, the path forward is clear—not in the sense that we know every step, but in the sense that we know how we must walk it: together, with courage, with imagination, and with the unwavering belief that resilience is not just about surviving the storm—but about learning how to dance in the rain, and then teaching others how to do the same.

EMOTIONAL INTELLIGENCE, EMPATHY, AND THE HUMAN SIDE OF LEADERSHIP

Leadership, at its core, is not about titles, authority, or decision-making prowess—it is about people. Especially in education, where the daily currency is human connection, leadership rises or falls on the leader's capacity to truly understand, feel, and respond to the emotional currents that flow beneath the surface of institutional life. It is in this subtle yet powerful realm that emotional intelligence and empathy become not just desirable traits, but necessary pillars of effective, human-centered leadership. It is tempting, in times of uncertainty or institutional pressure, to default to strategy, metrics, and structure. These offer the comfort of clarity and control. They allow leaders to create action plans, monitor performance, and generate reports that suggest things are moving forward. But beneath the numbers and above the noise, what people often remember most—what lingers long after policies are forgotten—is how they were made to feel. In moments of difficulty, the tone of a leader's voice, the patience in their response, the decision to pause and ask "How are you, really?"—these seemingly small acts often speak louder than any announcement or directive.

Emotional intelligence, then, is not a leadership accessory. It is the very language of connection. It enables leaders to recognize their own emotions without being ruled by them, to sense the emotional atmosphere in a room, to respond with nuance rather than reactivity. It allows them to notice when a teacher is burning out even if they haven't said a word, when a student's disengagement masks something deeper, or when a team's silence in a meeting signals more than just agreement. Leaders with emotional intelligence see more, not because they are superhuman, but because they are fully human—and paying attention (Drigas et al, 2023). Empathy goes even further. While emotional intelligence gives leaders the tools to recognize and manage emotions, empathy invites them to step into the emotional reality of others. It's not just understanding that someone is frustrated or afraid—it's feeling with them, sitting beside them in their discomfort without immediately trying to fix it or minimize it. In educational settings, this matters immensely. Students are not just learners; they are people navigating complex lives, often carrying invisible burdens. Teachers are not just deliverers of content; they are mentors, role models, caretakers, often stretched thin between expectations and limitations. Staff are not just support; they are the glue that holds systems together, frequently overlooked until something goes wrong.

When leaders lead with empathy, they foster cultures of safety and belonging. People open up. Ideas surface. Tensions ease. Mistakes become moments of growth rather than shame. And trust deepens—not because perfection has been achieved, but because humanity has been honored. In a time when burnout, anxiety, and disconnection are running high across educational systems, this kind of leadership is not a luxury. It is a lifeline. But let's not romanticize it: leading with emotional intelligence and empathy is hard. It demands vulnerability. It requires a leader to stay present in emotionally charged situations without retreating into defensiveness or control. It means holding space for pain without becoming paralyzed by it. It means listening to criticism not as a threat, but as an offering. And it means doing all of this while also managing responsibilities, expectations, and the ever-present push for measurable outcomes. This balancing act—between results and relationships, structure and sensitivity—is one of the quiet heroics of educational leadership. It is rarely recognized in performance reviews or celebrated in press releases. Yet it is here, in the emotional labor of leadership, that real transformation takes place. Because no innovation, policy change, or strategic plan will succeed in a culture where people feel unseen, unheard, or unsafe. Culture is shaped by what leaders prioritize, but also by how they show up—especially when things are difficult (Jerab & Mabrouk, 2023).

Consider a principal walking into a school the morning after a local tragedy. There are protocols to follow, announcements to make, logistics to arrange. But before any of that, they walk the halls. They look people in the eye. They ask, "How are you holding up?" They pause when a staff member begins to cry. They allow silence. They are not trying to be strong in the traditional sense—they are being present. And that presence speaks volumes. It says: You matter. We will get through this. Not because I have the answers, but because we're facing it together. Or take a university dean navigating faculty tensions after a divisive policy change. Instead of enforcing compliance or avoiding conflict, they create a space for dialogue. They begin the conversation not with rules but with a story—a personal one, about a moment when they too felt uncertain or unheard. They do not pretend neutrality, nor do they impose a solution. They listen. They reflect back what they hear. They admit their own discomfort. And slowly, something shifts. Not necessarily the policy itself, but the way people relate to it—and to each other. This is leadership not as performance, but as presence. There is also a deeply personal dimension to this kind of leadership (Nicholson & Kurucz, 2019). Leaders who consistently practice emotional intelligence and empathy often do so not because they've been trained to, but because they've lived it. They know what it's like to be overlooked, to feel pressure, to make hard decisions with imperfect information. They carry their own stories, and they use those stories not to dominate, but to relate. They understand that their emotional awareness is not separate from their leadership—it *is* their leadership.

And yet, this kind of leadership can take a toll. Because being emotionally attuned means carrying more—more of other people's fears, frustrations, hopes, and grief. It means absorbing tension, defusing conflict, and holding ambiguity without letting it unravel you. That's why emotionally intelligent leaders need care too. They need mentors, peers, spaces where they can speak without filtering. They need time to rest, to reflect, to reconnect with the values that brought them to this work in the first place. Empathy, when not supported by self-awareness and boundaries, can become depletion. But when nurtured wisely, it becomes an engine for resilience. In many ways, emotionally intelligent leadership is countercultural. It resists the myth of the invulnerable leader. It challenges the assumption that logic should always trump emotion, that productivity is more important than people, that control is a sign of strength. It replaces these notions with something more truthful: that leadership is relational, that emotion is not weakness but data, and that the human side of leadership is not the "soft" side—it is the heart of the work. And perhaps most beautifully, this kind of leadership is contagious (Kayyali, 2025 B). When leaders model emotional presence, others begin to mirror it. When they name their feelings, others feel safe to do the same (Wasylyshyn & Masterpasqua, 2018). When they show care, it ripples outward—through teams, classrooms, communities. Culture shifts. Walls come down. And in the spaces left behind, something new can emerge: not just better morale, but better ideas, better collaboration, deeper commitment.

The real measure of emotionally intelligent leadership is not in how composed a leader appears, but in how safe people feel around them (Olawoyin, 2018). It is in the quiet sigh of relief when someone realizes they don't have to pretend to be okay. It is in the honest feedback shared because there is trust. It is in the moment when someone feels seen for who they truly are—not just their role, but their humanity. In an age where uncertainty is a constant and pressure is high, educational institutions need more than strategic thinkers. They need compassionate presences. They need leaders who can hold paradox—who can be both decisive and empathetic, visionary and grounded, courageous and kind. They need people who understand that leading others begins with knowing oneself—not perfectly, but honestly. So yes, emotional intelligence and empathy may be hard to measure. They may not fit neatly into evaluation frameworks. But they are the very qualities that hold institutions together when everything else is falling apart. They are the glue that connects people to each other and to a shared sense of purpose. They are the thread that weaves resilience, innovation, and adaptability into something meaningful. And in the end, they are what will be remembered—not the plans, not the policies, but the moments of connection, the conversations that mattered, the leaders who made people feel seen. This is the human side of leadership. And it is, perhaps, the most powerful side of all.

LEADING WITH VISION: STRATEGIC FORESIGHT IN AN AGE OF VOLATILITY

In an age where change seems to come faster than we can process it, the role of a leader becomes more critical than ever. But leadership, especially in volatile times, is not about knowing every answer or having a clear map to the future. It's about vision—the ability to see beyond the immediate chaos, to grasp the patterns that lie beneath the surface of disruption, and to guide others through uncertainty with clarity and purpose. In these moments, the leader becomes both a beacon and a compass: someone who illuminates a possible future while also helping others navigate the stormy seas of the present. Strategic foresight is often a phrase used in corporate boardrooms or political circles, but it holds particular relevance in educational leadership (Burke, 2021). The challenge today is not simply managing the crises that come our way—it's about anticipating what's ahead. It's about understanding that we are living in a world defined by volatility, complexity, and unpredictability, where the old ways of doing things no longer suffice. If educational leaders are to thrive in this environment, they must cultivate the ability to think beyond the present moment—to see the long-term trends that will shape their institutions and prepare their communities for what lies ahead. This is not to say that foresight is about predicting the future with certainty—such a thing is impossible. Instead, it is about developing the ability to perceive emerging patterns, to recognize potential disruptions before they unfold, and to act on them in ways that align with the institution's core values and long-term goals. It's about shaping the future, rather than simply reacting to it.

In volatile times, this kind of vision is particularly important because it anchors an institution in a shared sense of purpose. When everything around us feels uncertain—when economic conditions fluctuate, when political landscapes shift, when social movements demand change—leaders who lead with vision provide something that cannot be easily taken away: a sense of direction. This is what allows a community to stay grounded, even when everything else seems in flux. Leaders with vision create a narrative that unites people—a vision that transcends individual challenges and connects everyone to something greater than themselves. The foundation of such leadership lies in the ability to ask the right questions, to engage in the kind of thinking that challenges assumptions and pushes boundaries (Kouzes & Posner, 2024). It involves a willingness to embrace ambiguity, to hold multiple perspectives simultaneously, and to acknowledge that certainty is no longer a given. Strategic foresight in education requires leaders to look beyond the immediate goals and consider how their decisions today will impact the institution tomorrow, next year, or even a decade from now. A leader with vision understands that the volatility of our age is not something to fear, but something to embrace. They recognize that disruption, while unsettling, brings with it the potential for reinvention. The turbulence that seems to shake the foundation of our institutions is not just a threat—it is an opportunity to rethink, reimagine, and rebuild. They understand that the future is not a fixed destination, but a landscape that is constantly shifting, shaped by the collective efforts of those who dare to envision a new path forward.

In education, this vision requires leaders to look not only at their own institutions but at the broader forces that shape the landscape of learning (Beerkens & van der Hoek, 2022). They must consider how technological advancements are transforming the way we teach and learn. They must understand the growing demand for personalized education, for inclusivity, for lifelong learning. They must anticipate how demographic shifts will impact their student populations, and how changing labor markets will affect the skills students need to thrive. They must be attuned to social movements and political changes that will influence policy and funding, and how global trends, from climate change to migration, will reshape the world in which their students live and work. The question that every leader must ask is: What kind of education do we need to prepare our students for this rapidly changing world? It's not enough to focus solely on the present, to ensure that the institution functions smoothly in the short term. Leaders must be proactive, thinking strategically about the future needs of their students, their faculty, and their communities. They must foster an environment where innovation is not just allowed, but expected. They must create systems that encourage risk-taking, experimentation, and collaboration. And they must be willing to make difficult decisions—sometimes unpopular ones—in order to ensure that their institutions remain relevant in the face of change. But even the most visionary leaders cannot succeed alone. Vision requires collaboration. In the age of volatility, no leader can see everything, anticipate every change, or navigate every challenge by themselves. The most effective leaders are those who can build networks of individuals who share a common purpose and a commitment to a collective vision. They must be able to communicate that vision clearly and inspire others to join in the journey. They must be able to translate abstract ideas into tangible actions, to turn dreams into reality.

Collaboration in this context is not just about working with others—it is about creating a shared sense of ownership and responsibility (Mattessich & Johnson, 2018). In an environment where volatility and uncertainty are the norms, the more people who are invested in the future of an institution, the stronger and more resilient that institution will be. This means leaders must empower others to contribute to the vision, to take ownership of the work, and to collaborate across departments, disciplines, and even sectors. They must build a culture of trust, where people feel safe to share ideas, to disagree, and to experiment. It is this collective energy that fuels resilience, and it is this resilience that allows institutions to navigate the challenges of a rapidly changing world. Leading with vision also means understanding that change is not linear. It does not happen in neat, predictable steps. Rather, it is messy, full of setbacks and surprises. It is in these moments, when things go off track, that the true strength of a leader's vision is revealed. A leader with strategic foresight is not dismayed by obstacles; they are energized by them (Zalman, 2019). They understand that disruption creates space for growth, for creativity, and for new opportunities. They are willing to adjust their course, to take risks, and to try new approaches—because they know that this is the way forward in an unpredictable world. One of the most critical aspects of visionary leadership is the ability to inspire others to share that vision. A leader's vision is only as strong as the people who believe in it. This requires not only strategic thinking, but emotional intelligence—the ability to connect with people on a human level, to understand their hopes and fears, and to communicate in a way that resonates deeply. It means being able to speak to the heart, not just the head, to create a sense of shared purpose that transcends individual goals and unites everyone in a common mission.

Strategic foresight is also about resilience—an ability to maintain focus on long-term goals, even in the face of short-term setbacks (Kohler, 2021). It is about seeing the bigger picture, even when immediate challenges feel overwhelming. It is about staying connected to the institution's mission, values, and purpose, and making decisions that align with these ideals, even when it is difficult. This kind of resilience is rooted in the understanding that leadership is a marathon, not a sprint. It requires patience, persistence, and a deep commitment to the long-term vision. Visionary leadership in an age of volatility requires humility. A leader with strategic foresight understands that they do not have all the answers (George, 2024). They know that the future is uncertain, and that the path forward is not always clear. This humility allows them to listen, to learn, and to adapt. It also allows them to be open to new ideas, to collaborate with others, and to engage in ongoing reflection and growth. A humble leader does not see themselves as the sole architect of the future, but as a guide—someone who can help others find their way, even when the road ahead is unclear. In a world that is increasingly volatile, complex, and unpredictable, the ability to lead with vision and strategic foresight is more important than ever. Visionary leaders are the ones who can navigate uncertainty, who can inspire others to join them in the journey, and who can shape the future of education in ways that align with the changing needs of society (Candrasari et al, 2023). They are the ones who understand that leadership is not about knowing the exact path forward—it's about having the courage to step into the unknown and to chart a course as they go. Education, at its best, has always been about preparing individuals not just for the world that is, but for the world that could be. And in an age of volatility, it is the visionary leaders who can see that future and inspire others to build it with them.

TRANSFORMATIVE LEGACIES: WHAT CRISIS-ERA LEADERSHIP TEACHES US FOR THE FUTURE

Crisis is often a word we associate with disruption, chaos, and breakdown, but if we look closely, we see that it can also be the breeding ground for transformation. It is in moments of profound upheaval—when the world seems to falter—that leadership becomes most meaningful. Not because it provides all the answers, but because it inspires others to rise, adapt, and create new solutions together. The legacies left by leaders in times of crisis are not just stories of survival, but of profound change that can reshape the very foundations of institutions, communities, and societies. What crisis-era leadership teaches us is that adversity does not simply test our limits; it expands them (Weatherington, 2024). It challenges us to rethink not only what we do, but who we are, and how we lead. In the past few decades, we have seen numerous crises unfold—financial collapses, political instability, social movements, and the most recent global pandemic. Each one has left an indelible mark on the institutions and communities it touched. But more than anything, these crises have demonstrated the power of leadership that goes beyond traditional management. They have shown us that crisis-era leadership is not about holding onto the status quo, but about evolving it. It is about not only surviving the storm but reimagining the world once the storm has passed. What makes crisis-era leadership so transformative is its ability to expose the hidden strengths and weaknesses within individuals and systems. When things are stable, it is easy to ignore the cracks, the blind spots, and the inefficiencies that lie just beneath the surface. But when a crisis strikes, these flaws are laid bare. This is where true leadership steps in—not to maintain control, but to build something stronger from the wreckage. This kind of leadership is characterized by its willingness to question assumptions, challenge existing structures, and confront uncomfortable truths. It is not about quick fixes or temporary measures. It is about long-term vision, integrity, and the courage to rebuild in new and better ways.

The legacy of crisis-era leadership is not simply measured by the decisions made in the heat of the moment, but by the lessons learned and the systems restructured in the aftermath. Leaders who rise to the challenge of a crisis leave behind more than policies and procedures—they leave behind a culture, a new way of thinking, and a set of values that will guide future generations. These legacies are transformative not because they are based on momentary achievements, but because they reflect a deeper commitment to growth, sustainability, and community. At the heart of these legacies is the ability of leaders to adapt. In the face of crisis, adaptation becomes not just a survival tactic, but a strategic imperative. Leaders who are able to adapt to new circumstances, who are open to rethinking old assumptions and embracing new ideas, set the foundation for an institution's long-term success. They understand that change is inevitable and that resistance to it often leads to stagnation. Instead of fearing change, they cultivate a culture of adaptability—encouraging flexibility, continuous learning, and innovation at every level of the organization. This is particularly relevant in education. In many ways, the pandemic has served as a crucible for educational leadership, testing the resilience and adaptability of systems that were already under strain. The transition to online learning, the rush to implement health and safety protocols, and the necessity of redefining how education is delivered have all forced leaders to rethink what education should look like in the future (Miller & Ives, 2023). But the most transformative legacy of this crisis is not the shift to digital learning platforms or hybrid models, but the deep reexamination of the purpose of education itself. It is in the question of what we value in education—what kind of learners we want to nurture, what skills we need to impart, and what kind of human beings we hope to shape—that crisis-era leadership has left its most lasting mark.

During this period, leaders in education have shown that they are capable of not only managing crisis but of shaping its outcomes. By embracing the challenge of online and hybrid learning, they have expanded the reach of education, breaking down geographic and socioeconomic barriers. By addressing the emotional and mental health needs of students and faculty, they have fostered a greater understanding of the human side of learning. And by questioning the long-standing assumptions of what education should be, they have opened the door to more inclusive, personalized, and adaptive models of learning (Kayyali, 2025 A). These leaders are not just responding to a crisis—they are actively shaping the future of education, creating a more equitable, flexible, and responsive system that can better meet the needs of future generations. This kind of leadership teaches us that crisis is not something to be feared, but something to be embraced as a catalyst for transformation (Fragouli & Tourlaki, 2020). It reveals the need for institutions to become more resilient, agile, and adaptable. It teaches us that leadership is not about maintaining control or preserving the status quo, but about moving with the currents of change, responding creatively to uncertainty, and engaging in constant reinvention. The leaders who rise to the challenge of a crisis are those who understand that in order to thrive, institutions must evolve—not just to survive, but to flourish. The transformative legacy of crisis-era leadership also lies in its emphasis on human connection (Kusumawati & Saputra, 2023). In times of uncertainty and fear, it is the human side of leadership that provides the stability and direction people need. Leaders who connect with their communities on a personal level, who listen with empathy and communicate with honesty, create a sense of trust and safety that allows others to follow. This is particularly evident in the way educational leaders have responded to the mental and emotional toll of the pandemic on students, faculty, and staff. Rather than simply focusing on logistical solutions, they have prioritized the well-being of their communities, recognizing that emotional resilience is just as important as academic achievement.

In many ways, the crisis has highlighted the need for a more compassionate, human-centered approach to leadership. Leaders who embrace this approach are not only more effective in times of crisis, but they also create a legacy that endures long after the immediate crisis has passed. They build institutions that are more resilient, more inclusive, and more supportive—places where people feel seen, heard, and valued. This is a leadership that recognizes the importance of community, of collaboration, and of empathy. It is leadership that acknowledges that success is not just measured in outputs or outcomes, but in the quality of relationships and the strength of the human bonds that hold an institution together. The crisis has also revealed the importance of vision in leadership (AlKnawy, 2018). In times of uncertainty, it is easy to become consumed by the immediacy of the moment, to focus only on the short-term problems that demand attention. But true leadership is about looking beyond the present moment, seeing the bigger picture, and charting a course for the future. Leaders who are able to maintain a long-term vision, who can inspire others to believe in that vision, and who are willing to make tough decisions in pursuit of that vision, leave behind a legacy that lasts. They create institutions that are not just prepared for the challenges of today, but equipped to meet the demands of tomorrow. The legacy of crisis-era leadership teaches us that leadership is not about being perfect—it is about being human (Williams et al, 2022). It is about embracing imperfection, learning from failure, and growing stronger through adversity. It is about having the courage to lead with empathy, to make decisions that are guided by a sense of purpose, and to inspire others to join in the work of transformation. The leaders who rise in times of crisis are the ones who understand that leadership is not a position of power, but a responsibility to serve, to listen, and to guide others through the storm. Ultimately, the legacy of crisis-era leadership is not something that can be measured in immediate results or short-term achievements. It is a legacy built over time—through the values that are upheld, the relationships that are cultivated, and the change that is made. It is a legacy that transcends individual accomplishments, and instead, reflects the collective strength and resilience of the entire community. It is a legacy that will continue to inspire future generations of leaders to rise to the challenge, to adapt to change, and to create a better, more inclusive world for all.

CONCLUSION

As we arrive at the close of this chapter, we are reminded that the true weight of leadership is not measured by the titles one holds or the policies one enacts, but by the courage it takes to remain human in the face of uncertainty. The most defining characteristic of innovative educational leadership in times of crisis is not mastery over chaos—but the willingness to move through it, with open eyes, steady hands, and a heart attuned to the people one serves. Throughout this exploration, we have seen that uncertainty does not simply test the resilience of institutions—it reveals the resilience of the people within them. Leaders in education are often pulled between competing demands: the need to assure stakeholders while managing risk, the urgency to act quickly while preserving thoughtfulness, the call to uphold tradition while daring to break it. In crisis, these tensions are heightened. Yet it is precisely in these tensions that the seeds of innovation are planted. It is in the pause between a decision and a doubt, in the hesitation before a bold step, that leadership finds its soul.

What we have learned from recent crises—pandemics, political unrest, climate emergencies, and more—is that there is no returning to the old normal. Nor should there be. Educational systems that were once designed for stability now require fluidity. Rigid hierarchies must give way to collaborative networks. Uniformity must make room for creativity. Leaders can no longer be distant figures delivering edicts from ivory towers; they must be facilitators, listeners, bridge-builders, and often, healers. Innovation in this context is not about chasing novelty. It is about responding authentically to complexity. It is about knowing when to lean into technology and when to lean into touch. When to act, and when to listen. When to redesign systems, and when to simply ask better questions. The leaders who can do this—who lead not by certainty, but by integrity—will be the ones who transform education from a reactive institution into a regenerative force.

Yet perhaps the most important takeaway is this: innovative leadership is not a solo endeavor. It does not reside in the heroic individual but in the collective capacity of communities to dream together, to take risks together, to imagine a better future together. Institutions that nurture this collective courage are the ones most likely to emerge not just intact from crisis, but renewed. The future of education will not be inherited—it must be created. And in this act of creation, it is leadership that holds the brush. In times of uncertainty, leaders have the rare opportunity to shape not only new responses, but new realities. May they lead with vision, with compassion, and with the unwavering belief that even in the darkest moments, learning—and the human spirit that drives it—can still light the way forward.

REFERENCES

Agosto, V., & Roland, E. (2018). Intersectionality and educational leadership: A critical review. *Review of Research in Education*, *42*(1), 255–285. DOI: 10.3102/0091732X18762433

Alkaher, I., & Gan, D. (2020). The role of school partnerships in promoting education for sustainability and social capital. *The Journal of Environmental Education*, *51*(6), 416–433. DOI: 10.1080/00958964.2020.1711499

AlKnawy, B. (2018). Leadership in times of crisis. BMJ Leader, leader-2018.

Bagga, S. K., Gera, S., & Haque, S. N. (2023). The mediating role of organizational culture: Transformational leadership and change management in virtual teams. *Asia Pacific Management Review*, *28*(2), 120–131. DOI: 10.1016/j.apmrv.2022.07.003

Bandyopadhyay, S., Bardhan, A., Dey, P., Bhattacharyya, S., Bandyopadhyay, S., Bardhan, A., ... & Bhattacharyya, S. (2021). Education Ecosystem: A Chronological Perspective. Bridging the Education Divide Using Social Technologies: Explorations in Rural India, 43-75.

Bataille, G. M., & Cordova, D. I. (Eds.). (2023). *Managing the unthinkable: Crisis preparation and response for campus leaders*. Taylor & Francis.

Beerkens, M., & van der Hoek, M. (2022). Academic leaders and leadership in the changing higher education landscape. In *Research Handbook on Academic Careers and Managing Academics* (pp. 121–136). Edward Elgar Publishing. DOI: 10.4337/9781839102639.00017

Buheji, M. (2018). *Understanding the power of resilience economy: An interdisciplinary perspective to change the world attitude to socio-economic crisis*. Mohamed Buheji.

Burke, R. (2021). Anticipatory action learning, leadership, strategy and foresight: Creating a successful future while enhancing results today. *Journal of Futures Studies*, *25*(3), 85–92.

Bush, T. (2020). Theories of educational leadership and management.

Cahill, H., & Leccardi, C. (2020). Reframing resilience. In *Youth and the new adulthood: Generations of change* (pp. 67–81). Springer Singapore.

Candrasari, R., Yorman, Y., Mayasari, N., Yulia, R., & Lake, F. (2023). Visionary leadership in education management: Leading toward optimal achievement in the era of independent learning. [INJOE]. *Indonesian Journal of Education*, *3*(3), 451–467.

D'Auria, G., & De Smet, A. (2020). Leadership in a crisis: Responding to the coronavirus outbreak and future challenges. *Psychology (Irvine, Calif.)*, *22*(2), 273–287.

Dalcher, D. (2020). Leadership in times of crisis: What's different now. *PM World Journal*, *9*(5), 1–18.

Darling-Hammond, L., Schachner, A., & Edgerton, A. K. (2020). *Restarting and Reinventing School: Learning in the Time of COVID and Beyond.* Learning Policy Institute.

Drigas, A., Papoutsi, C., & Skianis, C. (2023). Being an emotionally intelligent leader through the nine-layer model of emotional intelligence—The supporting role of new technologies. *Sustainability (Basel)*, *15*(10), 8103. DOI: 10.3390/su15108103

Dunn, R. (2020). Adaptive leadership: Leading through complexity. *International Studies in Educational Administration*, *48*(1), 31–38.

Fragouli, E., & Tourlaki, T. (2020). Leading change and crisis: How transformational changes can be successful nowadays? *Global Research in Higher Education*, *3*(3), 22–41. DOI: 10.22158/grhe.v3n3p22

Frei, F. X., & Morriss, A. (2020). Begin with trust. *Harvard Business Review*, *98*(3), 112–121.

George, A. S. (2024). The Symbiotic Relationship Between Visionary and Pragmatic Leadership in Propelling Organizational Success. *Partners Universal International Innovation Journal*, *2*(3), 57–79.

Jerab, D., & Mabrouk, T. (2023). The role of leadership in changing organizational culture. Available at *SSRN* 4574324. DOI: 10.2139/ssrn.4574324

Johnston, K., & Berger, J. G. (2021). Guide to adaptive challenges and action learning.

Kayyali, M. (2020). Post COVID-19: New era for higher education systems. *International Journal of Applied Science and Engineering*, *8*(2), 131–145. DOI: 10.30954/2322-0465.2.2020.6

Kayyali, M. (2022). Post-COVID-19 Pandemic Main Challenges with e-Learning.

Kayyali, M. (2024a). Crisis Response and Higher Education: Social Innovations in Times of Global Challenges. In Social Innovations in Education, Environment, and Healthcare (pp. 1-17). IGI Global.

Kayyali, M. (Ed.). (2024b). Rebuilding Higher Education Systems Impacted by Crises: Navigating Traumatic Events, Disasters, and More: Navigating Traumatic Events, Disasters, and More. IGI Global.

Kayyali, M. (Ed.). (2024c). Building Resiliency in Higher Education: Globalization, Digital Skills, and Student Wellness: Globalization, Digital Skills, and Student Wellness. IGI Global.

Kayyali, M. (2025). A). The Impact of Personalized Learning on Student Engagement and Achievement in STEAM. In *Integrating Personalized Learning Methods Into STEAM Education* (pp. 51–78). IGI Global Scientific Publishing. DOI: 10.4018/979-8-3693-7718-5.ch003

Kayyali, M. (2025). B). Leadership's Role in Fostering a Resilient Organizational Culture. In *Enhancing Resilience in Business Continuity Management* (pp. 267–294). IGI Global Scientific Publishing. DOI: 10.4018/979-8-3693-8809-9.ch009

Kohler, K. (2021). *Strategic foresight: Knowledge, tools, and methods for the future*. CSS Risk and Resilience Reports.

Kouzes, J. M., & Posner, B. Z. (2024). *The student leadership challenge: Five practices for becoming an exemplary leader*. John Wiley & Sons.

Kusumawati, R., & Saputra, A. D. (2023, August). FASCA Leadership: Alternative Leadership Styles in an Era of Change. In International Conference on Intelligent Networking and Collaborative Systems (pp. 295-305). Cham: Springer Nature Switzerland. DOI: 10.1007/978-3-031-40971-4_28

London, M. (2023). Causes and consequences of adaptive leadership: A model of leaders' rapid responses to unexpected events. Psychology of leaders and leadership, 26(1), 22.

Mattessich, P. W., & Johnson, K. M. (2018). Collaboration: What makes it work.

Miller, G. E., & Ives, K. S. (Eds.). (2023). *Leading the eLearning transformation of higher education: Leadership strategies for the next generation*. Taylor & Francis.

Mitchell, M. (2022). Building resilient organizations: Toward joy and durable power in a time of crisis. *Convergence.*

Nicholson, J., & Kurucz, E. (2019). Relational leadership for sustainability: Building an ethical framework from the moral theory of 'ethics of care'. *Journal of Business Ethics, 156*(1), 25–43. DOI: 10.1007/s10551-017-3593-4

Olawoyin, R. (2018). Emotional intelligence: Assessing its importance in safety leadership. *Professional Safety, 63*(08), 41–47.

Radfar, A., Ferreira, M. M., Sosa, J. P., & Filip, I. (2021). Emergent crisis of COVID-19 pandemic: Mental health challenges and opportunities. *Frontiers in Psychiatry, 12*, 631008. DOI: 10.3389/fpsyt.2021.631008 PMID: 34349675

Rice, C. (2023). *From Pandemic to Renewal: Practices for a World Shaken by Crisis*. InterVarsity Press.

Sayed, Y., Cooper, A., & John, V. M. (2021). Crises and disruptions: Educational reflections,(re) imaginings, and (re) vitalization. Journal of Education (University of KwaZulu-Natal), (84), 9-30.

Simula, B. L., & Scott, T. L. (2021). Disciplining academic identities: Boundaries and identity work among arts and sciences faculty. *Social Currents*, *8*(4), 378–397. DOI: 10.1177/23294965211001401

Sott, M. K., & Bender, M. S. (2025). The Role of Adaptive Leadership in Times of Crisis: A Systematic Review and Conceptual Framework. *Merits*, *5*(1), 2. DOI: 10.3390/merits5010002

Wasylyshyn, K. M., & Masterpasqua, F. (2018). Developing self-compassion in leadership development coaching: A practice model and case study analysis. *International Coaching Psychology Review*, *13*(1), 21–34. DOI: 10.53841/bpsicpr.2018.13.1.21

Weatherington, I. (2024). Adversity to Action: A Case Study Analysis of Crisis Leadership.

Williams, E., Armistead, J., & Rude, D. A. (2022). Leading for Innovation: A New Model for 21st–century Leadership. *New Horizons in Adult Education and Human Resource Development*, *34*(4), 3–13. DOI: 10.1002/nha3.20366

Zalman, A. (2019). Maximizing the power of strategic foresight. *Joint Force Quarterly*, *95*(4), 14–21.

Chapter 5
Voices of Resilience:
Student-Centered Learning During Disruption

Mustafa Kayyali
http://orcid.org/0000-0003-3300-262X
Maaref University of Applied Sciences, Syria

ABSTRACT

In times of educational upheaval, the prevailing systems are often stripped down to their bare essentials, exposing not just their strengths but also their deepest vulnerabilities. "Voices of Resilience: Student-Centered Learning During Educational Disruption" offers an intimate exploration of how learners themselves navigated these turbulent shifts—rising, adapting, and at times, resisting the changing tides of education. This chapter centers the experiences of students who were thrust into remote, hybrid, or reimagined learning environments during global crises such as the COVID-19 pandemic. By foregrounding the voices of students, this chapter argues for a reimagined educational future where learner agency is not a luxury but a foundational principle. It concludes with insights and implications for educators, policymakers, and institutions seeking to build more inclusive, empathetic, and resilient learning environments in the face of future uncertainties.

INTRODUCTION

There are moments in history when time fractures—when the normal rhythm of our daily lives is thrown off course and the systems we rely on are forced to recalibrate. For the global education community, the COVID-19 pandemic was one such rupture. Almost overnight, classrooms emptied, school gates closed, and learning—once bound to physical spaces and shared routines—was thrust into an ambiguous, digital

DOI: 10.4018/979-8-3373-1350-4.ch005

ether. But even before this pandemic, education systems had been grappling with chronic inequities, rigid structures, and growing calls for learner-centered reform. The pandemic did not invent the challenges; it exposed them (Kayyali, 2020). And in doing so, it offered something rare—a mirror. It is within this mirror that we begin to see not just the chaos and confusion, but the quiet persistence of learners, the emergence of unheard voices, and the resilience of those who had no choice but to adapt. This chapter is about those voices—the students who, amid systemic disruption, found ways to continue learning, questioning, creating, and dreaming. It is about the profound and often overlooked shift in educational dynamics: from institutions driving the learning experience, to learners themselves becoming central architects of their journeys.

In many ways, the disruption functioned as both a collapse and a catalyst. It collapsed what was already brittle: over-standardized curricula, one-size-fits-all assessment models, inflexible teaching methods, and limited emotional scaffolding. At the same time, it catalyzed innovation—sometimes clumsily, sometimes creatively. Schools experimented with asynchronous instruction, teachers tried to reinvent their roles, and students, from a variety of backgrounds, adapted in ways that defied expectations. The shift to digital platforms may have been born out of necessity, but for many, it became a space of possibility. And yet, not all stories were of triumph. For every tale of adaptation, there were stories of disconnection, exclusion, and loss. Students in underserved communities, those without stable internet, personal devices, or supportive home environments, were disproportionately affected. Mental health deteriorated. Motivation waned. The "digital divide" became more than a buzzword—it became a lived reality (Voinow, 2024). But amid the silence of empty classrooms and muted microphones, some students began to find their voice—not always loud, but honest and urgent. They spoke of what learning meant to them, what they needed from educators, and what it felt like to be part of a system not designed for such volatility.

The purpose of this chapter is not to romanticize struggle or frame resilience as an effortless triumph. Rather, it seeks to understand how student-centered learning evolved during a moment of intense disruption, and what that evolution reveals about the potential and limitations of our educational frameworks. By foregrounding student experiences—gathered from interviews, reflective journals, and participatory research—we aim to uncover the subtle ways learners reclaimed agency during a time when so much was out of their control. Crucially, this is not just a retrospective analysis. The relevance of this inquiry extends far beyond the pandemic. Climate emergencies, geopolitical instability, economic crises, and technological upheaval continue to threaten the stability of educational systems worldwide. If we are to build learning environments that are resilient, inclusive, and humane, we must learn from the voices of those who lived through uncertainty—not only as survivors but as active co-creators of meaning. This chapter, then, is a testament—not just to the endurance of students, but to their insights, innovations, and insistence on being more than just passive recipients of knowledge. It is a call to educators, leaders, and policymakers to listen differently, to design boldly, and to reimagine education as something fundamentally rooted in the lived experiences and aspirations of learners themselves.

THE DISRUPTION THAT REDEFINED LEARNING: CONTEXTUALIZING THE CRISIS

When we speak of disruption in education, we often imagine policy shifts, curriculum reforms, or technological upgrades (Lingard et al, 2021). But the disruption that unfolded globally in recent years was of a different nature—more abrupt, more disorienting, and far more revealing than anything the sector had previously experienced. It did not simply tinker with the structures of schooling; it dismantled them. It did not gradually challenge old assumptions; it forcefully laid them bare. The global pandemic arrived not as a distant possibility but as a blunt reality, collapsing the very premise that schools were immovable constants in children's lives. And in doing so, it became the greatest stress test modern education has ever faced.

In the early days of the crisis, the silence in school corridors was deafening. Classrooms, once alive with chatter, energy, and the soft friction of learning, became ghostly images flickering through screens (Kelly, 2025). Overnight, students were asked to stay home, teachers to reinvent their entire pedagogical repertoire, and parents to become makeshift facilitators of learning—all without preparation, without precedent, and often without the most basic tools needed to do so. What we faced was not just an educational shift, but a rupture in the social fabric that had long defined how, when, and with whom learning takes place. And yet, amidst this rupture, something extraordinary began to take shape. As school buildings closed, the meaning of school itself became a subject of urgent questioning (Rosén Rasmussen, 2021). What is school, if not a physical place? What is learning, if not scheduled instruction? Who is a teacher, when teaching must happen at a distance? These were not merely logistical questions. They were existential ones. They forced educators, students, and entire communities to re-examine long-held assumptions and values. And in that uncomfortable space of uncertainty, a window opened—a moment to reimagine education, not around systems, but around people.

What became immediately apparent was that learning, at its core, is not dependent on walls, bells, or desks. It is a deeply human process, driven by connection, curiosity, and care. And yet, our existing models had too often neglected these truths, focusing instead on measurable outcomes, compliance, and control. The crisis laid this bare. It exposed the fragility of educational systems that had been operating under the illusion of permanence and predictability. Timetables and testing structures, rigid hierarchies and standardized instruction—so much of it, we came to realize, could not withstand the unpredictable. But it wasn't just the systems that were unprepared. It was also our way of thinking about students. For decades, students had been positioned within education systems as passive recipients—consumers of knowledge, rather than co-creators of meaning. The sudden shift to online and home-based learning revealed how limited and limiting this framing was (Kayyali, 2023). In the absence of teachers physically present to guide each moment, students were asked to self-direct, self-regulate, and self-motivate. And while some struggled understandably, others flourished in unexpected ways. They began to organize their time, seek out resources independently, form peer support groups, and even question the relevance of what they were being taught. For some, the disruption was a setback. For others, it was a liberation.

Still, it is important not to romanticize this moment. The disruption was not equal in its impact. It magnified existing inequalities with unrelenting clarity. Students in rural areas, those from economically marginalized communities, or those with special needs faced barriers that went far beyond internet access. They encountered systems that were never truly designed for them in the first place—systems that now, under pressure, revealed their bias, rigidity, and neglect. Some students simply disappeared from the radar of schooling. They were not absent because they didn't care; they were absent because no one asked how to keep them present. The pandemic became, for them, not just an educational crisis but a crisis of recognition and belonging. And yet, for all its devastation, the disruption also became a mirror—an unflinching one. It reflected not only what was broken, but also what was possible. Educators found themselves reaching for empathy over efficiency, improvisation over perfection. They called students just to ask how they were doing, recorded voice notes late into the night, and allowed themselves to be vulnerable. There was an opening for trust, for experimentation, for humanity to enter the classroom—albeit virtually—in a way that had long been crowded out by institutional demands.

In this space, student-centered learning began to find new ground (Hoidn & Reusser, 2020). Not as a theoretical model, but as a lived necessity. With the traditional controls removed, students were invited—sometimes unintentionally—to take ownership. Some began customizing how they learned, choosing to engage through video, audio, or written formats, based on what suited them best. Others questioned why certain subjects mattered while others were left out. The crisis became, for many, a kind of curriculum in itself—a course in adaptability, in emotional intelligence, in learning how to learn. It redefined the "hidden curriculum" not as unspoken norms, but as the very skills and insights required to survive and make meaning in a time of chaos. This redefining of learning also invited a redefinition of the learner. Students were no longer only pupils; they became caregivers, translators, breadwinners, tech troubleshooters, and emotional anchors in their families. Their roles expanded beyond academic identity, revealing just how multidimensional young people's lives are. And in recognizing this, many educators began to shift—from expecting compliance to offering compassion, from measuring outputs to witnessing experiences. The teacher-student relationship became less about control and more about co-navigation. Learning, in this new terrain, was no longer linear or standardized (Garg et al, 2022). It was messy, fragmented, and deeply personal. Of course, this shift did not occur uniformly. For every student who discovered a new sense of agency, there were others left behind. And for every teacher who leaned into creativity, there were others paralyzed by the sheer scale of change. Institutions, too, varied widely in their responses—some doubling down on surveillance and discipline through screen monitoring, while others embraced flexibility and autonomy. The inconsistency was stark. But so too was the underlying truth: when pressed by crisis, every choice made by an institution reveals what it truly values. The pandemic did not just disrupt education; it clarified its priorities—or lack thereof.

The chapter title, "The Disruption That Redefined Learning," is not meant to imply that this redefinition was complete or even wholly positive. Rather, it suggests that the disruption forced a collective reckoning with what learning has been and what it could be. It challenged the illusion of neutrality in education, revealing how every structure and policy carries assumptions—about who learners are, what they need, and what they are capable of. And in this process, a quiet, yet powerful, transformation began: a re-centering of the learner not as an object of intervention, but as a subject with voice, agency, and complexity. As we reflect on this moment, it becomes clear that the crisis was never just about logistics or digital tools. It was about meaning. It was about how education responds—or fails to respond—to the lived realities of those it claims to serve. It was about whether schools could become not just sites of instruction, but sanctuaries of relevance, care, and possibility. And perhaps, most importantly, it was about whether we are willing to learn from those who had no choice but to learn in the hardest of ways. In listening to students, we hear stories of struggle, yes, but also of ingenuity, strength, and vision. These stories should not be footnotes to educational policy; they should be its foundation. If we take them seriously—if we see them not as anomalies but as guides—then we may finally begin to design learning systems that are not only resilient in the face of crisis but transformative in times of calm. The disruption was not the end of learning. It was the beginning of something far more honest.

LISTENING TO LEARNERS: NARRATIVES OF ADAPTATION AND STRUGGLE

During times of crisis, it's easy to prioritize logistics—how to keep systems running, how to replicate structures in unfamiliar contexts, how to preserve some version of "normal" for the sake of continuity. In the urgency to adapt, institutions often focus on delivery, access, infrastructure, and metrics. But in the noise of adaptation, something softer—and arguably more essential—can get lost: the voices of learners themselves (Kayyali, 2024 A). Not the survey responses or performance data, but their stories. Their frustrations, confessions, anxieties, moments of joy, moments of resignation. Their truths. Listening to learners during educational disruption is not just an act of empathy—it is a form of inquiry, a way of making sense of a broken world through the eyes of those most affected by its rupture. The narratives students carry are not simply accounts of remote schooling or internet issues. They are layered reflections of how young people experienced a world that suddenly became unfamiliar, sometimes unkind, and often overwhelming. These stories are not neat. They don't always resolve. But they matter. Because in them lie the insights we so often overlook in policy documents and pedagogical frameworks. There was Sarah, a sixteen-year-old who used to thrive in the structure of her school day but fell into silence once the pandemic began. "At school," she said, "I knew where I was supposed to be, what I was supposed to do. At home, everything blurred. I started feeling like nothing had a purpose anymore." Her teachers marked her as disengaged, but no one asked what was happening behind her blank Zoom screen. In reality, Sarah was caring for her younger siblings while her parents worked double shifts. She wasn't failing. She was surviving.

Then there was Ali, a high-achieving student from a rural town where electricity was unreliable and internet access inconsistent. "Sometimes I'd walk to the top of the hill at night," he said, "because that was the only place I could get a signal. I'd record voice notes to my teachers and hope they got through." His classmates in urban areas spoke of feeling trapped inside. For Ali, education became an outdoor ritual of persistence—one shaped by nature, distance, and a profound sense of being on the margins. "I used to feel proud of how well I did in school," he confessed. "Now I just hope people remember I tried." And we must not forget the students like Leila, who had always been soft-spoken in class but found her voice in digital learning spaces. "It's strange," she wrote in her journal, "but I finally felt like I could be myself. In the chat box, I didn't feel watched. I could think before I spoke. I could ask questions without feeling stupid." For Leila, the disruption opened a window she never knew she needed. She began participating more, not less. She grew into her own learning style. She found a rhythm of thinking that the in-person classroom had never made space for. These are not isolated experiences. They represent just a sliver of the diverse, complicated realities that students navigated during this global disruption. And within them, we find a tension: adaptation and struggle are not mutually exclusive. In fact, they often coexist. Students adapted while struggling. They showed resilience while carrying invisible burdens. They completed assignments while grieving lost family members. They showed up while being nowhere near okay. And still, much of the educational discourse during this period reduced them to performance graphs and attendance sheets, missing the soul of what they were enduring.

The phrase "student-centered learning" took on new meaning during this time—not because it was universally applied, but because students had no choice but to center themselves in their learning (Kaput, 2018). With teachers distanced, schedules irregular, and resources inconsistent, students began to construct meaning in highly personal and unpredictable ways. Some organized virtual study groups. Others created their own learning goals. A few rejected the school system altogether and found knowledge in unexpected places—through YouTube tutorials, community forums, books inherited from older siblings. Learning became less institutional and more instinctual, driven by relevance, survival, and sheer necessity. Yet, these acts of agency often went unnoticed. Schools and teachers, many of whom were themselves overwhelmed, tried to maintain structure by replicating the classroom online (Lemov, 2020). But for students, the world had already changed. Many expressed feeling unheard—obligated to follow plans that no longer fit their context. "It felt like I was being asked to pretend nothing had happened," said Ricardo, a high school junior. "But everything had happened. My dad lost his job. My grandma got sick. I was scared all the time. And they still wanted me to write essays like before." His words echo a broader truth: learning cannot be separated from life. And when life is disrupted, so too is the capacity for conventional learning. What students needed was not perfection from educators, but presence. Not polished slideshows or strict schedules, but authentic connection. The teachers who left a mark during this time were not necessarily the most tech-savvy, but the most human. They were the ones who began class by asking, "How are you—really?" The ones who allowed for silence, for late submissions, for vulnerability. They didn't abandon rigor, but they redefined it. For them, learning became less about covering content and more about uncovering meaning.

In spaces where student voices were genuinely heard, something remarkable happened. Education became reciprocal. Students offered feedback on lesson formats. They suggested changes to deadlines based on household responsibilities. They co-created projects rooted in their lived experiences—documenting the lockdown through poetry, analyzing misinformation in the media, exploring mental health in their communities. When given space, students did not demand less. They demanded better—more relevant, more compassionate, more flexible education. And in doing so, they reminded us what it means to be engaged: not compliance, but connection. Of course, not every story was hopeful. For many, the struggle became too heavy. Some students disappeared from school entirely. Others showed up with cameras off, voices muted—not out of disrespect, but out of exhaustion. The trauma was layered. Some were navigating unstable homes. Others were confronting racism, isolation, or anxiety at unprecedented levels. The screen became both a portal and a barrier. It offered access, but also anonymity. And within that anonymity, some students lost their sense of self. They began to ask: Do I matter here? Is anyone seeing me?

These questions must not be dismissed as emotional afterthoughts. They strike at the heart of what education is supposed to be: a space where young people are seen, heard, and valued. The crisis revealed that many had never truly felt this way—not before, not during, and unless something changes, not after. Listening to learners, then, is not just a matter of documenting what went wrong. It is about witnessing the full scope of their humanity—their confusion, their courage, their contradictions—and using those insights to build something better. As we move forward, the temptation will be to "return to normal." To restore the schedules, assessments, and classroom routines that defined pre-crisis education. But normal was never neutral. And for many students, normal was never equitable, empowering, or even tolerable. If we have truly listened, we cannot un-hear what was shared. We cannot un-see what was revealed. The stories of adaptation and struggle are not detours—they are the map. They show us where the system broke, but also where the learners broke through. Education must honor these stories—not by memorializing the crisis, but by transforming its lessons into lasting change (Bonds, 2022). This means designing systems that accommodate different ways of learning, teaching, and being. It means elevating student voices not as token gestures, but as essential elements of decision-making. It means training educators not only in digital tools, but in emotional intelligence, trauma-informed practice, and culturally responsive pedagogy. It means seeing every learner not as a statistic, but as a story in progress—one that deserves to be heard, supported, and celebrated. Because in the end, adaptation without acknowledgment is incomplete. Struggle without validation is unjust. If we want to build educational environments that are truly student-centered, we must begin with the radical act of listening. Not to judge, not to fix, but to understand. To hear the tremble in a voice, the hesitation in a message, the brilliance in a quiet insight. To recognize that behind every muted screen is a world—complex, fragile, resilient, real. And that world has something to teach us. If only we are willing to listen.

REDESIGNING PEDAGOGY: FLEXIBILITY, EMPATHY, AND AGENCY

For much of modern history, pedagogy has been defined by structure. It has been the blueprint through which learning unfolds—the timing of lessons, the order of subjects, the pacing of units, the delivery of content. It has promised efficiency, predictability, and outcomes that can be measured and reported. But for all its discipline and logic, traditional pedagogy often left something essential behind: the humanity of learners. In the pursuit of order, it lost touch with emotion. In the name of uniformity, it overlooked diversity. In valuing performance, it neglected meaning. And then came the disruption—the sudden unraveling of what was once stable. Schools closed. Schedules dissolved. Classrooms migrated into living rooms, bedrooms, shared kitchens. The routines and rituals that had long governed how we teach and learn vanished almost overnight. What remained was raw: a chaotic mix of digital tools, frayed nerves, shifting roles, and the quiet realization that pedagogy, as we knew it, was no longer enough. It could no longer serve the world in front of us, a world reeling from loss, fear, and uncertainty. What was needed was not just a digital adaptation of existing systems, but a full reimagining. A redesign.

This redesign did not begin in boardrooms or curriculum planning meetings. It began in the quiet gestures of overwhelmed teachers who chose to pause a lesson and check in with their students. In the improvised changes to assignments that allowed room for context and care. In the decision to extend a deadline not because policy allowed it, but because a student was grieving. These moments—small, often undocumented—signaled the emergence of a new kind of pedagogy: one rooted in flexibility, empathy, and agency. A pedagogy that did not just aim to teach, but to honor the full complexity of being human in a time of crisis (Eschenbacher & Fleming, 2021). Flexibility became more than just a practical necessity. It became an ethic. Teachers who had long been instructed to "stick to the plan" began to realize that the plan no longer applied. Students were joining classes from shared devices, caring for siblings, working jobs, managing anxiety. Under such conditions, learning could not be rigid. It had to breathe. And so, schedules became more forgiving. Assignments became more open-ended. Participation began to be measured not by how often a student spoke in class, but by how they engaged in ways that made sense for their context. Some submitted essays. Others created podcasts. Some turned in handwritten notes. Others sent voice memos recorded on their phones. The diversity of expression was not just tolerated—it was welcomed.

This kind of flexibility demanded trust. It required teachers to believe that students were doing their best, even when it didn't look like it. It also required a willingness to relinquish control, to move from being the sole authority in the room to being a co-facilitator of learning. For many educators, this was not easy. Letting go of control can feel like a loss. But what many found instead was a deepening of connection. When students were given room to make decisions—about how they learned, how they demonstrated understanding, how they managed their time—they began to show up differently (Veldkamp et al., 2020). Not always perfectly. But more honestly. More fully. Empathy, too, became a cornerstone of the new pedagogy. Not the performative kind, where we offer a kind word before moving on to the lesson. But the kind that lingers, that reshapes the way we teach. Empathy meant designing learning that acknowledged grief. That made room for silence. That didn't assume everyone was okay. It meant seeing the student who hasn't turned in any work not as lazy, but perhaps as lost. It meant realizing that the camera-off student might be hiding more than just their face—they might be hiding fear, sadness, shame. Empathy required educators to move beyond assumptions and ask, "What might this student be carrying that I cannot see?" But perhaps most transformative was the elevation of student agency. Agency has often been misunderstood in educational spaces. It is not about letting students do whatever they want, nor is it about abandoning guidance or standards. Agency is about choice, voice, and ownership. It is about students feeling that their learning belongs to them—that it reflects who they are, what they care about, and how they grow. During the disruption, agency was no longer a theoretical ideal. It became essential. Students had to decide when to log in, how to manage their time, what to prioritize. They had to learn how to learn without the constant physical presence of a teacher.

This forced autonomy exposed gaps. Many students had never been taught to think independently, to manage their time, to self-regulate (Nilson & Zimmerman, 2023). The system had trained them to follow instructions, not to explore. But alongside these gaps emerged growth. Some students, once disengaged in traditional classrooms, began to thrive when given freedom. Others discovered learning styles that had never been accommodated before. A student who struggled to take notes in class now found success by watching recorded lessons multiple times. Another who feared speaking in class now contributed confidently through written discussion forums. In schools and classrooms where this shift was embraced, the results were profound. Students became collaborators. Teachers became listeners. The curriculum became more alive, shaped by student interest and lived experience. This was not a perfect transformation, and it certainly was not universal. Many schools clung tightly to old models, trying to replicate brick-and-mortar schooling through digital means (Kingsbury, 2021). But those who leaned into the messiness of change found something rare: an education that felt more real. Redesigning pedagogy in this way also meant acknowledging that not all students begin from the same place. Equity became central—not just as a goal, but as a lens through which every decision was filtered. Who has access to technology? Who has a quiet place to learn? Who has support at home? Who is being left out of our assumptions? A truly empathetic pedagogy recognizes that fairness is not sameness. It offers different paths for different needs. It adapts. Importantly, this kind of pedagogical redesign does not end with the crisis. If anything, the crisis revealed how urgently we needed this change all along. The systems we once trusted to educate our children were, in many ways, brittle and unkind. They prioritized conformity over creativity, obedience over curiosity, standardization over selfhood. And they worked best for those who already had the resources to succeed. For the rest, they demanded that students bend themselves to fit a mold that was never made for them.

Now, we know better. We have seen what becomes possible when pedagogy centers humanity. We have witnessed what happens when students are given space to shape their learning and when teachers are trusted to teach with heart, not just with standards. We have learned that flexibility is not the enemy of rigor—that empathy does not dilute learning but deepens it. And we have seen, perhaps most importantly, that students are not passive vessels waiting to be filled. They are thinkers, creators, feelers, questioners. They are already learning—even when they are not in school. The question is whether our schools are willing to learn from them. As we look ahead, the challenge is not just to recover what was lost, but to preserve what was found. To hold on to the spirit of experimentation, the openness to difference, the courage to depart from scripts that no longer serve us. We must design learning environments that are nimble enough to change, compassionate enough to care, and brave enough to trust the learner (Taylor & Dearybury, 2021). Redesigning pedagogy is not about replacing one model with another. It is about remembering that education is, at its core, a relationship—a dynamic, evolving relationship between people, between ideas, between the world as it is and the world as it could be. That relationship deserves our attention, our imagination, and our love.

RESILIENCE AS PRAXIS: LESSONS FOR A POST-CRISIS EDUCATION FUTURE

In times of upheaval, when the foundations of life and society tremble, resilience often emerges as the defining trait that allows individuals and systems to navigate the storm. Resilience is not just about surviving; it is about adapting, growing, and finding strength in the most unexpected places. For education, resilience has been more than just a reactive response to crisis. It has been a transformative force, reshaping the ways we think about teaching, learning, and the very nature of what it means to educate in the face of uncertainty. But resilience in education is not just a concept—it must be lived, practiced, and embedded within the structures of our educational systems. It must be resilience as praxis, a sustained, deliberate effort to foster a culture of adaptability, care, and innovation that endures well beyond the crisis.

The disruptions of the past few years have forced us to confront the vulnerabilities of our education systems. Schools, which had long been seen as pillars of stability, were suddenly exposed as fragile and unprepared for a global crisis (De Anda, 2024). Technology that was once a luxury or an afterthought became an essential lifeline. Pedagogical models that had been taken for granted—teacher-centered, one-size-fits-all—began to crack under the weight of sudden shifts to remote and hybrid learning. But in the midst of these challenges, a new understanding of resilience began to emerge. It wasn't just about how schools could continue to function under pressure. It was about how education itself could evolve, becoming more adaptive, more inclusive, and more deeply connected to the lived experiences of students.

To think about resilience in education is to acknowledge that it is not just about the capacity to bounce back from adversity. It is about how we reshape our understanding of education to ensure that it is not only capable of withstanding crises but also capable of thriving through them. Resilience as praxis means making resilience not just a response to emergencies but a central guiding principle of educational design. It is about integrating flexibility, care, and innovation into the very fabric of how we teach and learn, ensuring that education is not a static system but a dynamic, evolving force that responds to the changing needs of society and the world. This means that the lessons we take from the crisis must not be limited to logistical adjustments or temporary fixes. They must extend to how we think about the purpose and practice of education itself. Resilience, in this sense, is not just about recovery but about re-imagining what is possible. It's about creating an educational future where resilience is inherent—not as an afterthought but as a core value woven through every aspect of the educational process. One of the most profound lessons of the past few years is that education, at its best, is about connection (Felten & Lambert, 2020). It's about the relationships between teachers and students, between students and their peers, and between the individual and the community. When schools were forced to move online, these connections became strained. The lack of physical proximity, the barriers of technology, and the emotional weight of the crisis all worked against the idea of connection. Yet, in this very moment of disruption, educators and students alike discovered the depth of these relationships in ways that hadn't been fully realized before. Teachers found new ways to connect with their students—through phone calls, through messages of encouragement, through the simple act of checking in. Students, in turn, discovered their own resilience, learning how to ask for help, how to support each other in virtual study groups, and how to advocate for their needs in a system that seemed to have forgotten them.

But the question remains: how can we preserve and amplify these connections in the post-crisis world? How can we ensure that the resilience that was born out of necessity becomes a permanent feature of our educational practices? The answer lies in the deliberate cultivation of an education system that values connection, adaptability, and compassion as essential components of learning. This means moving beyond the transactional model of education, where learning is simply about the transfer of knowledge from teacher to student, and towards a model that recognizes education as a deeply human process—one that is shaped by the experiences, emotions, and realities of those involved (Aguilar, 2018). At the heart of this transformation is the recognition that resilience cannot be built through top-down mandates or standardized solutions. It must be built from the ground up, by listening to the voices of students, teachers, and communities. In the past, education was often seen as something done to students. But the pandemic forced us to rethink that assumption. Students were not passive recipients; they were active participants in their own learning journeys, shaping how they adapted to the new reality, and in many cases, even helping to guide the direction of the educational process. This student-centered approach, born out of necessity, has the potential to redefine what it means to be a learner in the twenty-first century. It calls for a shift in how we design curricula, how we assess progress, and how we foster engagement.

Resilience as praxis means acknowledging that each student is unique, with their own experiences, struggles, and strengths. It requires flexibility—not just in terms of deadlines or learning environments but in the way we think about how learning happens. Flexibility, in this sense, means creating space for students to pursue their own interests, to learn in ways that align with their individual needs, and to engage with content in ways that feel relevant to their lives (Kariippanon et al, 2018). It also means being adaptable in the face of setbacks, whether they come in the form of a technical failure, a personal crisis, or a broader societal disruption. Resilience is not about being perfect. It's about being open to change and willing to embrace the messiness of learning. Empathy is another cornerstone of resilience as praxis. The pandemic revealed that when we fail to acknowledge the emotional, psychological, and social dimensions of learning, we miss the opportunity to truly connect with students. Education cannot be reduced to a set of tasks or outcomes—it must be rooted in the emotional experiences of learners. Teachers, too, have their own struggles, their own challenges, and their own vulnerabilities. In a resilient system, empathy must flow in both directions: teachers must empathize with their students, but students must also feel that their teachers see and understand them. The connection between educator and student is not just transactional—it is relational, built on mutual respect, trust, and care (Pranjić, 2021). But empathy alone is not enough. It must be paired with agency—the ability for students to take ownership of their learning, to make choices, and to direct the path they follow. Agency is the engine of resilience because it gives students a sense of control in an otherwise uncontrollable world. When students are empowered to make decisions about how they learn, they become not just survivors of a crisis but active agents in shaping their own futures. This requires a shift in how we view students, from passive recipients of knowledge to active contributors to the learning environment. It is through agency that students develop not only academic skills but life skills—the ability to adapt, to problem-solve, and to persist in the face of adversity.

To foster resilience as praxis, we must also embrace a broader vision of what education is for. The crisis revealed that the traditional focus on standardized testing, rigid curriculums, and predetermined outcomes fails to account for the full range of human experience. Education must be about more than just preparing students for exams; it must be about preparing them for life (Daniel, 2020). It must be about helping them develop the skills, values, and mindsets that will allow them to navigate an uncertain world. This means emphasizing not just academic knowledge but emotional intelligence, social responsibility, and a commitment to lifelong learning. As we move forward, it is essential to recognize that resilience is not a one-time fix, nor is it something that can be measured by a test score or graduation rate (Hazlett, 2025). Resilience is a process—a continuous effort to adapt, grow, and evolve in the face of challenges. It is about building systems that are flexible enough to respond to the needs of all students, systems that prioritize connection, empathy, and agency (Kayyali, 2024 B). It is about creating educational spaces where students feel valued, where they are seen not just as test-takers but as whole, complex human beings with unique experiences, dreams, and potential. The lessons of the crisis are clear: education must be resilient. But resilience in education is not just about surviving a crisis—it is about thriving in the face of it. It is about embracing the uncertainty, the messiness, and the complexity of the human experience, and using those qualities to build a better future for all learners. If we are to create a truly resilient educational system, we must move beyond the old paradigms and build a new vision—one that is grounded in flexibility, empathy, and agency, and that is committed to ensuring that every student has the opportunity to succeed, no matter what challenges lie ahead.

CONCLUSION

If disruption was the crucible, then resilience emerged as the most vital—and perhaps most misunderstood—outcome of this educational crisis. But resilience, as it revealed itself through the voices of learners, was not about stoicism or silent perseverance. It was a deeply relational, active process of negotiating identity, finding purpose, and asserting agency in an ever-shifting educational landscape. For many students, learning during disruption became more than the pursuit of grades or diplomas—it became a form of self-preservation, a declaration of dignity, and an act of hope. What this period taught us, perhaps more than anything else, is that learning does not cease when the classroom walls fall away. In fact, for many students, real learning began only when traditional structures crumbled. Stripped of rigid schedules, standardized testing regimes, and passive content delivery, learners were left with a fundamental question: What does learning mean to me? The answers varied, but they all pointed to a deeper truth—that education, when centered around students rather than systems, becomes a transformative experience.

This shift—from institutional control to learner agency—was not seamless. It was messy, uneven, and often painful. But it also carried within it the seeds of something radically hopeful: the possibility of an education system that is more responsive, more compassionate, and more just. Students did not merely adapt to new tools or platforms; they adapted in ways that forced educators to rethink pedagogical assumptions. Some began designing their own schedules, negotiating assignment formats, and co-creating content with teachers. Others found solace in peer networks, online communities, or in reflective practices that helped them make sense of their fractured realities. In the absence of certainty, students cultivated creativity. In the absence of traditional support systems, they built new ones. And in the absence of in-person validation, they turned inward, developing self-motivation and self-regulation skills that no curriculum could have taught so effectively. These are not just survival strategies; they are lifelong competencies that deserve to be recognized, nurtured, and integrated into the future of learning. The implications are profound. Educational institutions must move beyond simply "recovering" from disruption. We must resist the urge to revert to pre-crisis norms as a measure of success. The goal should not be to return to business as usual, but to evolve—deliberately and democratically—towards models of learning that honor the diverse realities and aspirations of students. This means embedding student voice in curriculum design, prioritizing mental health and emotional literacy, embracing flexibility in assessment, and challenging the hierarchies that have long governed teacher-student relationships.

Equity must no longer be treated as an optional goal but as the foundation of any legitimate learning system. Technology, while offering tremendous promise, must be critically examined for the access gaps it can widen. And most of all, we must continue to ask students—not occasionally, but consistently—what they need, what they see, and how they learn best. Their responses, as this chapter has shown, are not only insightful—they are essential. In amplifying the voices of learners, we are reminded that education is not merely a system of instruction. It is a space of becoming. During the darkest moments of disruption, it was not standardized protocols that sustained learning—it was the human spirit, the capacity to adapt, to resist, and to reimagine. As we face an increasingly unpredictable world, let us not forget this lesson. Let us listen to our students—not just with sympathy, but with reverence. Let us make space for their resilience, not as a background trait, but as the very heartbeat of education's future. Because in the end, if we are to build a learning system that endures the storms to come, we must begin by centering the ones who have already weathered them.

REFERENCES

Aguilar, E. (2018). *Onward: Cultivating emotional resilience in educators*. John Wiley & Sons. DOI: 10.1002/9781119441731

Bonds, M. D. S. (2022). *Navigating Memorialization and Commemoration on US Campuses: Approaches to Crisis Recovery*. Routledge. DOI: 10.4324/9781003165491

Daniel, S. J. (2020). Education and the COVID-19 pandemic. *Prospects*, *49*(1), 91–96. DOI: 10.1007/s11125-020-09464-3 PMID: 32313309

De Anda, A. (2024). Experiences with Crisis and Resilience in Education in Emergencies: Voices and Lenses of Teachers in Nicaragua (Doctoral dissertation, Concordia University, St. Paul).

Eschenbacher, S., & Fleming, T. (2021). Toward a critical pedagogy of crisis. European journal for Research on the Education and Learning of Adults, 12(3), 295-309.

Felten, P., & Lambert, L. M. (2020). *Relationship-rich education: How human connections drive success in college*. Jhu Press. DOI: 10.1353/book.78561

Garg, S., Tsipras, D., Liang, P. S., & Valiant, G. (2022). What can transformers learn in-context? a case study of simple function classes. *Advances in Neural Information Processing Systems*, *35*, 30583–30598. DOI: 10.52202/068431-2217

Hazlett, B. (2025). Resiliency: An Exploration of Non-Cognitive Characteristics of First-Generation College Students as It Relates to Academic Success (Doctoral dissertation, Northeastern University).

Hoidn, S., & Reusser, K. (2020). Foundations of student-centered learning and teaching. In *The Routledge international handbook of student-centered learning and teaching in higher education* (pp. 17–46). Routledge. DOI: 10.4324/9780429259371-3

Kaput, K. (2018). Evidence for Student-Centered Learning. Education evolving.

Kariippanon, K. E., Cliff, D. P., Lancaster, S. L., Okely, A. D., & Parrish, A. M. (2018). Perceived interplay between flexible learning spaces and teaching, learning and student wellbeing. *Learning Environments Research*, *21*(3), 301–320. DOI: 10.1007/s10984-017-9254-9

Kayyali, M. (2020). Post COVID-19: New era for higher education systems. *International Journal of Applied Science and Engineering*, *8*(2), 131–145. DOI: 10.30954/2322-0465.2.2020.6

Kayyali, M. (2023). *Virtual Universities: An Overview & Trends*. Online Submission.

Kayyali, M. (2024a). Crisis Response and Higher Education: Social Innovations in Times of Global Challenges. In Social Innovations in Education, Environment, and Healthcare (pp. 1-17). IGI Global.

Kayyali, M. (Ed.). (2024b). Building Resiliency in Higher Education: Globalization, Digital Skills, and Student Wellness: Globalization, Digital Skills, and Student Wellness. IGI Global.

Kelly, J. M. (2025). The Glitch Is the Machine: Technological Mediation and Identity in The Collected Works of Rowan Black (Doctoral dissertation, Drew University).

Kingsbury, I. (2021). Online learning: How do brick and mortar schools stack up to virtual schools? *Education and Information Technologies*, *26*(6), 6567–6588. DOI: 10.1007/s10639-021-10450-1 PMID: 33551662

Lemov, D. (2020). *Teaching in the online classroom: Surviving and thriving in the new normal*. John Wiley & Sons.

Lingard, B., Wyatt-Smith, C., & Heck, E. (2021). Transforming schooling through digital disruption: Big data, policy, teaching, and assessment 1. In *Digital disruption in teaching and testing* (pp. 1–44). Routledge.

Nilson, L. B., & Zimmerman, B. J. (2023). *Creating self-regulated learners: Strategies to strengthen students' self-awareness and learning skills*. Routledge. DOI: 10.4324/9781003443803

Pranjić, S. S. (2021). Development of a caring teacher-student realationship in higher education. *The Journal of Education, Culture, and Society*, *12*(1), 151–163. DOI: 10.15503/jecs2021.1.151.163

Rosén Rasmussen, L. (2021). Building Pedagogies. A historical study of teachers' spatial work in new school architecture. *Education Inquiry*, *12*(3), 225–248. DOI: 10.1080/20004508.2020.1857495

Taylor, J., & Dearybury, J. (2021). *The Courageous Classroom: Creating a Culture of Safety for Students to Learn and Thrive*. John Wiley & Sons.

Veldkamp, A., Van De Grint, L., Knippels, M. C. P., & Van Joolingen, W. R. (2020). Escape education: A systematic review on escape rooms in education. *Educational Research Review*, *31*, 100364. DOI: 10.1016/j.edurev.2020.100364

Voinow, A. V. (2024). Holistic Strategies for a Connected Future: Can we Bridge the Digital Divide by Redefining Smart City Agendas?.

Chapter 6
Leading Through Change: Empowering Educators in Times of Crisis

Satya Subrahmanyam
http://orcid.org/0000-0003-0441-2742
Holy Spirit University of Kaslik, Lebanon

ABSTRACT

Periods of uncertainty and crisis, such as the COVID-19 pandemic and geopolitical conflicts, challenge traditional educational systems, necessitating a shift in teaching and learning methods. This chapter, Navigating the Unknown: Strategies for Educators in Times of Change, explores strategies for educators to adapt, foster resilience, and maintain continuity in education during turbulent times. It highlights key challenges, including rapid technology adoption, equity, mental health concerns, and the psychological toll on both educators and students. Practical strategies such as fostering a growth mindset, collaboration, and innovative pedagogical approaches are discussed. The chapter emphasizes the importance of leadership, well-being, and proactive planning, offering case studies and actionable insights to equip educators for future uncertainties.

1. INTRODUCTION

Periods of uncertainty and crisis, ranging from global pandemics and geopolitical unrest to rapid technological shifts, pose significant challenges for educational systems worldwide. These disruptions often manifest as sudden transitions to remote learning, institutional instability, and heightened emotional and cognitive stress among both educators and learners (Bozkurt & Sharma, 2020). In such contexts, the concept of "uncertainty" refers to unpredictable conditions with limited foresight, while

DOI: 10.4018/979-8-3373-1350-4.ch006

"crisis" denotes high-stakes, often urgent situations that disrupt normal functioning (Capano, 2020). For educators, these conditions can upend pedagogical routines, strain resources, and necessitate swift adaptation to unfamiliar environments, placing a heavy burden on their professional and personal well-being.

The purpose of this chapter is to explore strategic responses that empower educators to navigate these unpredictable terrains. By identifying adaptive strategies, core competencies, and institutional support, the chapter seeks to provide a comprehensive framework for responding effectively to disruption. Central to this exploration is the recognition that educators are not merely transmitters of knowledge but also leaders, caregivers, and change agents during crises. As such, understanding their evolving role and equipping them with the necessary tools is vital to ensuring continuity, resilience, and innovation in education during times of change.

1.1 Context and Rationale

The 21st century has been marked by increasingly frequent and complex global disruptions that profoundly affect all societal sectors, including education. From the COVID-19 pandemic to geopolitical instability, economic uncertainties, environmental crises, and digital transformation, educational institutions have found themselves at the intersection of profound and unpredictable change. Within this context, the terms *uncertainty* and *crisis* in education refer to situations characterized by ambiguity, volatility, and a lack of reliable information or precedents for decision-making. These can be global in nature - such as health pandemics or climate emergencies - or more localized events, such as civil unrest or abrupt policy changes, all of which can destabilize learning environments and disrupt the continuity of educational delivery (Bozkurt et al., 2020).

Crises in education are not solely defined by the nature of the external event but also by the level of preparedness and resilience of the institution and its members. While the COVID-19 pandemic posed a global threat, its impact on schools and universities varied widely depending on existing digital infrastructure, educator competencies, and leadership support (Reimers & Schleicher, 2020). In some regions, educational activities came to a complete halt, whereas others quickly transitioned to online modalities. The experience underscored that uncertainty is not merely a state of not knowing, but a condition that tests the agility, creativity, and moral commitment of educators.

The repercussions of uncertainty are multidimensional. For educators, crises often mean rethinking pedagogical practices, managing personal stress while supporting students, and navigating new technological environments with limited training or support. Students, on the other hand, face challenges related to access, equity, motivation, and mental health. At the institutional level, unpredictability can hinder strategic planning, affect funding and enrollment, and expose systemic vulnerabilities (Daniel, 2020). Thus, the need for informed, adaptable, and proactive strategies to manage educational uncertainty is both immediate and essential.

1.2 Purpose of the Chapter

This chapter is situated within the broader framework of understanding and responding to the *dynamics of educational change in periods of uncertainty and crisis*. It specifically aims to explore and present strategic responses that empower educators to navigate such periods with resilience and efficacy. By analyzing the roles educators play in times of upheaval and outlining practical strategies derived from research and case studies, the chapter seeks to serve as a guide for practitioners, policymakers, and educational leaders committed to ensuring continuity, quality, and equity in education under adverse conditions.

One central aim of this chapter is to challenge the notion that crisis must inherently result in chaos or regression. Instead, it presents uncertainty as an opportunity for transformation, a catalyst for reimagining educational practices and structures in ways that better serve diverse learners. Educators are not only knowledge transmitters but also frontline change agents whose actions can significantly influence student outcomes and institutional resilience. As such, equipping them with the necessary tools, mindsets, and frameworks to act decisively during crises is paramount.

This chapter seeks to provide a coherent framework for navigating unpredictability in educational contexts. The framework will include components such as fostering personal resilience, leveraging reflective practice, utilizing technology effectively, and engaging in collaborative problem-solving. Each of these elements contributes to a holistic understanding of how educators can maintain instructional continuity, promote student well-being, and support institutional adaptability.

The insights presented in this chapter are informed by interdisciplinary research in education, psychology, and organizational change, as well as by the lived experiences of educators who have navigated uncertainty. The goal is not to present a one-size-fits-all solution but to offer adaptable strategies that can be customized to various contexts and levels of education, from primary and secondary schools to universities and adult learning environments.

This chapter aspires to encourage a shift in how we perceive and prepare for uncertainty in education. Rather than reacting with fear or resistance, educators and institutions should approach the unknown with a strategic mindset that values flexibility, inclusivity, and innovation. In doing so, they can not only withstand the pressures of a crisis but emerge stronger, more cohesive, and more aligned with the evolving needs of learners in an unpredictable world.

2. UNDERSTANDING THE NATURE OF UNCERTAINTY IN EDUCATION

The education sector, long perceived as a bastion of continuity and tradition, has increasingly become subject to a multitude of unpredictable and destabilizing forces. Whether emanating from global pandemics, regional conflicts, economic downturns, or internal institutional disruptions, uncertainty has emerged as a pervasive theme in the modern educational landscape. To effectively navigate and respond to such unpredictability, it is imperative to first understand the types of crises that impact education, the unique characteristics that define educational uncertainty, and the broad spectrum of effects these crises have on all educational stakeholders.

2.1 Types of Crises Impacting Education

When crises occur on a **global scale**, they affect educational systems all across the globe. Among the most significant worldwide disruptions of the 21st century, the COVID-19 epidemic had a major impact on the delivery and perception of education. School closures impacted nearly 1.5 billion pupils in over 190 countries during the height of the epidemic, according to UNESCO (2020). Instructors were forced into online classes unprepared, while parents and students struggled with concerns related to digital access, motivation, and safety.

Beyond pandemics, other global crises such as climate change, international economic instability, and cybersecurity threats also pose significant challenges to educational continuity. Climate-induced natural disasters such as floods, hurricanes, and wildfires have led to the displacement of school populations and destruction of infrastructure, disrupting learning trajectories and community cohesion (Venegas Marin et al., 2024).

Regional Crises are geographically contained yet can be equally devastating to affected communities. These include political unrest, armed conflicts, economic recessions, and public health emergencies specific to certain areas. Prolonged conflict in regions such as Syria and Yemen has led to the destruction of schools, displacement of students, and the loss of qualified teachers (UNICEF, 2024). In economically distressed regions, austerity measures often lead to reduced educational budgets, resulting in teacher layoffs, increased class sizes, and deteriorating school conditions.

Economic collapse at the regional level can be particularly damaging, as it undermines both the physical infrastructure of education and the psychological stability of the stakeholders involved. In countries facing hyperinflation or currency devaluation, the purchasing power of families and institutions diminishes, limiting access to quality education, learning resources, and up-to-date technologies (Al Khalili, 2025).

Institutional Crises, while less visible on the global stage, can cause profound disruption at the micro level. These crises often involve internal challenges such as abrupt leadership changes, accreditation losses, governance failures, or funding cuts. The sudden departure of a university president or school principal during a strategic overhaul can disrupt leadership continuity and stakeholder confidence. Funding crises in public education systems frequently result in the elimination of programs, delayed salaries, and an erosion of morale among staff and students alike (Reimers et al., 2022).

Institutional crises may also stem from internal conflicts, such as labor disputes, mismanagement, or cases of misconduct. These can create a toxic environment, leading to attrition, declining student performance, and reputational damage. Unlike global and regional crises, institutional crises often require localized and context-sensitive interventions, placing significant responsibility on the competence and preparedness of school leaders.

Table 1. Types of crises impacting education

Type of Crisis	Scope / Examples	Impact on Education
Global Crises	COVID-19 pandemic, climate change, international economic instability, cybersecurity threats	- School closures affecting ~1.5 billion students (COVID-19) - Shift to online learning unpreparedly - Disruption of learning continuity due to natural disasters, displacement, infrastructure damage
Regional Crises	Political unrest, armed conflicts, regional economic recessions, localized public health emergencies	- Destruction of schools and displacement of students (Syria, Yemen) - Loss of qualified teachers - Reduced educational budgets, larger class sizes, deteriorating school conditions - Limited access to quality education and learning resources during economic collapse
Institutional Crises	Abrupt leadership changes, accreditation loss, governance failures, funding cuts, internal conflicts	- Disruption of leadership continuity and stakeholder confidence - Program eliminations, delayed salaries, staff/student morale decline - Attrition, declining performance, reputational damage

2.2 Characteristics of Uncertainty in Educational Contexts

Uncertainty in education is distinguished by several key characteristics that collectively challenge traditional approaches to planning, instruction, and evaluation.

Rapid Change

The first hallmark of educational uncertainty is the speed with which events unfold and demand response. The sudden shift to online learning during the COVID-19 pandemic is a prime example of rapid change. Many institutions were forced to adopt new pedagogical technologies, revise curricula, and train faculty within weeks. Such accelerated transformations often outpace the ability of educators and institutions to adapt, resulting in temporary or permanent disruptions to learning (Bozkurt et al., 2020).

Limited Information

Another defining feature of educational uncertainty is the paucity of reliable, actionable information. During times of crisis, decisions must often be made with incomplete data, ambiguous guidelines, or conflicting expert opinions. During the early months of the COVID-19 pandemic, debates around the effectiveness of online learning tools, student assessment methods, and the safety of reopening schools highlighted the difficulties of decision-making under uncertainty (Subrahmanyam, 2021).

Disrupted Routines

Educational systems thrive on structure - daily schedules, syllabi, semesters, and academic calendars provide predictability and coherence. In periods of uncertainty, these routines are frequently disrupted. Students and teachers may face abrupt shifts in mode of instruction, school closures, or cancellation of exams and graduations. The breakdown of these structures can lead to confusion, anxiety, and disengagement, particularly among younger learners and vulnerable populations (Hodges et al., 2020).

Disruption is not confined to academic routines; it extends to social and emotional support systems embedded within schools. The lack of face-to-face interaction, extracurricular activities, and mentoring opportunities can leave students feeling isolated and unmotivated. Teachers who rely on collegial collaboration may struggle with the isolation of remote teaching environments.

2.3 Effects on Educational Stakeholders

Emotional and Psychological Toll

Crisis-induced uncertainty takes a substantial emotional and psychological toll on both educators and learners. Educators may experience heightened stress, burnout, and a sense of helplessness, particularly when they are expected to fulfill professional obligations without adequate support or resources. According to a survey conducted by the RAND Corporation, nearly one in four teachers in the U.S.

reported symptoms of depression during the pandemic, a rate significantly higher than the national average (Steiner & Woo, 2021).

Anxiety, dread, and emotional exhaustion are common among students as well. Important cognitive processes for learning, including memory, concentration, and problem-solving, might be hindered by these emotional states. Academic failure and high dropout rates have been associated with chronic stress and trauma, particularly in areas hit hard by war or by economic hardship (Cerna-Turoff et al., 2021).

Cognitive Challenges

Cognitive development and academic progress are often disrupted during crises. Remote learning, for instance, poses challenges in maintaining attention, comprehension, and critical thinking, especially for younger students or those without prior exposure to digital platforms. The lack of immediate feedback and peer interaction can hinder the development of higher-order thinking skills and reduce the effectiveness of collaborative learning strategies (Means et al., 2020).

Educators, too, must rapidly assimilate new information, technologies, and teaching methods, often with minimal guidance. This cognitive overload can reduce their ability to deliver engaging and effective instruction, leading to lower student achievement and satisfaction.

Professional and Institutional Strain

On a professional level, uncertainty can derail career progression, disrupt professional development, and challenge educators' sense of identity and purpose. In higher education, adjunct faculty and contract-based educators often face job insecurity during financial crises. Professional development opportunities may be postponed or eliminated, depriving educators of vital skills needed to navigate new teaching environments (American Federation of Teachers, 2022).

Institutions also face considerable strain, including reputational risks, diminished student enrollment, and increased operational costs. Universities that fail to adapt quickly to remote learning may lose competitive standing, while those that invest in expensive technological solutions may struggle with long-term financial sustainability.

Understanding the nature of uncertainty in education is a prerequisite for developing effective strategies to mitigate its impact. Crises - whether global, regional, or institutional - disrupt educational continuity, affect the well-being of stakeholders, and challenge the resilience of systems. These disruptions are characterized by rapid change, limited information, and the breakdown of established routines. The consequences are profound, encompassing emotional distress, cognitive challenges, and professional setbacks. By acknowledging these realities and examining the multifaceted impacts of uncertainty, educators and policymakers can begin to craft responses that not only address immediate needs but also lay the foundation for more adaptive, resilient, and inclusive educational systems in the future.

3. THE ROLE OF EDUCATORS IN TIMES OF CHANGE

The 21st-century education landscape has been dramatically transformed by unprecedented disruptions, ranging from pandemics and technological revolutions to shifting socio-political dynamics. These changes have redefined not only the structure of educational institutions but also the role of educators within them. The contemporary educator is no longer a mere dispenser of knowledge but a multifaceted leader who must adapt, innovate, and provide emotional and strategic guidance amid continuous changes. This section explores how educators have transitioned into adaptive leaders, outlines the essential competencies they require, and offers illustrative vignettes that capture their evolving roles in real-world contexts.

3.1 From Instructor to Adaptive Leader

Historically, educators have functioned primarily as content experts and classroom managers. In times of systemic and sudden change - such as during the COVID-19 pandemic, political unrest, or digital transformation - this conventional role is no longer sufficient. Educators are now expected to assume dynamic responsibilities that extend beyond pedagogy, encompassing leadership, emotional support, and innovation.

Facilitator: Modern educators must serve as facilitators of learning rather than authoritative figures. This entails fostering environments that encourage inquiry, collaboration, and self-directed learning. During remote learning phases in the COVID-19 pandemic, teachers had to facilitate discussions in virtual settings, guiding students in navigating digital tools and sustaining engagement despite physical separation (Ribeiro & Subrahmanyam, 2021).

Counselor: Amid crises, students often face emotional and psychological challenges. Educators, therefore, must take on a supportive role, recognizing signs of distress and offering empathy and stability. The ability to counsel does not require professional therapeutic expertise but demands awareness and responsive action, particularly during widespread crises such as war, economic instability, or health emergencies.

Innovator: In rapidly evolving educational environments, innovation becomes indispensable. Educators must be able to design creative solutions to overcome new teaching challenges, whether it be through flipped classrooms, gamification, or integrating artificial intelligence tools. Innovation also includes rethinking assessments and creating inclusive learning strategies that cater to diverse learners, particularly in under-resourced or crisis-affected regions (Bozkurt & Sharma, 2020).

Thus, the educator's transition from instructor to adaptive leader is not a matter of choice but a necessity to ensure continuity and relevance in education during uncertain times.

3.2 Essential Competencies for Educators

In their expanded roles, educators must cultivate a diverse set of competencies. These competencies not only equip them to manage immediate disruptions but also empower them to contribute meaningfully to long-term educational transformation.

Emotional Intelligence: Emotional intelligence (EI) is critical for building relationships, managing stress, and responding to students' emotional needs. Goleman (1995) outlines five dimensions of EI - self-awareness, self-regulation, motivation, empathy, and social skills - that are particularly relevant to education. An educator with high EI can create a supportive classroom climate, navigate interpersonal conflicts, and maintain a sense of calm in turbulent times.

Flexibility and Adaptability: Unpredictable educational contexts demand educators who are flexible in their teaching approaches and responsive to change. Adaptability enables educators to pivot from traditional classroom settings to remote learning environments, adjust lesson plans in real-time, and respond effectively to institutional shifts, such as curriculum overhauls or policy changes (Reimers & Schleicher, 2020).

Digital Literacy: The proliferation of digital tools in education means that educators must possess digital literacy, not only in using technology but in leveraging it for pedagogical purposes. Competence in learning management systems (e.g., Moodle, Canvas), video conferencing platforms (e.g., Zoom, MS Teams), and emerging educational technologies (e.g., AI-based tutors, simulations) is now integral to teaching effectiveness.

Strategic Thinking: Strategic thinking enables educators to make informed decisions aligned with long-term educational goals. It involves anticipating potential challenges, assessing risks, and designing interventions that consider both immediate outcomes and future impacts. In times of institutional crises such as funding cuts or leadership changes, educators with strategic insight can advocate for sustainable solutions and institutional resilience (Subrahmanyam, 2024).

3.3 Educator Experiences and Case Vignettes

Real-world cases and educator narratives vividly illustrate how the roles and competencies discussed are operationalized in times of change.

Vignette 1: Ms. Amina – The Facilitator During Conflict

In a conflict-affected region of Syria, Ms. Amina, a secondary school teacher, found herself teaching displaced students in a makeshift school camp. With minimal resources, she adopted storytelling and role-playing techniques to engage students traumatized by war. She encouraged group projects that allowed children to express their emotions and build peer support systems. Her approach demonstrated the facilitator role, promoting learning despite the limitations of the environment and the psychological burdens on students.

Vignette 2: Mr. James – Digital Innovator Amidst COVID-19

Mr. James, a mathematics teacher in South Africa, struggled initially with online teaching during the pandemic. He soon mastered tools like GeoGebra and Kahoot to create interactive lessons. He also recorded short video tutorials and formed WhatsApp study groups for students without consistent internet access. His journey from digital novice to innovator exemplified how adaptability and digital literacy can transform pedagogical effectiveness under pressure (Mpungose, 2020).

Vignette 3: Dr. Lin – Strategic Leader in a Time of Institutional Change

At a major university in Taiwan undergoing leadership transition and budget reductions, Dr. Lin, a senior lecturer, spearheaded a faculty initiative to redesign curricula for blended learning. She consulted with students and staff, analyzed enrolment trends, and coordinated with administration to secure funding for pilot programs. Her actions reflect strategic thinking and leadership, ensuring both academic integrity and institutional adaptability.

Vignette 4: Ms. Clara – The Counselor in Post-Pandemic Recovery

Returning to in-person classes after the pandemic, Ms. Clara, an elementary school teacher in Italy, noticed signs of anxiety and disengagement among her students. She incorporated mindfulness exercises and weekly check-in circles into her routine. By providing a safe emotional space, she nurtured trust and helped students regain a sense of normalcy. Her role as a counselor highlighted the importance of emotional intelligence in education recovery efforts.

Figure 1. The role of educators in times of change

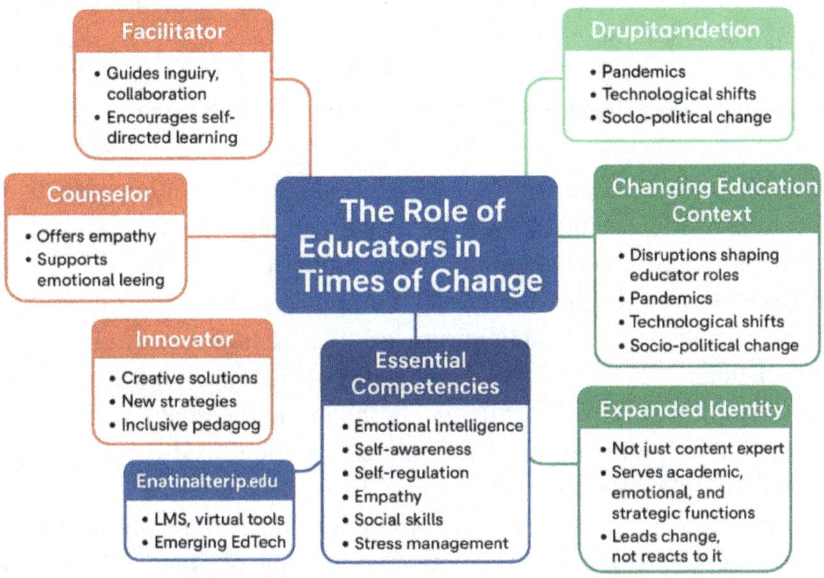

(Author, 2025)

In times of disruption and uncertainty, educators are not merely conveyors of knowledge but pivotal agents of transformation. Their roles must evolve to meet the multifaceted demands of changing educational landscapes. This shift requires the cultivation of competencies such as emotional intelligence, adaptability, digital literacy, and strategic thinking. Real-world experiences affirm that educators who embrace these roles and skills can not only sustain learning during crises but also lead systemic educational improvements. As education systems worldwide face ongoing and future disruptions, investing in educational development is both urgent and essential.

4. CORE STRATEGIES FOR NAVIGATING EDUCATIONAL UNCERTAINTY

Periods of uncertainty and crisis in education - whether precipitated by global pandemics, sociopolitical unrest, or digital disruptions - demand intentional, adaptive responses from educators. To remain effective and sustainable in such contexts, educators must be equipped not only with tools and knowledge but also with resilient mindsets and community support systems. This section delineates seven core strategies that empower educators to navigate uncertainty while continuing to inspire, support, and guide learners. Each strategy offers both theoretical grounding and practical application, aiming to foster resilience and adaptability at individual, institutional, and systemic levels.

4.1 Building Personal and Professional Resilience

Resilience is the cornerstone of sustainability in the face of ongoing disruption. For educators, resilience entails the capacity to maintain psychological well-being, professional efficacy, and adaptability despite stress or adversity.

Self-care and mental health: Self-care practices, including mindfulness, exercise, adequate sleep, and maintaining work-life balance, are vital for emotional resilience (Kuntz, 2021). Burnout among educators has escalated during periods of crisis, such as the COVID-19 pandemic, highlighting the urgent need to prioritize mental health through institutional support and personal well-being strategies.

Professional development: Engaging in continuous professional learning can significantly enhance educators' confidence and coping skills. Participation in webinars, certification programs, and communities of practice helps educators stay abreast of pedagogical innovations and build a sense of purpose and control (Subrahmanyam, 2024).

By investing in self-care and professional growth, educators can better withstand the emotional toll of uncertainty and emerge as more effective practitioners.

4.2 Embracing Reflective Practice

Reflective practice fosters a deeper understanding of teaching processes and enhances decision-making during periods of uncertainty. It involves examining one's experiences, beliefs, and assumptions to improve professional performance and emotional awareness.

Critical reflection: Critical reflection, as proposed by Cornish (2024), encourages educators to think deeply about their actions, particularly when standard practices no longer apply. In a crisis, this reflection becomes essential to adapt quickly and make informed pedagogical choices.

Journaling: Keeping a reflective journal helps educators process emotions and assess the effectiveness of their responses to challenges. It offers a safe space for self-exploration and growth.

Peer feedback: Structured opportunities for feedback from colleagues can also enhance reflection. Collaborative review sessions and teaching circles foster mutual learning and the co-construction of adaptive strategies (Brookfield, 2017).

Embracing reflective practice enables educators to respond more thoughtfully and less reactively, supporting both personal and pedagogical growth during times of flux.

4.3 Cultivating a Growth Mindset

The ability to adapt to new situations depends on having a growth mindset, which is the conviction that one's skills may be enhanced with hard work and focus. Viewing problems as chances for learning rather than threats may greatly boost resilience and performance, according to Ba and others' (2025) research.

Openness to change: Educators with a growth mindset are more likely to experiment with new tools and methods, take risks in curriculum design, and accept feedback constructively. They foster the same mindset in students, creating a culture of continuous improvement.

Reframing challenges: Challenges such as shifting to online teaching or managing diverse learner needs can be reframed as opportunities to innovate, improve inclusivity, or reach broader audiences. This reframing transforms adversity into a catalyst for development.

A growth mindset not only equips educators to survive periods of uncertainty but also empowers them to lead transformative change in their institutions.

4.4 Leveraging Technology for Continuity

Technology is an indispensable tool for maintaining educational continuity amid disruption. From virtual classrooms to learning management systems, digital tools enable flexibility and inclusivity.

Digital tools and platforms: Educators must be competent in using video conferencing (e.g., Zoom, Microsoft Teams), asynchronous platforms (e.g., Moodle, Google Classroom), and interactive tools (e.g., Padlet, Kahoot) to engage students remotely or in hybrid settings (Bozkurt & Sharma, 2020).

Hybrid and asynchronous models: Blended learning approaches allow for greater flexibility in instruction. Asynchronous models are especially beneficial during crises, enabling learners with connectivity or scheduling challenges to access content at their own pace (Hodges et al., 2020).

Digital equity: Leveraging technology also involves addressing access disparities. Educators should consider low-bandwidth tools and mobile-accessible content to ensure inclusivity.

Through strategic and inclusive use of technology, educators can mitigate learning disruptions and reach students in diverse and challenging contexts.

4.5 Establishing Flexible and Responsive Pedagogy

Rigid instructional approaches are ill-suited to volatile educational environments. Flexibility in pedagogy allows educators to adapt to student needs and changing conditions while maintaining academic rigor.

Universal Design for Learning (UDL): UDL is a framework that promotes inclusive instruction by offering multiple means of representation, engagement, and expression (CAST, 2018). It accommodates diverse learning needs, especially relevant during crises when students face heightened personal and academic challenges.

Trauma-informed teaching: Educators must recognize the psychological impacts of crises on learners. Trauma-informed pedagogy emphasizes safety, trust, empowerment, and empathy in classroom interactions (Norrish & Brunzell, 2023). This approach helps re-establish a sense of normalcy and support.

Assessment flexibility: Offering varied assessment formats - project-based, reflective, formative - ensures that students can demonstrate learning even when traditional exams are not feasible.

By adopting responsive pedagogical practices, educators create inclusive, equitable environments that remain effective despite uncertainty.

4.6 Strengthening Collaboration and Community

Collaboration is essential during times of uncertainty, both for professional support and for collective problem-solving. Building networks and communities enhances resource sharing, morale, and innovation.

Peer support networks: Informal and formal support systems among educators provide emotional and intellectual resources. These networks facilitate mentoring, team teaching, and co-reflection, reducing isolation and increasing collective efficacy (Hendrickx et al., 2025).

Interdisciplinary collaboration: Working across disciplines allows educators to integrate diverse perspectives and design holistic learning experiences. Interdisciplinary teams are especially effective in responding to complex challenges such as curriculum redesigning during institutional transitions.

Shared decision-making: Empowering educators to participate in institutional decisions regarding crisis response strategies fosters trust and ownership. Participatory leadership enhances resilience by tapping into collective intelligence and shared values (de La Hoz-Ruiz et al., 2024).

Strong professional communities foster mutual growth and ensure that no educator faces uncertainty alone.

4.7 Scenario Planning and Contingency Thinking

Educators must not only respond to crises but also anticipate and prepare for them. Scenario planning and contingency thinking provide frameworks for proactive strategy development.

Preparing for multiple futures: Scenario planning involves identifying potential future developments and crafting responses to each. An educator may plan differentiated teaching strategies for full in-person, hybrid, or entirely online modes of instruction. This anticipatory mindset increases agility (Burt, 2023).

Strategic foresight and agile planning: Strategic foresight combines trend analysis, horizon scanning, and risk assessment to inform long-term planning. Agile educational planning incorporates flexibility, rapid feedback, and iterative improvement, ensuring plans can evolve as conditions change (Salza et al., 2019).

Incorporating scenario-based thinking transforms uncertainty from a reactive burden into a manageable and even strategic opportunity.

Figure 2. Core strategies for navigating educational uncertainty

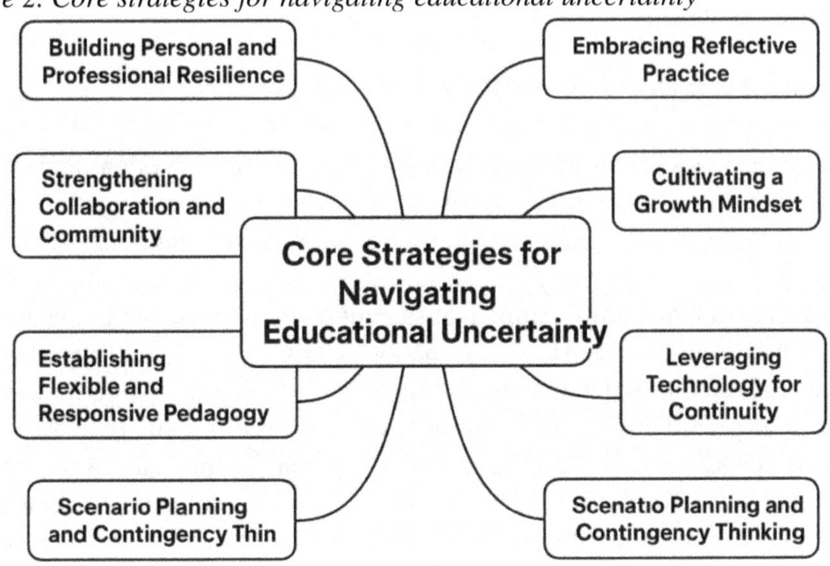

(Author, 2025)

Uncertainty in education is no longer an occasional disruption but an enduring reality of the modern era. To navigate this complex landscape, educators must adopt multifaceted strategies that encompass personal well-being, pedagogical agility, technological proficiency, and collaborative engagement. Building resilience, embracing reflective practices, cultivating growth mindsets, and developing adaptive teaching frameworks all serve to equip educators for sustained success. Strategic foresight and strong professional communities provide structural support to weather the unknown. By intentionally developing these competencies and approaches, educators can transform uncertainty into a space for innovation, inclusivity, and transformative learning.

5. INSTITUTIONAL SUPPORT SYSTEMS AND POLICY IMPLICATIONS

The rapidly changing landscape of education, influenced by global crises, technological disruption, and evolving learner needs, calls for robust institutional support systems and responsive policy frameworks. Institutions that effectively support educators not only enhance teaching and learning outcomes but also ensure long-term resilience and adaptability. This section explores the vital role of leadership in supporting educators, offers targeted policy recommendations, and highlights examples of institutional best practices that demonstrate how strategic planning and innovative responses can help navigate educational uncertainty.

5.1 Role of Leadership in Supporting Educators

Leadership plays a central role in establishing a supportive, resilient, and adaptive educational ecosystem. In times of uncertainty, the effectiveness of institutional responses often depends on how leadership communicates, engages stakeholders, and fosters trust.

Transparent Communication: Transparent communication is foundational in times of crisis. Leaders must share information clearly, promptly, and consistently to mitigate confusion and reduce anxiety among educators. Transparent communication helps align expectations, maintains morale, and ensures that faculty feel valued and informed. According to Fullan (2020), leadership transparency during crises builds relational trust, which is critical for the collective efficacy of educational teams.

Inclusive Planning and Shared Governance: Educational leaders must involve educators in institutional decision-making processes to ensure that policies and strategies reflect on-the-ground realities. Shared governance not only democratizes decision-making but also fosters ownership and accountability. Inclusive planning enables leaders to tap into the diverse perspectives and experiences of faculty, creating more responsive and context-sensitive strategies (Bryk et al., 2015). This approach is especially crucial when decisions affect classroom instruction, workload distribution, or assessment standards.

Empathetic and Transformational Leadership: In volatile times, leaders who demonstrate empathy and adopt transformational leadership practices tend to be more effective. They inspire innovation, provide emotional support, and cultivate a vision that empowers educators. Bass and Riggio (2006) suggest that transformational leaders create environments where faculty are motivated to surpass traditional goals and embrace change as an opportunity for growth.

5.2 Policy Recommendations

In order to institutionalize support for educators and ensure agility in response to future challenges, educational policies must be forward-looking, flexible, and equity-driven. The following policy recommendations are essential for strengthening institutional resilience.

Flexible and Adaptive Policies: Rigid policies often hamper the ability of institutions to respond to unexpected events. Policies must allow flexibility in curriculum delivery, assessment modalities, and faculty workload management. For example, offering hybrid teaching options and flexible deadlines for research outputs can alleviate pressure on faculty during emergencies. These adaptive measures must be formalized within institutional frameworks to enable a proactive rather than reactive approach (Bozkurt & Sharma, 2020).

Emergency Preparedness and Crisis Response Planning: Educational institutions should develop comprehensive emergency preparedness plans that encompass not only infrastructural concerns but also pedagogical and human resource considerations. Policies must outline protocols for remote teaching, data protection, communication hierarchies, and mental health support. Establishing a crisis response task force within each institution can facilitate faster and more coordinated action when disruptions occur (OECD, 2020).

Investment in Professional Development and Technology Training: Continuous professional development (CPD) is key to equipping educators with the skills required to navigate digital learning environments and pedagogical innovations. Institutions should allocate budgetary resources for training in digital literacy, online pedagogical methods, and emerging educational technologies. Structured training programs - especially in asynchronous learning, inclusive teaching practices, and data analytics - enable faculty to adapt to varied learning needs and contexts (Trust & Whalen, 2020).

Equity-Focused Policy Frameworks: Policies must be designed with an emphasis on equity and inclusion. This includes provisions for supporting underrepresented or marginalized faculty, ensuring accessible digital tools, and addressing the digital divide among students and staff. Institutions must promote Universal Design for Learning (UDL) principles and invest in tools that support diverse learning needs (Saborío-Taylor & Rojas-Ramírez, 2024).

5.3 Examples of Institutional Best Practices

Several educational institutions worldwide have demonstrated exemplary approaches in managing change and uncertainty. The following case studies showcase the best practices that can serve as models for others.

The University of Cape Town (South Africa): A Multi-Pronged Crisis Response - In response to the COVID-19 pandemic, the University of Cape Town (UCT) adopted a multi-pronged strategy that combined leadership agility with community-centered decision-making. The university launched an Emergency Remote Teaching (ERT) plan that included free data bundles for staff and students, asynchronous learning modules, and an online teaching helpline. UCT initiated mental health programs for educators facing burnout and stress (Czerniewicz et al., 2020). Their approach demonstrated how contextual awareness, inclusive communication, and resource equity can support institutional resilience.

Arizona State University (USA): Investment in Technological Infrastructure and Faculty Training - Arizona State University (ASU) has long been recognized for its strategic investments in digital innovation. Even prior to the pandemic, ASU had developed an extensive infrastructure for online learning. During the crisis, ASU scaled its digital platforms and provided on-demand faculty training in virtual pedagogy, assessment design, and student engagement techniques. Their faculty development model, centered on micro-credentials and peer mentoring, has been instrumental in building teaching capacity (Crow & Dabars, 2020).

Monash University (Australia): Fostering Collaboration and Leadership Empowerment - Monash University implemented a decentralized model of leadership that empowered departmental heads to make context-specific decisions during times of disruption. This autonomy allowed for faster response and greater alignment with local realities. The institution also prioritized collaboration, forming interdisciplinary teams to develop remote learning resources and mental health programs. Their success was underpinned by a strategic commitment to adaptive governance and collaborative decision-making (Eva et al., 2021).

Zayed University (UAE): Hybrid Learning Models and Faculty Wellness Programs - Zayed University introduced an innovative hybrid model that blended asynchronous and synchronous delivery, catering to both faculty and student needs. The university also launched a comprehensive Faculty Wellness Program that included mindfulness sessions, peer coaching, and access to mental health professionals. This holistic approach to educator support underscores the significance of integrating wellness into institutional strategy (Zayed University, 2020).

Institutional support systems and policy frameworks are critical for enhancing the resilience, adaptability, and well-being of educators. The role of leadership is pivotal in establishing transparent communication, inclusive planning, and empathetic engagement. Effective policy design - characterized by flexibility, preparedness, and equity - can transform institutions into agile and responsive learning environments. Real-world examples from institutions across the globe illustrate that thoughtful investment in faculty development, technology infrastructure, and community building leads to sustained educational quality amid uncertainty. Moving forward, educational institutions must continue to build on these strategies, fostering ecosystems where educators are empowered to thrive in dynamic and often unpredictable contexts.

6. INSIGHTS GAINED AND FUTURE DIRECTIONS

Crises such as the COVID-19 pandemic have served as inflection points for education systems worldwide, exposing structural weaknesses while simultaneously highlighting areas of strength and resilience. Beyond the immediate disruptions, these experiences have presented an opportunity to reevaluate educational priorities and chart new directions for sustainable reform. This section explores the insights gained from recent educational crises, the need to cultivate a culture of adaptability, and the implications for long-term transformation of curricula, assessments, and institutional design.

6.1 What Crises Reveal About Educational Systems

The COVID-19 pandemic, among other global crises, has acted as a magnifying lens, bringing long-standing inequities and structural challenges to the forefront. At the same time, it has revealed the resilience and creativity of educators and institutions under pressure.

Exposure of Inequities: Perhaps the most striking revelation during the pandemic was the vast inequity embedded within educational systems. Differences in access to technology, reliable internet, safe learning environments, and parental support created stark learning gaps. According to UNESCO (2020), nearly 1.6 billion learners were affected by school closures, with those from marginalized communities bearing the greatest burden. The digital divide not only separated students but also placed disproportionate stress on educators in under-resourced settings who lacked training or tools to adapt quickly to remote instruction.

Further, disparities in policy implementation and emergency response capabilities revealed institutional fragility in low-income contexts. In many cases, national or regional governments failed to provide coherent strategies, leaving schools to improvise responses, thereby exacerbating inconsistencies across regions (Reimers & Schleicher, 2020).

Recognition of Strengths: Despite the challenges, the pandemic also revealed the extraordinary adaptability and dedication of teachers, administrators, and students. Educators reimagined instruction with limited preparation time, developed creative pedagogical approaches, and supported learners emotionally and academically. Collaborative networks emerged, sharing resources and best practices through digital communities (Trust & Whalen, 2020). Institutions that had invested in digital infrastructure and professional development before the crisis were notably more agile in navigating disruptions.

A Potential for Transformation: The crisis acted as a catalyst, pushing institutions toward innovation and modernization at a pace previously thought impossible. What emerged was the recognition that traditional, rigid models of education are no longer sufficient. There is now widespread acknowledgment of the need for flexible learning pathways, technology integration, and learner-centric pedagogies. As Schleicher (2020) notes, education systems must move from reactive responses to proactive transformation, building future-ready institutions that can withstand volatility.

6.2 Towards a Culture of Adaptability

A major takeaway from recent crises is the necessity of embedding adaptability into the very fabric of education systems. Cultivating adaptability not only prepares systems for future disruptions but also enhances overall teaching and learning quality.

Encouraging Innovation and Risk-Taking: Educational systems have historically been slow to change, often characterized by bureaucratic inertia and resistance to innovation. The pandemic demonstrated the need for risk-taking, experimentation, and localized innovation. Teachers who were previously confined to rigid curricula began designing context-specific learning modules, blending synchronous and asynchronous modalities. Institutions that allowed for pedagogical autonomy saw greater innovation and responsiveness (Bozkurt & Sharma, 2020).

Encouraging a culture of innovation requires more than episodic change - it demands structural support through funding, time allocation for experimentation, and recognition mechanisms that reward creativity. Creating "innovation incubators" within schools and universities can empower educators to test new ideas and share successful models across institutions.

Embedding Adaptability in Teacher Education: Teacher education must evolve from a focus on content delivery to a broader emphasis on flexibility, digital competence, emotional intelligence, and systems thinking. Future educators must be equipped to navigate uncertainty, integrate technology meaningfully, and personalize learning. According to Darling-Hammond et al. (2020), adaptive expertise should be a core competency in teacher training programs.

This shift requires redesigning initial teacher education curricula and professional development to include modules on crisis pedagogy, blended learning strategies, data-informed instruction, and digital well-being. Partnerships between teacher education institutions and schools can create practicum experiences that simulate real-world complexities, preparing future teachers to manage dynamic classroom environments.

6.3 Implications for Long-Term Educational Reform

Moving forward, the insights gleaned from crisis experiences should inform ambitious, long-term reforms that strengthen the relevance, equity, and adaptability of educational systems. These reforms must address curriculum design, assessment practices, and institutional frameworks.

Redesigning Curriculum for the 21st Century: Traditional curricula, heavily weighted toward rote learning and standardized content, have shown their limitations in preparing students for real-world challenges. The post-pandemic era calls for curricula that prioritize critical thinking, problem-solving, collaboration, digital literacy, and global citizenship. These competencies must be integrated across disciplines and grade levels, promoting interdisciplinary learning and real-world application.

Curricula should be designed with flexibility in mind, allowing for modular learning paths, differentiated instruction, and responsiveness to local contexts. The integration of project-based and experiential learning approaches can foster deeper engagement and skill development (OECD, 2020).

Rethinking Assessment Practices: The pandemic prompted a reevaluation of high-stakes, standardized testing as the dominant mode of assessment. With in-person exams disrupted, institutions turned to alternative forms of assessment such as portfolios, formative assessments, and peer evaluations. These alternatives highlighted the potential of assessment for learning rather than assessment of learning.

Future assessments must be designed to capture a broader range of competencies, including creativity, collaboration, and resilience. Digital tools and AI-driven analytics can support real-time, personalized assessment feedback, while also enhancing accessibility for diverse learners (Subrahmanyam, 2025). Assessments must be culturally sensitive and inclusive, reflecting the diverse backgrounds and experiences of learners.

Redefining Institutional Frameworks: The crisis underscored the importance of institutional agility, decentralization, and collaboration. Rigid, hierarchical structures often delayed responses, while institutions with flexible governance models adapted more efficiently. Long-term reform should promote distributed leadership, collaborative planning, and stakeholder engagement.

Educational institutions must also invest in robust digital infrastructure, cloud-based learning environments, and data systems that enable real-time decision-making. Institutional resilience depends on the integration of physical and digital learning spaces, supported by continuous professional development and well-being initiatives for staff and students alike.

Institutions must develop mechanisms for regular reflection and feedback, ensuring that reforms remain dynamic and responsive. The use of learning analytics, stakeholder surveys, and action research can inform ongoing improvements and strategic planning.

Crises such as the COVID-19 pandemic have exposed systemic inequities and challenges in education but have also illuminated areas of resilience, creativity, and potential transformation. The future of education depends on our ability to learn from these disruptions and embed adaptability, equity, and innovation into the core of our systems. This calls for a new paradigm of teacher education, curriculum design, assessment practices, and institutional governance.

By fostering a culture of adaptability, encouraging innovation, and designing flexible and learner-centered frameworks, educational institutions can not only prepare for future uncertainties but also enhance the quality and relevance of education for all learners. Long-term educational reform should no longer be viewed as a periodic undertaking but as a continuous, iterative process driven by evidence, collaboration, and a shared vision for inclusive and future-ready learning.

7. CONCLUSION

The unprecedented educational disruptions brought about by global crises such as the COVID-19 pandemic have served as a wake-up call, underscoring the vulnerabilities and strengths of our educational systems. This chapter has explored the multifaceted challenges faced by educators and institutions - including inequities in access, technological limitations, mental health concerns, and the inertia of traditional policies - and highlighted the adaptive strategies that emerged in response. From leadership-driven institutional support and flexible policy formulation to innovations in pedagogy and the embedding of adaptability within teacher training, these experiences have provided invaluable insights into how education systems can become more resilient and inclusive.

A key takeaway from these turbulent times is the urgent necessity of embracing uncertainty not as a threat, but as a catalyst for transformation. Crises have exposed deep systemic inequities but have also unveiled untapped potential for creativity, collaboration, and reform. Educators demonstrated remarkable agility, often redesigning pedagogical approaches overnight. Institutions that fostered transparent communication, empowered decision-making, and ongoing professional development were better able to manage transitions and ensure continuity in learning.

As the education sector moves forward, there is a moral and strategic imperative to shift from reactive to proactive planning. This involves embedding flexibility into curricula, assessments, and institutional frameworks and cultivating a mindset of lifelong learning and continuous improvement among educators. Building a culture of adaptability - where innovation is encouraged, risk-taking is supported, and inclusivity is prioritized—is not merely desirable but essential for long-term sustainability.

Thus, this chapter concludes with a call to action: educators, policymakers, and institutional leaders must seize the lessons learned from past disruptions to future-proof education. Proactive preparation for change, through investment in digital infrastructure, robust teacher training, and inclusive policies, should become a cornerstone of educational planning. Institutions must move beyond short-term crisis responses to build adaptive systems capable of thriving in uncertainty.

Only by doing so can we ensure that education remains a pillar of equity, empowerment, and resilience in a rapidly changing world.

REFERENCES

Al Khalili, T. (2025). Parental strategies and educational inequalities during severe economic, political and refugee crises in conflict-affected Lebanon. *Asia Pacific Education Review*, *26*(2), 553–566. DOI: 10.1007/s12564-025-10040-7

American Federation of Teachers. (2022). *AFT report: Pandemic intensified job insecurity for adjuncts.* https://www.aft.org/news/aft-report-pandemic-intensified-job-insecurity-adjuncts

Ba, Y., Ming, W., & Zhang, H. (2025). Unlocking academic success: How growth mindset interventions enhance student performance through self-belief and effort regulation. *Acta Psychologica*, *256*, 104977. DOI: 10.1016/j.actpsy.2025.104977 PMID: 40209651

Bass, B. M., & Riggio, R. E. (2006). *Transformational leadership* (2nd ed.). Psychology Press. DOI: 10.4324/9781410617095

Bozkurt, A., Jung, I., Xiao, J., Vladimirschi, V., Schuwer, R., Egorov, G., & Paskevicius, M. (2020). A global outlook to the interruption of education due to COVID-19 pandemic: Navigating in a time of uncertainty and crisis. *Asian Journal of Distance Education*, *15*(1), 1–126. DOI: 10.5281/zenodo.3878572

Bozkurt, A., & Sharma, R. C. (2020). Emergency remote teaching in a time of global crisis due to CoronaVirus pandemic. *Asian Journal of Distance Education*, *15*(1), i–vi. https://www.asianjde.com/ojs/index.php/AsianJDE/article/view/447

Brookfield, S. D. (2017). *Becoming a critically reflective teacher* (2nd ed.). Jossey-Bass.

Bryk, A. S., Gomez, L. M., Grunow, A., & LeMahieu, P. G. (2015). *Learning to improve: How America's schools can get better at getting better*. Harvard Education Press.

Burt, G. (2023). *Evolution of scenario planning: Theory and practice from disorder to order*. De Gruyter., DOI: 10.1515/9783110792065

Capano, G. (2020). Policy design and state capacity in the COVID-19 emergency in Italy: If you are not prepared for the (un)expected, you can be only what you already are. *Policy and Society*, *39*(3), 326–344. DOI: 10.1080/14494035.2020.1783790 PMID: 35039724

CAST. (2018). *Universal design for learning guidelines (Version 2.2).* http://udlguidelines.cast.org

Cerna-Turoff, I., Fischer, H. T., Mansourian, H., & Mayhew, S. (2021). The pathways between natural disasters and violence against children: A systematic review. *BMC Public Health*, *21*(1), 1249. DOI: 10.1186/s12889-021-11252-3 PMID: 34247619

Cornish, L. (Ed.). (2024). *(Re)learning as Reflective Practitioners: Insights for Other Professions from Reflective Practice in Teacher Education*. Springer., DOI: 10.1007/978-3-031-60211-5

Crow, M. M., & Dabars, W. B. (2020). *The fifth wave: The evolution of American higher education*. Johns Hopkins University Press. DOI: 10.1353/book.73164

Czerniewicz, L., Agherdien, N., Badenhorst, J., Belluigi, D., Chambers, T., Chili, M., de Villiers, M., Felix, A., Gachago, D., Gokhale, C., Ivala, E., Kramm, N., Madiba, M., Mistri, G., Mgqwashu, E., Pallitt, N., Prinsloo, P., Solomon, K., Strydom, S., & Wissing, G. (2020). A wake-up call: Equity, inequality and COVID-19 emergency remote teaching and learning. *Postdigital Science and Education*, *2*(3), 946–967. DOI: 10.1007/s42438-020-00187-4 PMID: 40477091

Daniel, S. J. (2020). Education and the COVID-19 pandemic. *Prospects*, *49*(1–2), 91–96. DOI: 10.1007/s11125-020-09464-3 PMID: 32313309

Darling-Hammond, L., Hyler, M. E., & Gardner, M. (2020). *Effective teacher professional development*. Learning Policy Institute. https://learningpolicyinstitute.org/product/effective-teacher-professional-development-report

de La Hoz-Ruiz, J., Khalil, M., Domingo Segovia, J., & Liu, Q. (2024). Learning analytics for enhanced professional capital development: A systematic review. *Frontiers in Psychology*, *15*, 1302658. DOI: 10.3389/fpsyg.2024.1302658 PMID: 38318080

Eva, N., Wolfram Cox, J., Tse, H., & Lowe, K. B. (2021). *The concept of collective leadership and how it could lead us out of a pandemic*. Monash Impact. https://www2.monash.edu/impact/articles/covid-19/the-concept-of-collective-leadership-and-how-it-could-lead-us-out-of-a-pandemic/

Fullan, M. (2020). *The new meaning of educational change* (5th ed.). Teachers College Press.

Goleman, D. (1995). *Emotional intelligence: Why it can matter more than IQ*. Bantam Books.

Hendrickx, M. M. H. G., Thurlings, M. C. G., & den Brok, P. (2025). Teachers' collaborative knowledge building in professional learning communities: Connecting interaction patterns to learning gains. *European Journal of Psychology of Education*, *40*(1), 39. Advance online publication. DOI: 10.1007/s10212-024-00938-y

Hodges, C., Moore, S., Lockee, B., Trust, T., & Bond, A. (2020). The difference between emergency remote teaching and online learning. *EDUCAUSE Review*, *27*, •••. https://er.educause.edu/articles/2020/3/the-difference-between-emergency-remote-teaching-and-online-learning

Kuntz, J. C. (2021). Resilience in times of global pandemic: Steering recovery and thriving trajectories. *Applied Psychology = Psychologie Appliquée*, *70*(1), 188–215. DOI: 10.1111/apps.12296 PMID: 33362330

Means, B., & Neisler, J. with Langer Research Associates. (2020). *Suddenly Online: A National Survey of Undergraduates During the COVID-19 Pandemic*. San Mateo, CA: Digital Promise.

Mpungose, C. B. (2020). Emergent transition from face-to-face to online learning in a South African University in the context of the Coronavirus pandemic. *Humanities & Social Sciences Communications*, *7*(1), 113. Advance online publication. DOI: 10.1057/s41599-020-00603-x

Norrish, J., & Brunzell, T. (2023). How is trauma-informed education implemented within classrooms? A synthesis of trauma-informed education programs. *The Australian Journal of Teacher Education*, *48*(3). Advance online publication. DOI: 10.14221/1835-517X.6159

OECD. (2020). *The impact of COVID-19 on education: Insights from Education at a Glance 2020*. Paris: OECD Publishing. https://web-archive.oecd.org/2020-09-08/562941-the-impact-of-covid-19-on-education-insights-education-at-a-glance-2020.pdf

Reimers, F. M., Amaechi, U., Banerji, A., & Wang, M. (Eds.). (2022). *Education to build back better: What can we learn from education reform for a post-pandemic world?* Springer., DOI: 10.1007/978-3-030-93951-9

Reimers, F. M., & Schleicher, A. (2020). *A framework to guide an education response to the COVID-19 pandemic of 2020*. OECD Publishing., DOI: 10.1787/6ae21003-

Ribeiro, N., & Subrahmanyam, S. (2021). Emotional challenges of teaching online during the pandemic. In Uddin, M., Sony, K. K., & Verma, N. (Eds.), *Psychosocial issues in COVID-19 pandemic* (pp. 244–255). SHINE Publication. RED.

Saborío-Taylor, S., & Rojas-Ramírez, F. (2024). Universal design for learning and artificial intelligence in the digital era: Fostering inclusion and autonomous learning. *International Journal of Professional Development. Learners and Learning*, *6*(2), ep2408. Advance online publication. DOI: 10.30935/ijpdll/14694

Salza, P., Musmarra, P., & Ferrucci, F. (2019). Agile methodologies in education: A review. In Parsons, D., & MacCallum, K. (Eds.), *Agile and lean concepts for teaching and learning: Bringing methodologies from industry to the classroom* (pp. 25–45). Springer Singapore., DOI: 10.1007/978-981-13-2751-3_2

Schleicher, A. (2020). *The impact of COVID-19 on education: Insights from Education at a Glance 2020.* OECD. https://www.improvingthestudentexperience.com/library/covid19/the-impact-of-covid-19-on-education-insights-education-at-a-glance-2020.pdf

Steiner, E. D., & Woo, A. (2021). *Job-related stress threatens the teacher supply: Key findings from the 2021 State of the U.S. Teacher Survey*. RAND Corporation. https://www.rand.org/pubs/research_reports/RRA1108-1.html

Subrahmanyam, S. (2021). Attitudes and beliefs of faculty on teaching online during pandemic. In Sony, K. K., Verma, N., & Uddin, M. (Eds.), *Psychosocial issues in Covid-19 pandemic* (pp. 256–268). SHINE Publication. RED.

Subrahmanyam, S. (2024). Implementing agile talent management practices to adapt to changing higher education landscapes. In Kayyali, M. (Ed.), *Cultivating creativity and navigating talent management in academia* (pp. 187–210). IGI Global Scientific Publishing., DOI: 10.4018/979-8-3693-6880-0.ch009

Subrahmanyam, S. (2024). The role of leadership in nurturing innovation and creativity among faculty and staff. In Kayyali, M. (Ed.), *Cultivating creativity and navigating talent management in academia* (pp. 449–468). IGI Global Scientific Publishing., DOI: 10.4018/979-8-3693-6880-0.ch021

Subrahmanyam, S. (2025). Open Educational Resources and Their Impact on Curriculum Flexibility and Accessibility. In Cruz, M., Queirós, R., & Mascarenhas, D. (Eds.), *Building Teaching Competencies for AI-Driven and Inclusive Learning* (pp. 69–102). IGI Global Scientific Publishing., DOI: 10.4018/979-8-3373-7729-2.ch002

Trust, T., & Whalen, J. (2020). Should teachers be trained in emergency remote teaching? Lessons learned from the COVID-19 pandemic. *Journal of Technology and Teacher Education*, *28*(2), 189–199. https://www.learntechlib.org/primary/p/215995/. DOI: 10.70725/307718pkpjuu

UNESCO. (2020). *Education: From disruption to recovery.* UNESCO. https://en.unesco.org/covid19/educationresponse

UNICEF. (2024). *Education in emergencies.* https://www.unicef.org/education/emergencies

Venegas Marin, S., Schwarz, L., & Sabarwal, S. (2024). Impacts of extreme weather events on education outcomes: A review of evidence. *The World Bank Research Observer, 39*(2), 177–226. DOI: 10.1093/wbro/lkae001

Zayed University. (2020). *Zayed University switches to hybrid model of teaching, learning for Fall 2020 semester.* ZU News. https://www.zu.ac.ae/main/en/news/2020/august/hybrid20

KEY TERMS AND DEFINITIONS

Adaptive Leadership: A style of leadership in which educators and administrators adjust their strategies, roles, and approaches in response to changing circumstances and complex challenges.

Collaborative Networks: Professional and peer support systems that encourage shared decision-making, interdisciplinary collaboration, and collective problem-solving during times of change.

Crisis: A critical event or period in education that threatens the stability, safety, or functioning of institutions and stakeholders, requiring rapid and strategic responses.

Digital Literacy: The ability of educators and learners to effectively use digital tools, platforms, and technologies to ensure continuity of learning in hybrid or fully online environments.

Growth Mindset: The belief that abilities and skills can be developed through effort, learning, and adaptation, fostering openness to change and innovation in educational contexts.

Reflective Practice: A systematic approach to self-evaluation and critical thinking, where educators analyze their experiences to improve teaching effectiveness and professional growth.

Resilience: The capacity of educators and learners to recover from setbacks, manage stress, and maintain professional effectiveness during periods of uncertainty.

Scenario Planning: A strategic foresight technique that involves anticipating multiple potential futures and preparing adaptable responses to manage uncertainty in educational settings.

Uncertainty in Education: A state in which educators, students, and institutions face unpredictable changes that disrupt normal routines, decision-making, and outcomes, often caused by crises such as pandemics, conflicts, or digital disruptions.

Universal Design for Learning (UDL): An educational framework that provides flexible learning environments to accommodate diverse learner needs, promoting inclusivity and responsiveness during disruptions.

Chapter 7
Modern Pedagogy and Smart Technologies:
Empowering Quality Education for Sustainable Development Goal 4

Vivek Ahuja
http://orcid.org/0000-0001-5073-4342
FLAME University, India

ABSTRACT

The pursuit of Sustainable Development Goal 4 (SDG4) – ensuring inclusive and equitable quality education for all – necessitates a shift in teaching and learning. This chapter explores the intersection of modern pedagogy and smart technologies in transforming education to enhance accessibility, engagement, and outcomes. It examines innovative teaching approaches like inquiry-based, active, personalized, and project-based learning, emphasizing their role in fostering critical thinking and problem-solving skills. Additionally, the chapter highlights smart technologies, including e-learning, gamification, augmented and virtual reality (AR/VR), and artificial intelligence (AI), in redefining education. These tools facilitate personalized learning, data-driven decision-making, and immersive experiences, ensuring adaptability in diverse environments. The discussion underscores how these advancements support SDG4 by promoting inclusivity, empowering educators, and improving monitoring and evaluation.

I. INTRODUCTION

The integration of modern pedagogical methods and smart technologies is a dynamic and transformative endeavor with profound implications for educational

DOI: 10.4018/979-8-3373-1350-4.ch007

quality (Chaudhry & Malik, 2014; Hosseini & Kinnunen, 2021; Sathishkumar, 2012; Visvizi, 2018). This chapter reviews existing literature to explore the intersection of innovative teaching methodologies and cutting-edge technologies, emphasizing their synergistic role in achieving Sustainable Development Goal 4 (SDG4) - Quality Education.

Modern pedagogy methods play a central role in enhancing educational quality, as highlighted by various approaches. *Inquiry-Based Learning*, for instance, encourages students to ask questions, developing curiosity and independent learning. *Active Learning Strategies*, including group discussions and problem-solving activities, promote engagement and collaborative learning, enhancing comprehension and retention. *Personalized Learning*, tailored to individual student needs, is facilitated through adaptive learning platforms, adjusting the learning experience based on progress and preferences. *Project-Based Learning* contributes to practical skill development and creativity, preparing students for real-world challenges.

Complementing these pedagogical advancements are smart technologies that further revolutionize education. E-Learning Platforms and Massive Open Online Courses (MOOCs) make education accessible globally, breaking down barriers of location and time. Gamification in Education enhances motivation and engagement, making learning enjoyable and effective. Augmented Reality (AR) and Virtual Reality (VR) create immersive learning environments, providing hands-on experiences. Artificial Intelligence (AI) automates administrative tasks, offers personalized learning experiences, and facilitates intelligent tutoring systems.

When the world faced the challenges posed by the COVID-19 pandemic, the need to integrate technology into pedagogy became even more apparent, with a focus on user-centric design (Hosseini & Kinnunen, 2021). This chapter also acknowledges the urgency brought about by the pandemic and discusses the significance of modern pedagogy methods and smart technologies in ensuring continued and enhanced educational delivery.

The chapter further explores how this integration contributes to the achievement of SDG4. The discussion emphasizes upon the role of these advancements in promoting inclusivity by bridging educational gaps for marginalized communities. Additionally, addresses the importance of enhancing teacher training to empower educators with the skills needed to effectively integrate smart technologies into their teaching methods. Moreover, the chapter explores the use of data analytics and AI-driven tools for monitoring and evaluation, underlining their role in improving teaching methodologies and student outcomes through data-driven decision-making.

In essence, this review paper aims to provide a comprehensive explanation of the interrelationship between modern pedagogy methods and smart technologies, shedding light on their collective potential to reshape education and contribute significantly to the global pursuit of quality education for all.

While this chapter focuses on modern pedagogical approaches, it is important to distinguish them from traditional pedagogies to better understand their transformative potential. Traditional pedagogies are largely teacher-centered, emphasizing lecture-based instruction, standardized curricula, and passive knowledge transmission. In contrast, modern pedagogies are learner-centered, focusing on active engagement, critical thinking, collaboration, and real-world application of knowledge. Approaches such as inquiry-based learning, project-based learning, and personalized learning shift the role of the educator from a knowledge provider to a facilitator of learning. This transition is particularly significant in the context of rapidly evolving technological environments, where learners are expected not only to acquire knowledge but also to apply, analyze, and create. The integration of smart technologies further strengthens modern pedagogies by enabling interactive, adaptive, and data-driven learning experiences.

II. MODERN PEDAGOGY METHODS: ENHANCING EDUCATIONAL QUALITY

Inquiry-Based Learning (IBL)

Inquiry-Based Learning (IBL) stands as a learner-focused pedagogical approach that has demonstrated its efficacy in enhancing the understanding and integration of educational content (Lower-Hoppe et al., 2021). At its core, IBL encourages students to actively engage with the learning process by posing meaningful questions, exploring topics, and applying critical thinking skills. This approach is designed to promote a sense of curiosity and independence in learners, developing a lifelong commitment to acquiring knowledge (Ismail, 2006).

The impact of IBL on students' cognitive development is substantial. IBL not only deepens students' comprehension of the subject matter but also stimulates their intellectual curiosity by placing an emphasis on the generation of meaningful questions. This process of inquiry goes beyond rote memorization, encouraging students to explore material, connect concepts, and draw conclusions. In doing so, IBL lays the foundation for a deeper understanding that extends beyond the immediate learning context.

Furthermore, the application of knowledge is a crucial component of IBL. Students are not merely passive recipients of information; instead, they actively use their acquired knowledge to explore real-world problems, scenarios, or applications. This hands-on engagement facilitates further internalization of concepts and enhances the transferability of knowledge to various contexts. As a result, IBL not only promotes a mastery of subject matter but also cultivates practical skills valuable in both academic and real-world settings.

One of the notable outcomes of IBL is its positive impact on critical thinking skills. Research indicates that students engaged in inquiry-based learning experiences demonstrate significant improvements in their ability to analyze information, evaluate arguments, and think critically (Ghaemi & Mirsaeed, 2017). The skills developed through IBL extend beyond the specific subject matter, empowering students to approach challenges with a more analytical and reflective mindset.

While IBL has found widespread application in general education, its integration into higher education presents both benefits and challenges (Kori, 2021). The benefits lie in the potential for deeper engagement, enhanced critical thinking, and the development of independent learning skills among college and university students. However, challenges may arise due to the need for resources, faculty training, and the adaptation of traditional educational structures.

Active Learning Strategies

Active Learning Strategies represent a dynamic and student-centered approach to education, utilizing methods such as group discussions, problem-solving activities, and debates to promote engagement, collaboration, and ultimately enhance comprehension and retention. Research findings throw light on the significance of these strategies, emphasizing their positive impact on the learning experience (Patil & Dharwadkar, 2020; Konopka et al., 2015).

The essence of active learning lies in its departure from traditional, passive instructional methods. Instead of being mere recipients of information, students actively participate in the learning process, becoming contributors to their own educational journey (Niemi, 2007). This shift in the role of students from passive observers to active participants has been linked to heightened levels of motivation, improved understanding of content, and increased retention of knowledge.

Group discussions serve as a prime example of active learning, enabling students to exchange ideas, challenge perspectives, and collaboratively construct knowledge. Through dialogue and interaction, students are not only exposed to diverse viewpoints but are also prompted to articulate and defend their own understanding, reinforcing their grasp of the subject matter. Moreover, group discussions cultivate communication skills and teamwork, valuable competencies in both academic and professional settings.

Problem-solving activities are another effective active learning strategy. This method encourages students to apply theoretical knowledge to practical situations by presenting them with real-world challenges or scenarios. This process reinforces conceptual understanding and cultivates problem-solving skills and the ability to adapt theoretical concepts to real-world contexts. The hands-on nature of problem-solving activities enhances the transferability of knowledge, preparing students for challenges of their future endeavors.

Debates, as a form of active learning, stimulate critical thinking and argumentation skills. Students engage in structured discussions, presenting and defending their positions on specific topics. This deepens their subject matter understanding and also hones their ability to analyze information critically and articulate persuasive arguments.

In the context of technology-integrated classrooms, active learning strategies can be further augmented through the strategic use of appropriate technology tools (Ntuli, 2019). Online platforms, collaborative software, and interactive multimedia resources provide avenues for extending active learning beyond the physical classroom, facilitating virtual group discussions, interactive problem-solving scenarios, and digital debates.

Personalized Learning

Personalized learning, a focal point in modern education, revolves around tailoring the educational experience to individual student needs and learning styles. This approach is particularly facilitated by the advent of adaptive learning platforms, which dynamically adjust the learning journey based on students' progress and preferences. Scholars such as Kem (2022), Popescu (2010), and Bachari, Abelwahed, and Adnani (2011) emphasize the critical role of these platforms in providing customized and adaptive learning experiences.

Adaptive learning platforms, as highlighted by Kem (2022) and Popescu (2010), serve as instrumental tools in the implementation of personalized learning. These platforms leverage advanced algorithms and data analytics to continuously assess students' performance and adapt instructional content accordingly. These platforms create a tailored learning path for each student, optimizing engagement and understanding by gauging individual strengths, weaknesses, and learning preferences.

Kem (2022) explores the practical applications of adaptive tools, emphasizing their capacity to enhance the learning process. These tools not only adjust the difficulty level of content but also give real-time feedback, helping students in tracking their progress and make informed decisions about their learning journey. This real-time feedback loop contributes to a more dynamic and responsive educational experience.

Popescu (2010) introduces an innovative approach to personalized learning by focusing on individual learning preferences. Recognizing that students have unique styles of acquiring and processing information, Popescu's framework tailors instructional strategies to align with these preferences. This approach acknowledges the diversity in learners and aims to create a more inclusive and effective learning environment.

Bachari, Abelwahed, and Adnani (2011) stress the importance of fitting e-learning into the diverse needs of learners, advocating for a framework that considers learner preferences. This perspective aligns with the broader concept of personalized learning, highlighting the necessity of accommodating different learning paces, styles, and preferences to maximize educational effectiveness.

Arapi et al. (2007) contribute to the discourse by introducing a pedagogy-driven personalization framework that considers the variety of learners and their individual needs. This framework emphasizes the alignment of instructional methods with pedagogical goals, ensuring that personalization enhances learning outcomes while maintaining a strong educational foundation.

Project-Based Learning

Project-Based Learning (PBL) stands out as an effective pedagogical approach that goes beyond traditional instructional methods, emphasizing hands-on, practical experiences. The benefits of PBL, as illustrated by research, extend to the development of practical skills and the enhancement of creativity in students (Blumenfeld et al., 1991). Furthermore, PBL plays a crucial role in creating a sense of ownership and teamwork, ultimately preparing students for real-world challenges (Ayas & Zeniuk, 2001).

One of the key strengths of project-based learning lies in its ability to develop practical skills in students. Students gain valuable hands-on experience by engaging in projects that require the application of theoretical knowledge to real-world scenarios. This experiential learning not only reinforces their understanding of academic concepts but also equips them with the skills necessary for success in practical, professional settings. Whether it's problem-solving, critical thinking, or technical proficiency, PBL provides students with a platform to develop and refine these practical skills.

Creativity is another dimension that is nurtured through project-based learning. The open-ended nature of projects allows students to explore novel solutions, think critically, and express themselves creatively. This freedom to innovate and create helps develop a mindset that is essential not only for academic success but also for adapting to the dynamic challenges of the real world. PBL encourages students to approach problems with a creative mindset, enhancing their ability to think outside the box.

A significant aspect of project-based learning is the sense of ownership it instills in students. Unlike traditional instructional methods, where students may passively receive information, PBL requires active participation and decision-making. Students take ownership of their projects, from conceptualization to completion, developing a sense of responsibility for their learning outcomes. This autonomy contributes to a deeper and more meaningful engagement with the subject matter.

Moreover, PBL promotes teamwork as students often collaborate on complex projects. Working in teams allows students to leverage diverse skills and perspectives, simulating real-world professional environments where collaboration is crucial. The ability to communicate effectively, delegate tasks, and deal with group dynamics are essential skills cultivated through collaborative project work.

While the benefits of PBL are evident, there is a recognized need for further research to explore its impact in post-secondary education (Helle et al., 2006). This suggests an ongoing interest in refining and optimizing the application of project-based learning at higher education levels.

Grant (2011) emphasizes the importance of understanding student perspectives in project-based learning. Internal and external influences, beliefs, and technology play crucial roles in shaping learning outcomes and products in the context of PBL. This insight highlights the need for a holistic understanding of the factors influencing the effectiveness of project-based learning, emphasizing a student-centric approach to instructional design and implementation.

In essence, this exploration into active learning strategies and project-based learning reveals two distinct yet complementary pedagogical approaches that prioritize student engagement and holistic skill development. Active learning strategies, manifested through activities like group discussions, problem-solving, and debates, propel students to the forefront of their educational experience. Beyond enhanced comprehension and knowledge retention, these strategies empower students with vital skills in effective communication, collaboration, and critical thinking. The role played by adaptive learning platforms stands out as pivotal in realizing the goals of personalized learning, offering tailored pathways for students to explore and grasp concepts at their own pace.

Simultaneously, project-based learning (PBL) emerges as a powerful pedagogical strategy with broad-ranging benefits. PBL not only develops practical skills but also enhances creativity, nurtures a sense of ownership, and promotes teamwork by actively involving students in real-world projects. This multifaceted preparation extends beyond academic success, equipping students to confront and navigate the challenges they will inevitably encounter in the real world.

The connection between modern pedagogies and educational quality lies in their ability to transform how learning occurs. Educational quality is not solely defined by content delivery but by learning outcomes such as critical thinking, problem-solving ability, creativity, and adaptability. Modern pedagogies directly contribute to these outcomes by engaging learners actively, encouraging inquiry, enabling collaboration, and personalizing learning pathways. Unlike traditional approaches that prioritize memorization, these methods promote deeper understanding and long-term retention. As a result, modern pedagogies play a crucial role in enhancing not only academic achievement but also the broader competencies required in the 21st century.

The upcoming sections of this chapter will explore the possibilities that arise when these pedagogical approaches converge with smart technologies.

III. SMART TECHNOLOGIES IN EDUCATION

The integration of smart technologies in education has ushered in a transformative era, redefining traditional learning paradigms and expanding educational opportunities. E-Learning Platforms and Massive Open Online Courses (MOOCs) have become instrumental in democratizing education, offering a diverse array of courses that transcend geographical boundaries. The flexibility and convenience inherent in these online platforms break down barriers of location and time, providing learners worldwide with unprecedented access to educational resources. Additionally, the infusion of gamification into education has revolutionized engagement and motivation. Learning experiences become enjoyable and effective, enhancing knowledge retention by incorporating game elements. Augmented Reality (AR) and Virtual Reality (VR) further elevate the educational landscape by creating immersive and interactive environments. These technologies, applied in fields such as science, history, and art, provide students with hands-on experiences that transcend traditional boundaries. Artificial Intelligence (AI) takes center stage in automating administrative tasks, personalizing learning experiences, and enabling intelligent tutoring systems. The ability of AI to analyze extensive datasets offers educators valuable insights, shaping a more adaptive and efficient educational ecosystem. Together, these smart technologies are shaping a future where education is not only accessible but also engaging, personalized, and enriched by immersive experiences.

E-Learning Platforms and Massive Open Online Courses (MOOCs)

The advent of E-Learning Platforms and Massive Open Online Courses (MOOCs) has ushered in a transformative era in education, reshaping the traditional learning landscape and democratizing access to knowledge on a global scale. This paradigm shift, as emphasized by research findings (Glusac et al., 2015; (Haron et al., 2019; Tsai & Wong, 2013; Baturay, 2015), has revolutionized how individuals engage with educational content.

MOOCs, exemplified by platforms like Coursera, Udacity, and edX, have emerged as powerful tools, offering a diverse array of courses from top universities worldwide.

This proliferation of online platforms has effectively dismantled geographical barriers, making education accessible to learners irrespective of their location. Learners no longer need to be in specific physical spaces to access high-quality educational content; instead, they can engage in learning experiences from the comfort of their homes or any location with internet connectivity.

The hallmark of E-Learning Platforms and MOOCs is the flexibility and convenience they bring to education. Learners can access courses at their own pace, facilitating a learner-centric approach that accommodates various schedules and learning styles. This flexibility is particularly beneficial for working professionals, students with diverse commitments, and individuals who may face geographical constraints. The asynchronous nature of many online courses allows learners to tailor their educational experiences to fit into their busy lives.

However, the proliferation of MOOCs has not been without challenges. Scholars have raised concerns about the pedagogical foundations and technological challenges associated with these platforms (Glusac et al., 2015; (Haron et al., 2019; Tsai & Wong, 2013). The rapid growth of MOOCs has prompted discussions about the need for open education standards to ensure quality and consistency in course delivery.

Nevertheless, MOOCs persist as vital contributors to global education, offering a platform for lifelong learning and knowledge sharing (Baturay, 2015). The vast array of courses available on these platforms spans diverse subjects, providing learners with the opportunity to acquire new skills, deepen existing knowledge, and explore areas of interest. In essence, E-Learning Platforms and MOOCs represent a dynamic force in education, embodying the principles of accessibility, flexibility, and lifelong learning that are essential for meeting the diverse needs of learners worldwide.

Gamification in Education

The integration of game elements into educational contexts, commonly referred to as gamification, represents a dynamic and innovative approach to learning, particularly in higher education. Numerous studies highlight its positive impact on student engagement, motivation, and performance (Langendahl et al., 2016; Subhash & Cudney, 2018). Gamified learning experiences go beyond traditional instructional methods, creating an environment that activates students and contributes significantly to their learning outcomes by incorporating elements such as narrative, challenge, progression, and feedback (Langendahl et al., 2016).

One of the key strengths of gamification is its ability to enhance student motivation. The inherent design of games, with their goal-oriented structure and rewards systems, taps into intrinsic motivators, encouraging students' active participation in the learning process. The narrative component of gamification, for example, can immerse students in a storyline, making the learning experience more compelling and personally meaningful. As a result, students tend to invest time and effort into their educational pursuits.

Engagement is another area where gamified learning experiences excel. The inclusion of challenges and progression mechanisms within the educational framework provides students with a sense of accomplishment as they overcome obstacles and advance through levels. This sense of achievement creates a positive learning environment, where students are continually motivated to tackle new challenges and explore the subject matter in depth.

Gamification also addresses the issue of knowledge retention. The interactive nature of gamified learning experiences requires students to actively apply what they have learned to solve problems or navigate challenges within the game. This application of knowledge in a practical context enhances retention, as students are more likely to remember information that they have actively used in problem-solving scenarios.

While the potential benefits of gamification are evident, it is essential to acknowledge that its effectiveness can vary. Some studies report mixed results, suggesting that the impact of gamified learning experiences depends on various factors such as the design of the gamification elements, the subject matter, and individual student preferences (Domínguez et al., 2013). Nevertheless, the overall potential of gamification in improving learning outcomes, student motivation, and engagement is significant (Yu et al., 2020).

Augmented Reality (AR) and Virtual Reality (VR)

The integration of Augmented Reality (AR) and Virtual Reality (VR) into education opens up huge possibilities, providing students with immersive and interactive learning experiences that transcend traditional instructional methods. Both AR and VR have demonstrated significant potential, each offering unique advantages in enhancing engagement and knowledge retention.

Virtual Reality (VR) stands out for its ability to transport students to simulated environments, creating a sense of presence and immersion that traditional learning methods cannot replicate. Research, such as that conducted by Shankar et al. (2023), highlights how VR can significantly enhance engagement and retention of knowledge by providing experiential learning opportunities to students. Whether exploring historical landmarks, conducting scientific experiments, or creating art in a virtual studio, VR allows students to engage with the subject matter in a way that goes beyond course material and lectures.

On the other hand, Augmented Reality (AR) enhances conventional educational materials by superimposing digital information onto the physical surroundings. Gupta and Rohil (2017) emphasize how AR aids in developing critical thinking by seamlessly integrating virtual elements into the physical world. In fields like science, history, and art, AR enhances learning experiences by providing additional contextual information, interactive elements, or three-dimensional models. For instance, AR can bring historical artifacts to life, allowing students to virtually explore ancient civilizations or witness historical events unfold in their physical surroundings.

In the context of science education, the potential of AR and VR becomes even more evident. Huang et al. (2019) highlight the effectiveness of AR in conveying auditory information, while VR excels in creating immersive and engaging experiences. Whether dissecting virtual specimens, exploring the human anatomy, or conducting complex experiments in a simulated laboratory, AR and VR offer unparalleled opportunities for students to interact with scientific concepts in a tangible and memorable way.

Furthermore, AR-VR integration has the potential to reduce cognitive overload by presenting information in a more visually intuitive manner (Bower et al., 2014). Complex concepts can be conveyed in a simplified yet comprehensive fashion, catering to various learning styles and facilitating a deeper understanding of the subject matter.

From exploring historical events to conducting scientific experiments and creating art in virtual spaces, AR and VR offer students hands-on experiences that enhance engagement, develop critical thinking, and significantly contribute to the retention of knowledge.

Artificial Intelligence (AI) in Education

The integration of Artificial Intelligence (AI) within education represents a groundbreaking paradigm shift, promising to revolutionize learning and teaching by offering personalized experiences and streamlining administrative tasks (Harry, 2023). AI's transformative potential extends beyond the traditional boundaries of education, enhancing access and addressing contemporary challenges (Ahmad et al., 2021). However, realizing the full benefits of AI in education requires absolute understanding of its effective implementation by educators (Zawacki-Richter et al., 2019).

One significant facet of AI in education is its capability to automate administrative tasks, freeing up valuable time for educators to focus on instructional activities. AI-powered tools can efficiently handle routine administrative responsibilities, such as grading, scheduling, and data management (Harry, 2023). This automation not only enhances the efficiency of educational institutions but also allows educators to allocate more time and attention to personalized teaching and mentoring.

The aspect of personalization is a key strength of AI in education. AI can analyze vast amounts of data related to students' learning preferences, performance, and engagement by leveraging machine learning algorithms. This data-driven approach enables the creation of personalized learning experiences tailored to individual needs and learning styles. Adaptive learning platforms powered by AI can dynamically adjust the difficulty level of content, suggest targeted resources, and provide real-time feedback, ensuring that each student receives a customized and optimized learning path.

AI's role in student evaluation is also noteworthy. Through continuous analysis of student performance data, AI can offer insights into individual strengths and weaknesses, facilitating a more accurate assessment of a student's progress. This data-driven evaluation can inform educators about areas that may require additional attention, allowing for timely interventions and personalized support.

Intelligent tutoring systems represent another dimension of AI's impact on education. Malik, Tayal, and Vij (2018) emphasize the role of AI in the development of these systems, which provide interactive and adaptive support to students. AI-driven tutoring systems can dynamically adjust the pacing and content of instruction based on individual student responses and performance, offering a personalized and responsive learning experience.

Despite the immense potential, there are challenges associated with the effective utilization of AI in education. Educators need to be equipped with the requisite skills and knowledge to integrate AI tools seamlessly into their teaching practices (Zawacki-Richter et al., 2019). Addressing ethical considerations, ensuring data privacy, and mitigating biases in AI algorithms are also crucial aspects that require careful attention in the deployment of AI in educational settings.

While the potential benefits are substantial, there is a parallel need for comprehensive teacher training, ethical considerations, and ongoing research to maximize the positive impact of AI on teaching and learning.

Recent discussions on AI in education also highlight critical concerns related to ethics, data privacy, and algorithmic bias. While AI offers significant opportunities for personalization and efficiency, its implementation must be guided by responsible and transparent practices. Scholars such as Holmes et al. (2019) and Zawacki-Richter et al. (2019) emphasize the need for educator readiness and ethical frameworks in deploying AI-driven tools. Emerging global guidelines, including those from Holmes and Miao (2023), further stress the importance of ensuring that AI enhances, rather than replaces, human-centered teaching and learning processes.

IV. ACHIEVING SDG4 THROUGH MODERN PEDAGOGY AND SMART TECHNOLOGIES

Sustainable Development Goal 4 (SDG 4), adopted as part of the United Nations' 2030 Agenda, aims to "ensure inclusive and equitable quality education and promote lifelong learning opportunities for all." SDG 4 extends beyond access to education and emphasizes key dimensions such as quality of learning outcomes, equity, inclusion, relevant skills development, and lifelong learning. It calls for improved teaching practices, well-trained educators, inclusive learning environments, and the effective integration of technology. In this context, the alignment of modern pedagogies with smart technologies becomes essential, as both contribute not only to expanding access but also to enhancing the effectiveness, inclusivity, and relevance of education systems globally.

The intersection of modern pedagogy and smart technologies thus emerges as a powerful catalyst for achieving SDG 4. This section explores how this integration promotes inclusivity, strengthens teacher capacity, and enables data-driven monitoring and evaluation, ultimately contributing to improved educational outcomes.

Promoting Inclusivity

The fusion of modern pedagogy with smart technologies holds immense promise in dismantling educational disparities for marginalized communities, thereby ensuring equitable access to quality education (Campbell, 1999; Treviranus & Coombs, 2000; Garcia & Lee, 2020; Messiou, 2012). By leveraging these technologies, educators can tailor instructional approaches to accommodate diverse learning styles and needs, effectively addressing barriers that have historically limited access to education. The integration of smart technologies within modern pedagogical frameworks further enables the creation of inclusive learning environments that are responsive to a wide spectrum of learners.

For example, MOOCs have enabled learners from remote and economically disadvantaged regions to access courses from globally recognized institutions at minimal or no cost. Mobile-based learning platforms have further expanded educational reach in areas where physical infrastructure remains limited. In addition, assistive technologies such as screen readers, speech-to-text tools, and captioned content have significantly improved accessibility for learners with disabilities. These developments illustrate how the integration of pedagogy and technology can actively reduce structural barriers and democratize access to education.

However, realizing the full potential of this integration requires moving beyond access toward equity-centered implementation. It is essential to address the structural and systemic causes of educational inequality and to ensure that technology adoption does not inadvertently reinforce existing disparities (Garcia & Lee, 2020). In this regard, educators and institutions must actively promote inclusive practices by amplifying diverse voices, fostering participatory learning environments, and embedding equity within pedagogical design (Messiou, 2012). This alignment between inclusivity, pedagogy, and technology is central to achieving the broader vision of SDG 4.

Enhancing Teacher Training

Enhancing teacher training stands out as a critical imperative in maximizing the effective utilization of smart technologies in the classroom. Numerous studies, including those by Estes, Dailey-Hebert, and Choi (2017) and Tan et al. (2003), emphasize the need for comprehensive training that goes beyond basic computer literacy to include pedagogical support and the development of teacher self-efficacy. This reflects the understanding that effective integration of smart technologies requires a multifaceted skill set, combining technical proficiency with pedagogical competence.

For instance, professional development programs incorporating blended learning models enable teachers to experience technology-enabled pedagogy firsthand. Micro-credentialing platforms and online certification courses further support continuous professional development without requiring educators to disengage from their ongoing responsibilities. Additionally, simulation-based training and virtual classroom environments provide opportunities for teachers to practice, experiment, and refine instructional strategies within technology-rich contexts.

Studies by Kode and Nori (2016) and Oigara (2013) further highlight structured initiatives aimed at empowering educators through the use of educational technologies and holistic training models. Such approaches move beyond tool-based training and focus on integrating technology meaningfully into teaching practices.

A key insight emerging from the literature is that successful technology integration is not merely about tool adoption but about pedagogical transformation. Educators must understand how to align technological tools with learning objectives, student needs, and instructional design principles. In this regard, teacher training must foster not only technical skills but also innovation, adaptability, and reflective practice, ensuring that educators are equipped to navigate evolving digital learning environments effectively. This, in turn, directly contributes to SDG 4 by strengthening teaching quality and improving learning outcomes.

Monitoring and Evaluation

The integration of data analytics and AI-driven tools in education holds immense promise for transforming the monitoring and evaluation of educational programs. As highlighted by Vilanova et al. (2019) and Ogor (2007), these tools play a critical role in identifying at-risk students and predicting academic performance. The ability to leverage data analytics enables educators to adopt a proactive approach, facilitating timely interventions and targeted support, thereby improving student retention and academic success.

A significant dimension of this transformation lies in the capacity of data analytics and AI to uncover complex educational dynamics. Maia, Bueno, and Sato (2023) emphasize the role of AI models in not only understanding these dynamics but also in informing and supporting evidence-based public policy. By analyzing large-scale datasets, these tools reveal patterns, trends, and correlations that can guide strategic decision-making related to curriculum design, resource allocation, and systemic reform.

Wayman (2005) further highlights the practical application of student data in driving school improvement, particularly through supporting teacher inquiry and reflective practice. The integration of data-driven decision-making into teaching methodologies enables educators to continuously refine their instructional approaches based on empirical evidence. This iterative cycle of analysis, reflection, and adjustment fosters a culture of continuous improvement in both teaching practices and learning outcomes.

For example, learning management systems (LMS) equipped with analytics dashboards allow educators to monitor student engagement, attendance, and performance in real time. AI-driven early warning systems can identify at-risk students and trigger timely interventions. At the institutional level, data analytics supports curriculum improvement, enhances teaching effectiveness, and informs policy decisions, thereby strengthening overall educational quality.

However, the effective use of these technologies requires careful consideration of ethical concerns, including data privacy, algorithmic bias, and responsible use of student data. Addressing these challenges is essential to ensure that data-driven approaches contribute to equitable and meaningful educational improvement. In this way, monitoring and evaluation become central to achieving both accountability and continuous improvement under SDG 4.

Quality Education as the Core of SDG 4

Collectively, these dimensions contribute to the broader goal of achieving quality education under SDG 4. Quality education extends beyond access to include meaningful learning outcomes, equity, relevance, and lifelong learning. It requires effective pedagogy, inclusive environments, continuous assessment, and adaptive learning systems.

The integration of modern pedagogies with smart technologies directly supports this vision by enabling interactive learning, personalized instruction, inclusive access, and data-driven improvements. These approaches shift education from a content-delivery model to a learner-centered, outcome-oriented process that fosters critical thinking, creativity, and real-world problem-solving abilities.

Ultimately, the convergence of modern pedagogy and smart technologies redefines quality education as an adaptive, inclusive, and future-ready system, capable of preparing learners not only for academic success but also for meaningful participation in an increasingly complex and interconnected world.

V. CONCLUSION

In summary, this chapter serves to illuminate the transformative power that arises from the harmonious integration of modern pedagogy methods with smart technologies, particularly in the context of achieving Sustainable Development Goal 4 (SDG4). The discourse within these pages not only emphasize upon the qualitative enhancements these tools bring to education but also positions them as catalysts for building a sustainable and equitable future. The amalgamation of modern pedagogy and smart technologies emerges as a potent force in reshaping the fabric of education by creating innovative and inclusive learning experiences.

Throughout the chapter, insightful analysis and practical examples have been presented, weaving a narrative that goes beyond theoretical considerations to showcase the tangible impact of this convergence. The intention is clear – to inspire and galvanize educators, policymakers, and researchers alike, encouraging them to dive further into the dynamic intersection of modern pedagogy and smart technologies. This will transform the prospect of positive change in education from a mere possibility into a concrete reality, aligning seamlessly with the worldwide commitment to guaranteeing quality education for everyone.

Further, the chapter calls for a collective commitment to continuous exploration and adaptation. It recognizes the potential for these integrated approaches to not only meet the immediate needs of learners and educators but also to anticipate and address future challenges. This forward-looking perspective positions education as a dynamic, evolving entity that evolves in tandem with technological advancements and societal changes.

In essence, this chapter serves as an invitation to all stakeholders in education – an invitation to embrace the transformative potential of modern pedagogy and smart technologies. It is a call to action for educators to reimagine their teaching methodologies, for policymakers to design inclusive policies, and for researchers to explore the dynamics of this integration. The ultimate goal is to collectively contribute to the realization of SDG4, creating an educational ecosystem that is not only of the highest quality but is also accessible, innovative, and equitable for learners around the globe. Through collaboration and sustained effort, the intersection of modern pedagogy and smart technologies stands as a guiding light, pointing the way toward a brighter, more inclusive, and globally impactful future in education.

REFERENCES

Ahmad, S. F., Rahmat, M. K., Mubarik, M. S., Alam, M. M., & Hyder, S. I. (2021). Artificial intelligence and its role in education. *Sustainability (Basel)*, *13*(22), 12902. DOI: 10.3390/su132212902

Arapi, P., Moumoutzis, N., Mylonakis, M., & Christodoulakis, S. (2007). A pedagogy-driven personalization framework to support adaptive learning experiences. In *Proceedings of the Seventh IEEE International Conference on Advanced Learning Technologies (ICALT 2007)* (pp. 96–97). DOI: 10.1109/ICALT.2007.27

Ayas, K., & Zeniuk, N. (2001). Project-based learning: Building communities of reflective practitioners. *Management Learning*, *32*(1), 61–76. DOI: 10.1177/1350507601321005

Bachari, E. E., Abelwahed, E. H., & Adnani, M. E. (2011). E-learning personalization based on dynamic learners' preference. *International Journal of Computer Science and Information Technologies*, *3*(3), 200–216. DOI: 10.5121/ijcsit.2011.3314

Baturay, M. H. (2015). An overview of the world of MOOCs. *Procedia: Social and Behavioral Sciences*, *174*, 427–433. DOI: 10.1016/j.sbspro.2015.01.685

Blumenfeld, P., Soloway, E., Marx, R., Krajcik, J. S., Guzdial, M., & Palincsar, A. S. (1991). Motivating project-based learning: Sustaining the doing, supporting the learning. *Educational Psychologist*, *26*(3-4), 369–398. DOI: 10.1080/00461520.1991.9653139

Bower, M., Howe, C., McCredie, N., Robinson, A., & Grover, D. (2014). Augmented reality in education: Cases, places and potentials. *Educational Media International*, *51*(1), 1–15. DOI: 10.1080/09523987.2014.889400

Campbell, K. (1999). The promise of computer-based learning: Designing for inclusivity. *IEEE Technology and Society Magazine*, *18*(4), 28–34. DOI: 10.1109/44.808846

Chaudhry, M. Y., & Malik, A. M. (2014). Intersection of pedagogy and emerging technologies to enhance student-centred learning in higher education.

de Souza Zanirato Maia, J., Bueno, A. P., & Sato, J. R. (2023). Applications of artificial intelligence models in educational analytics and decision making: A systematic review. SSRN *Electronic Journal*. DOI: 10.3390/world4020019

Domínguez, A., Saenz-de-Navarrete, J., Marcos, L. D., Sanz, L. F., Pagés-Arévalo, C., & Martínez, J. (2013). Gamifying learning experiences: Practical implications and outcomes. *Computers & Education*, *63*, 380–392. DOI: 10.1016/j.compedu.2012.12.020

Estes, J. S., Dailey-Hebert, A., & Choi, D. H. (2017). Integrating technological innovations to enhance the teaching-learning process.

Garcia, A., & Lee, C. H. (2020). Equity-centered approaches to educational technology.

Ghaemi, F., & Mirsaeed, S. J. (2017). The impact of inquiry-based learning approach on critical thinking skill of EFL students.

Glusac, D., Karuović, D., & Milanov, D. (2015). Massive open online courses: Pedagogical overview. In *Proceedings of the 16th International Carpathian Control Conference (ICCC)* (pp. 142–146).

Grant, M. M. (2011). Learning, beliefs, and products: Students' perspectives with project-based learning. *The Interdisciplinary Journal of Problem-Based Learning*, *5*(2), 37–69. DOI: 10.7771/1541-5015.1254

Gupta, N., & Rohil, M. K. (2017). Exploring possible applications of augmented reality in education. In *Proceedings of the 4th International Conference on Signal Processing and Integrated Networks (SPIN)* (pp. 437–441). DOI: 10.1109/SPIN.2017.8049989

Haron, H., Yusof, A. R., Samad, H., Ismail, N., Juanita, A., & Yusof, H. (2019). The platform of MOOC (massive open online course) on open learning: Issues and challenges.

Harry, A. (2023). *Role of AI in education. Interdisciplinary Journal and Hummanity*. INJURITY.

Helle, L., Tynjälä, P., & Olkinuora, E. (2006). Project-based learning in post-secondary education: Theory, practice and rubber sling shots. *Higher Education*, *51*(2), 287–314. DOI: 10.1007/s10734-004-6386-5

Holmes, W., Bialik, M., & Fadel, C. (2019). *Artificial intelligence in education: Promises and implications for teaching and learning*. Center for Curriculum Redesign.

Holmes, W., & Miao, F. (2023). *Guidance for generative AI in education and research*. UNESCO Publishing.

Hosseini, Z., & Kinnunen, J. (2021). Integration of pedagogy into technology: A practical paradigm. In *Education and New Developments 2021*. DOI: 10.36315/2021end086

Huang, K., Ball, C., Francis, J., Ratan, R. A., Boumis, J. K., & Fordham, J. A. (2019). Augmented versus virtual reality in education: An exploratory study examining science knowledge retention when using augmented reality/virtual reality mobile applications. *Cyberpsychology, Behavior and Social Networking*, *22*(2), 105–110. DOI: 10.1089/cyber.2018.0150 PMID: 30657334

Ismail, N. (2006). Inquiry-based learning: A new approach to classroom learning.

Kem, D. (2022). Personalised and adaptive learning: Emerging learning platforms in the era of digital and smart learning. *International Journal of Social Science and Human Research*, *5*(2). Advance online publication. DOI: 10.47191/ijsshr/v5-i2-02

Kode, S., & Nori, K. V. (2016). Enhancing IT education: Education technology for teacher training. In *Proceedings of the International Conference on Learning and Teaching in Computing and Engineering (LaTICE)* (pp. 94–98). DOI: 10.1109/LaTiCE.2016.4

Konopka, C. L., Adaime, M. B., & Mosele, P. H. (2015). Active teaching and learning methodologies: Some considerations. *Creative Education*, *6*(14), 1536–1545. DOI: 10.4236/ce.2015.614154

Kori, K. (2021). Inquiry-based learning in higher education. In *Technology Supported Active Learning*. DOI: 10.1007/978-981-16-2082-9_4

Langendahl, P., Cook, M., & Mark-Herbert, C. (2016). Gamification in higher education.

Lower-Hoppe, L. M., Brgoch, S. M., Chen, Y., & Sutherland, S. (2021). Inquiry-based learning in action.

Malik, G., Tayal, D. K., & Vij, S. (2018). An analysis of the role of artificial intelligence in education and teaching. In *Advances in Intelligent Systems and Computing*.

Messiou, K. (2012). *Confronting marginalisation in education: A framework for promoting inclusion*.

Niemi, A. M. (2007). What are effective strategies to support student engagement and learning.

Ntuli, E. (2019). *Active learning strategies in technology integrated K-12 classrooms*. Early Childhood Development. DOI: 10.4018/978-1-5225-7507-8.ch035

Ogor, E. N. (2007). Student academic performance monitoring and evaluation using data mining techniques. In *Electronics, Robotics and Automotive Mechanics Conference (CERMA 2007)* (pp. 354–359). DOI: 10.1109/CERMA.2007.4367712

Oigara, J. N. (2013). Integrating technology in teacher education programs.

Patil, S., & Dharwadkar, D. N. (2020). Improving students' engagement through active learning strategies: Case study based active review sessions and skillathon. *Journal of Engineering Education Transformations*.

Popescu, E. (2010). Adaptation provisioning with respect to learning styles in a web-based educational system: An experimental study. *Journal of Computer Assisted Learning*, *26*(4), 243–257. DOI: 10.1111/j.1365-2729.2010.00364.x

Sathishkumar, A. S. (2012). Emerging technology of smart class teaching in school education: A literature review. *International Journal of Scientific Research*, *3*, 446–449. DOI: 10.15373/22778179/August2014/137

Shankar, D. A., Tewari, D. V., Rahman, M., Mishra, D. A., & Bajaj, M. K. (2023). Impact of virtual reality (VR) and augmented reality (AR) in education. *Tuijin Jishu/ Journal of Propulsion Technology*.

Subhash, S., & Cudney, E. A. (2018). Gamified learning in higher education: A systematic review of the literature. *Computers in Human Behavior*, *87*, 192–206. DOI: 10.1016/j.chb.2018.05.028

Tan, S. C., Hu, C., Wong, S. K., & Wettasinghe, C. M. (2003). Teacher training on technology-enhanced instruction: A holistic approach. *Journal of Educational Technology & Society*, •••, 6.

Treviranus, J., & Coombs, N. (2000). Bridging the digital divide in higher education.

Tsai, F. S., & Wong, K. (2013). The state of massive open online courses (MOOCs) in engineering education: Where do we go from here?

Vilanova, R., Domínguez, M., Vicario, J. L., Prada, M. A., Barbu, M., Varanda, M. J., Alves, P., Podpora, M., Spagnolini, U., & Paganoni, A. M. (2019). Data-driven tool for monitoring of students' performance. *IFAC-PapersOnLine*, *52*(9), 165–170. DOI: 10.1016/j.ifacol.2019.08.188

Visvizi, A. (2018). *The future of innovation and technology in education: Policies and practices for teaching and learning excellence*.

Wayman, J. C. (2005). Involving teachers in data-driven decision making: Using computer data systems to support teacher inquiry and reflection. *Journal of Education for Students Placed at Risk*, *10*(3), 295–308. DOI: 10.1207/s15327671espr1003_5

Yu, Z., Gao, M., & Wang, L. (2020). The effect of educational games on learning outcomes, student motivation, engagement and satisfaction. *Journal of Educational Computing Research*, *59*(3), 522–546. DOI: 10.1177/0735633120969214

Zawacki-Richter, O., Marín, V. I., Bond, M., & Gouverneur, F. (2019). Systematic review of research on artificial intelligence applications in higher education: Where are the educators? *International Journal of Educational Technology in Higher Education*, *16*(1), 16. DOI: 10.1186/s41239-019-0171-0

Chapter 8
Reimagining School Climate With Filipino Core Values:
Lessons for Effective School Leadership

Marie Grace Avelino Gomez
http://orcid.org/0000-0003-0180-9875
University of the Philippines, Philippines

Ethelbert G. Galang
http://orcid.org/0009-0001-7631-970X
University of the Philippines, Philippines

Susan Labartinos Cobarrubias
http://orcid.org/0009-0009-7420-3007
Jose Rizal University, Philippines

Joel C. Javiniar
http://orcid.org/0009-0001-0230-2307
University of the Philippines-Diliman, Philippines

ABSTRACT

A supportive school climate boosts student achievement, resilience, and overall well-being, especially in the face of social, economic, and environmental challenges. In the Philippines, fostering such an engaged atmosphere involves embracing key Filipino values—kapwa (shared identity), kalinga (care), malasakit (compassion), and bayanihan (collective effort). These values cultivate inclusive, nurturing, and resilient school communities that emphasize mental health and emotional safety.

DOI: 10.4018/979-8-3373-1350-4.ch008

As students navigate a VUCA (volatile, uncertain, complex, ambiguous) world, it is essential for school leaders to practice culturally grounded leadership. By redefining school environments through these values, they can create caring spaces where students and educators can flourish.

INTRODUCTION

A positive school climate is a cornerstone of academic success and social development. It can help students feel a sense of safety and belonging, where relational trust prevails. It improves academic achievement, test scores, grades, and engagement, while also reducing the negative effects of poverty on academic achievement (Darling-Hammond & DePaoli, 2020). The combination of safety, nurturing relationships, and effective teaching and learning contributes to the well-being of students, teachers, and staff (Hammar et al., 2024). However, in the Philippines, challenges such as resource limitations, overcrowding, and cultural diversity make fostering an ideal school climate particularly demanding (Malbas et al., 2023). These realities call for a culturally grounded approach, leveraging Filipino values to redefine school climate and leadership practices.

Cultivating a positive school climate is crucial for navigating the volatile, uncertain, complex, and ambiguous (VUCA) world. Although the COVID-19 virus has become less virulent, challenges such as climate change, natural disasters, economic turmoil, and emerging viruses continue to afflict our world. This requires identifying all possible interventions within the school system to mitigate the impacts of these challenges. These global disruptions require schools to be not only sites of instruction but also spaces of adaptability, emotional strength, and civic resilience. A strong and values-driven school climate becomes essential in this environment, providing both grounding and flexibility.

Filipino core values—*kapwa* (shared identity), *kalinga* (care), *malasakit* (compassion), and *bayanihan* (collective effort)—offer transformative potential in shaping school environments. The Philippines has a collectivist culture. Kapwa emphasizes shared humanity and interconnectedness, cultivating a sense of belonging in schools. Kalinga highlights the importance of genuine care and nurturing relationships, promoting the well-being of both students and teachers. Malasakit encourages empathy, urging members of the school community to act with understanding and kindness. Lastly, bayanihan embodies the spirit of teamwork, enabling collective action to overcome challenges. These values align seamlessly with the goals of a thriving school climate.

In the Philippine context, the school climate can be influenced by close familial ties, respect for authority, and resilience, which are culturally inherent. Leaders can harness these traits to create inclusive, safe, and equitable learning environments.

The Filipino concept of kapwa can guide leaders in building inclusive communities. Addressing discrimination, fostering equity, and ensuring every member of the school community feels valued are key actions aligned with this value. Similarly, practicing Kalinga means prioritizing mental health and well-being through programs and support systems that address students' and teachers' emotional and psychological needs. Leaders embody a culture of care by ensuring that vulnerable groups are protected.

Malasakit transforms leadership into an empathy-driven process. School leaders can model compassionate decision-making by actively listening to the needs of their communities and creating systems for conflict resolution rooted in understanding. Meanwhile, Bayanihan underscores the power of collective effort. Leaders can inspire collaboration among teachers, parents, and community members, organizing initiatives to address issues such as overcrowding, infrastructure gaps, or resource shortages.

Despite its potential, integrating Filipino values into school leadership faces several challenges. Resistance to change, cultural inertia, and systemic constraints, such as a lack of funding, often hinder progress. Balancing traditional values with modern leadership practices necessitates intentionality, creativity, and persistence. However, examples of schools that successfully incorporate these values into their climate demonstrate that these obstacles can be overcome with the right strategies.

School leaders must integrate Filipino values into daily practices, professional development, and policies to institutionalize this approach. Training programs can emphasize values-driven leadership, while collaborative initiatives can foster community involvement. These efforts transform individual schools and provide a foundation for systemic change across the Philippine education sector.

School climate is a multifaceted construct that incorporates issues related to safety, student engagement, and the school environment (Bradshaw et al., 2021). It plays a pivotal role in shaping the everyday experiences of learners, educators, and school leaders. It reveals the overall atmosphere of a school—how people interact, how safe they feel, and how engaged they are in the process of teaching and learning (Grazia & Molinari, 2021). A positive school climate contributes significantly to students' academic achievement, emotional regulation, social adjustment, and long-term success (Monsillion et al., 2023). In schools where respect, inclusion, and shared responsibility thrive, learners are more motivated and less likely to drop out or engage in harmful behaviors.

In collectivist cultures like the Philippines, the role of school climate is crucial. Considering the country's socio-economic diversity, regional disparities, and varying levels of access to resources, school climate can serve as either a protective or a risk factor for students' educational outcomes. Many learners in public schools, particularly in remote and marginalized areas, face additional challenges such as poverty, malnutrition, armed conflict, and climate-related disruptions like strong typhoons, earthquakes, volcanic eruptions, and landslides. In such environments, a strong and nurturing school climate provides stability, hope, and a sense of belonging. Issues like large class sizes, teacher burnout, limited facilities, and bureaucratic constraints continue to affect the daily realities of both learners and teachers.

Despite these difficulties, the Philippine cultural context offers unique resources for fostering a climate of connection, care, and resilience. Filipino values have long emphasized interdependence, relational harmony, and collective well-being. When these values are intentionally infused into school leadership and daily school practices, they possess the power to transform educational spaces into environments of mutual support, emotional safety, and shared growth.

In this context, Filipinizing the school climate goes beyond simply adapting Western models; it represents a strategic and culturally significant method for transforming education. By integrating Filipino core values into school practices, we affirm students' identities rather than diminish them and respect their backgrounds rather than overlook them. This enhances learning, making it not just more effective but also more compassionate and inclusive, particularly for students from historically marginalized communities.

A positive school climate equips students with the psychological safety and social-emotional skills needed to face the uncertainties of the future. It cultivates empathy, collaboration, and ethical decision-making, essential for both survival and success in a rapidly changing world. In a VUCA environment, simply having academic knowledge is not enough; students must develop character, confidence, and a profound sense of connection with others—attributes that thrive in supportive school climates.

Ultimately, reimagining school climate through Filipino core values presents a path toward more inclusive, resilient, and thriving school communities. School leaders can foster environments where students and educators flourish by embracing values such as kapwa, kalinga, malasakit, and bayanihan. The future of Philippine education hinges on integrating these values into leadership, ensuring that schools are not just places of learning but also hubs of care, collaboration, and shared success.

This chapter posits that enhancing Philippine education requires more than just curriculum revisions or infrastructure upgrades; it necessitates a focus on the human relationships and values fundamental to our schools. By reevaluating and embracing Filipino cultural values as foundational elements of school culture, educators and leaders can cultivate institutions that not only educate the mind but also nurture the heart. This approach equips our youth not only for academic achievement but also for the duty of fostering a compassionate, just, and unified nation.

OBJECTIVES OF THE CHAPTER

This chapter aims to explore how Filipino core values can meaningfully redefine the concept and practice of school climate in the local educational context. While global frameworks emphasize elements like safety, relationships, teaching and learning, and institutional support, these concepts gain deeper relevance when viewed through the lens of Filipino cultural values. Core values such as *kapwa*, *kalinga*, *malasakit*, and *bayanihan* are not mere cultural footnotes but living principles that shape how Filipinos relate, lead, and learn. By integrating these values, schools can move beyond compliance-based structures and instead foster climates rooted in shared identity, relational harmony, and mutual responsibility.

Redefining school climate through Filipino values involves more than surface-level inclusion of cultural symbols or traditions. It requires a transformation of mindsets and practices at every level of school life—from classroom interactions and discipline approaches to leadership styles and community engagement. For instance, *malasakit* reframes teacher-student dynamics to emphasize empathy and genuine concern. *Kapwa* challenges hierarchical barriers, promoting inclusive participation and mutual respect. These values offer culturally attuned ways to address systemic issues such as student disengagement, teacher burnout, and community-school disconnects.

To support this redefinition, the chapter presents practical examples for school leaders on integrating Filipino values into everyday leadership practices. School heads and administrators play a critical role in shaping school climate, setting the tone for building relationships and making decisions. This chapter draws from grounded examples, reflective insights, and contextual leadership strategies to illustrate how to practice values-driven leadership. Whether through compassionate discipline, participatory governance, or inclusive communication, the emphasis is on leading with cultural integrity and moral purpose.

These lessons are not prescriptive checklists but guiding principles that can be adapted to diverse school contexts across the Philippines. Recognizing the geographic, linguistic, and socio-economic diversity of the country, the chapter encourages school leaders to adapt these values creatively and reflectively. Urban or rural, public or private—every school has its unique climate challenges and cultural strengths. Thus, the goal is not to impose a singular model but to empower leaders to co-create school environments that reflect the best of Filipino heritage and contemporary educational goals.

Ultimately, the objectives of this chapter converge toward one broader aspiration: to build Filipino schools that are not only academically excellent but also culturally grounded and responsive to the needs of both students and faculty. In a VUCA world, schools need to become safe havens where students and teachers can flourish as whole persons. Filipino values offer timeless wisdom and social capital that, if embraced intentionally, can revolutionize school climate and leadership from within. By grounding our educational spaces in what is deeply human and authentically Filipino, we create schools that nurture not only smart students—but also kind, connected, and compassionate citizens.

SCHOOL CLIMATE IN THE PHILIPPINE CONTEXT

School climate refers to the quality and character of school life as experienced by students, teachers, and staff. It includes norms, goals, interpersonal relationships, teaching and learning practices, and organizational structures (Konishi et al., 2022). In the Philippines, this concept is enriched by cultural and social nuances that shape how people interact within the school environment. A Filipinized perspective of school climate not only incorporates universal dimensions such as safety and relationships but also emphasizes *pakikisama* (smooth interpersonal relationships), *malasakit* (compassion), and *bayanihan* (community spirit) as key factors in creating a nurturing school culture (Lim, 2024).

The Impact of Positive School Climate

A positive school climate has profound effects on everyone in the school community. For students, it increases engagement, academic performance, and emotional well-being (Manla, 2021). For teachers, it leads to job satisfaction, reduced burnout, and higher motivation. Staff and administrators also benefit through improved collaboration and professional growth (Rizzotto & França, 2022). In the Philippines, where many schools serve as second homes for students, especially in low-income areas, a nurturing school climate can compensate for external hardships and contribute significantly to holistic development (Jabar et al., 2023).

Core Elements of School Climate

School climate is the collective atmosphere of a school as experienced by students, teachers, staff, and even the wider community. It reflects how people feel, interact, and function within the school setting. In the Philippines, where schools are often seen as extensions of the home and community, the climate of a school can either nurture or hinder the holistic development of learners. The core elements of school climate are safety, relationships, teaching and learning, and the institutional environment. Safety includes physical security and emotional safety, especially in areas prone to bullying or political tension (Charlton et al., 2021). The current Philippine Educational Commission reports an increase in the incidence of bullying in public school (Second Congressional Commission on Education, 2024).

Hierarchical respect, familial dynamics, and collectivist values shape relationships. The Philippines is strongly rooted on the notion of *Kapwa* (*sense of unity with others*) wherein other significant people are given much respect and recognition.

Lastly, the *institutional environment* refers to infrastructure, resources, and leadership styles that support effective learning. The institutional environment sets the foundation for how schools operate and prioritize student and staff well-being. When institutional leaders promote inclusivity, support professional development, and uphold transparency, these values trickle down and are mirrored within individual schools.

Safety: Beyond Physical Protection in Schools

Safety in schools traditionally refers to the absence of violence, bullying, or physical threats. In the Philippines, where some schools face risks due to natural disasters, overcrowding, or political conflict, physical safety must be constantly addressed through disaster preparedness, clear evacuation protocols, and safe infrastructure. However, safety also includes emotional security. Filipino students, especially those from marginalized backgrounds or conflict-affected areas, need assurance that their voices are heard, their identities respected, and their struggles acknowledged within the school space.

The study of Delos Reyes (2019) focuses on evaluating the safety measures at Bauan Technical High School in Batangas, Philippines, as perceived by senior high school students. It examines various aspects of school safety, including policies, emergency procedures, building and grounds security, visitor handling, communication systems, transportation protocols, and the prevalence of illegal activities. The findings indicate that procedures for handling visitors were highly effective, as perceived by the students as there were guards and safety forms accomplished. The school is also safe from individuals distributing harmful substances and all forms of bullying. Issues related to illegal activities were less effectively addressed as the school is not ready for bomb threats, fires, and earthquakes (Delos Reyes, 2019).

The study of Madrid et al. (2020) focuses on empowering teachers and students to combat sexual abuse. It introduces the “4R's” training, namely, Recognizing, Recording, Reporting, and Referral. This is aimed at enhancing teachers' abilities to identify and address signs of abuse effectively. Additionally, the program includes life skills education for students, fostering self-awareness and assertiveness to help them resist and report abusive situations. The study highlights the importance of a supportive school environment and active community involvement in reinforcing these preventive measures. Overall, the initiative demonstrates that comprehensive training and education can significantly strengthen the capacity of both educators and students to stand up against sexual abuse (Madrid et al., 2020).

The study of Delaliarte et al. (2024) investigates the fire safety knowledge and behaviors of Grade 12 STEM students in the central Philippines. Utilizing the Knowledge-Attitude-Practice (KAP) framework, the research reveals that students generally possess a strong awareness of fire safety and can demonstrate appropriate preventive and responsive action. A significant correlation exists between students' awareness and their safety practices, suggesting that increased knowledge leads to better safety behaviors. The authors emphasize the need for continuous enhancement of teacher instruction further to strengthen students' preparedness and resilience in fire-related emergencies.

Cruz & Ormilla's (2022) study examines how public elementary schools in the Alfonso Lista District, Ifugao, implement Disaster Risk Reduction Management (DRRM) programs. The study highlights that schools have established safe learning environments, formed school-based DRRM teams, and integrated DRRM into the curriculum, benefiting students, staff, parents, and visitors. While school personnel demonstrate awareness and understanding of DRRM activities, the study notes that disaster preparedness activities are still in the process of improvement. A significant finding is the direct association between the level of DRRM implementation and the extent of disaster preparedness in schools, emphasizing the importance of comprehensive DRRM strategies in enhancing disaster management capacity (Cruz & Ormilla, 2022).

Emotional safety is particularly important in a culture where *hiya* (shame) and *pakikisama* (getting along) influence student behavior. A student might stay silent even when bullied or confused during lessons to avoid standing out or being perceived as difficult. Teachers must be trained to recognize signs of distress, create emotionally inclusive classrooms, and use non-threatening discipline strategies. School climate improves when learners feel they can make mistakes, seek help, and express themselves without fear of ridicule or punishment.

Fernandez et al.'s (2022) study assessed the effectiveness of the *Mindfulness for Safe Schools* program, a mindfulness-based intervention aimed at preventing sexual abuse among Filipino public school students in Grades 7 and 8. The program sought to enhance students' emotional regulation and reduce impulsive reactions during emotionally charged situations, particularly in peer-to-peer dating contexts. Utilizing a mixed-methods design, the evaluation included quantitative surveys, focus group discussions, and video content analysis to gauge both teacher and student responses. Results indicated that teachers responded positively to the training, applying mindfulness techniques to improve patience and understanding in their roles. Students demonstrated significant improvements in emotion regulation and decreased susceptibility to peer pressure following the intervention. However, the study also noted challenges in program implementation, such as conflicts with teachers' existing workloads, highlighting the need for coordinated efforts among educators, schools, and government bodies to ensure the program's success (Fernandez et al., 2022). In addition to this, there are child protection policies in all schools to ensure safety of all the students.

Relationships

At the core of Filipino culture is the deep value for relationships, extending to schools. Teacher-student relationships are often modeled after familial dynamics, where the teacher is seen as a second parent. This dynamic creates opportunities for trust and care but may also lead to overly authoritarian practices if not balanced with respect for student agency. Peer relationships, meanwhile, thrive when *bayanihan* (community spirit) and *kapwa* (shared identity) are promoted. Strong, respectful relationships lead to better collaboration, less bullying, and greater student motivation.

In creating a Filipinized school climate, it is essential to establish relationships grounded in mutual respect and cultural awareness. Students from diverse ethnic backgrounds must be accorded equal respect and representation in school activities. Educators who demonstrate *kalinga* and *malasakit* inspire students to cultivate prosocial behaviors, thus reinforcing the social fabric of the school. Additionally, fostering relationships involves maintaining open and consistent communication with families, acknowledging parents as partners in the educational process.

School administrators take the lead or support in-service training aimed at equipping and motivating teachers. These training sessions enhance teachers' knowledge of both subject matter and teaching techniques. Proficient school administrators also consistently share updates on new programs and initiatives from different school divisions with principals and teachers. The effectiveness of these programs is periodically evaluated with input from teachers and other stakeholders. Administrators frequently observe teachers- not to intimidate, but to assist in refining their skills. Compassionate administrators step in when teachers encounter overwhelming deadlines and responsibilities. Alongside supporting teachers, training is provided for parents of children with disabilities to maintain a consistent learning experience at home.

Teaching and Learning: Establishing a Culturally Responsive Pedagogy

The classroom serves as the core of the school climate. Effective teaching involves not just academic rigor, but also how lessons are communicated, how students are engaged, and how varied learning needs are met. In the Philippines, educators must consider students' socioeconomic backgrounds, linguistic diversity, and prior experiences influenced by their community and family. Approaches that are participatory, inclusive, and relevant to everyday experiences create a positive environment where learning becomes meaningful.

Teaching and learning thrive when the curriculum and instruction reflect the local context. For example, integrating Philippine history, indigenous knowledge systems, and local literature into lessons can make students feel seen and validated. A school climate that supports differentiated instruction and values multiple intelligences creates a space where every learner can succeed.

Reyteran's (2021) study examines the experiences of teachers working in Indigenous Peoples (IP) schools in Occidental Mindoro, Philippines, focusing on how care is provided to IP learners. Findings reveal that teachers often engage in implementing the curriculum with IP content, monitoring student progress, and integrating into the IP community. which fosters a deeper understanding and respect for indigenous cultures. Teachers highlighted the need for pre-service education programs to include IP education, cultural awareness initiatives, incorporation of IP languages, training for multigrade classes, and immersion activities in IP communities. These recommendations aim to enhance the capacity of future educators to deliver culturally responsive and caring education to IP learners (Reyteran, 2021).

Institutional Environment: Leadership and Infrastructure

A school's climate is deeply influenced by its physical and organizational environment. Many public schools in the Philippines face challenges like overcrowded classrooms, fading facilities, and insufficient resources, which can adversely affect morale and academic performance (Chua-Reyes et al., 2022). Nonetheless, schools can improve their climate by strategically leveraging existing resources through effective leadership. Principals and administrators who demonstrate *malasakit* towards both staff and students can foster a culture of resilience and creativity, even in environments with limited resources.

The institutional environment is also influenced by how school leaders manage their teams, address conflicts, and enact policies. Various committees for grievance handling and policy development exist within Philippine schools. These institutions benefit from leaders who are empathetic, inclusive, and community-focused. Implementing support systems such as guidance counseling, student welfare programs, and mental health initiatives can further enhance the school climate. When teachers and students feel supported by the system instead of overwhelmed by it, they are more inclined to thrive and fully engage in school life.

The study by Najera & Gearlan (2021) explores justice and compassion in leadership. Leaders at Filipino Catholic schools are encouraged to integrate these values into their decision-making and relationship-building within the school community, reflecting the servant-leadership model inspired by Jesus Christ. This model emphasizes that school leaders should focus on serving others rather than seeking personal gain. All leaders are expected to showcase empathy and fairness. By embodying these ideals, Filipino school leaders aim to foster supportive educational environments that enhance the growth of both students and staff (Najera & Gearlan, 2021).

Fostering a positive school climate involves more than just separate initiatives; it requires a deep understanding of cultural values, structural realities, and community involvement. When educators and policymakers recognize the interconnections among safety, relationships, teaching and learning, and the institutional environment, they can create schools that are safe, effective, and genuinely reflective of Filipino values. A school climate shaped by Filipino culture values both individuals and the community as a whole, forming a solid foundation for lifelong learning and aiding in nation-building.

Challenges in Philippine School Climate

While it has great potential, developing and sustaining a positive school climate in the Philippines encounters many challenges. Public schools frequently deal with overcrowded classrooms, inadequate facilities, and restricted access to technology. The imbalance in the teacher-student ratio negatively impacts classroom management and personalized instruction. Additionally, natural disasters, poverty, and political instability in certain areas disrupt both learning and safety. These systemic problems obstruct attempts to foster environments where learners feel both supported and empowered.

Cultural Diversity and Inclusion

The Philippines is a multicultural country with a tapestry of languages, religions, and traditions. While diversity is a strength, it can also pose challenges in school climate when cultural misunderstandings or biases occur. For example, indigenous students may face exclusion or stereotyping. Teachers must be culturally responsive and inclusive in their teaching strategies and discipline approaches. Filipinizing school climate means recognizing, respecting, and celebrating these differences rather than overlooking them.

Filipino Values that Shape School Climate

Filipino cultural values, including *hiya* (sense of shame), *utang na loob* (debt of gratitude), and *pakikisama* (smooth interpersonal relationships), significantly shape school interactions (Liboon et al., 2024). These values impact communication, discipline, and authority within educational settings. For instance, teachers may refrain from confrontation to avoid embarrassment or disrespect, while students often choose silence even when they require assistance. Understanding these cultural dynamics is vital for aligning school policies and practices with students' actual experiences rather than creating conflicts with them (Sruthi & Seema Menon, 2024).

Filipinizing the school climate also involves integrating indigenous knowledge, local languages, and cultural rituals into curricula and school life. Initiatives such as culturally responsive teacher training, classroom rituals that reflect Filipino heritage, and strong parent-teacher partnerships based on *pakikipagkapwa-tao* contribute to creating culturally safe, inclusive environments. Policies and programs must consider both formal institutional structures and the informal support systems embedded in Filipino communities.

LESSONS FOR EFFECTIVE SCHOOL LEADERSHIP

Effective school leadership in the Philippines is culturally anchored in core Filipino values—*Kapwa*, *Kalinga*, *Malasakit*, and *Bayanihan*. These values guide leadership behavior and provide practical tools for fostering inclusive, resilient, and compassionate school communities. By weaving these values into leadership practices, school leaders can nurture student well-being, empower teachers, and engage communities.

Leadership with Kapwa: Building Inclusive Communities

The Filipino value of *Kapwa*—a shared identity that sees others as extensions of oneself—plays a foundational role in educational leadership. Rooted in Filipino psychology, *Kapwa* promotes empathy, inclusivity, and social responsibility (Enriquez, 1992; Solitario, 2022). In schools, it encourages participatory decision-making and relationship-centered leadership, fostering belonging among diverse students and staff.

As *Kapwa* emphasizes mutual respect and collective identity, school leaders who practice it strive to dismantle systemic inequalities and foster equitable environments. This can be operationalized through anti-discrimination policies, cultural sensitivity training, and equity-focused student councils (Reyes, 2015; Shields, 2010). Despite the challenges posed by Western-centric leadership models, *Kapwa* remains crucial for establishing socially responsive, community-oriented schools (Dayson et al., 2024).

Leadership with Kalinga: Prioritizing Well-Being

Kalinga represents care, compassion, and attentiveness to others' needs. Educational leadership requires leaders to prioritize the emotional and psychological well-being of students and staff. According to Creasman (2022), care-driven leadership involves self-awareness, support systems, and culturally sensitive engagement.

Policies from the Philippine Department of Education, such as Department of Education Order 72, series of 2009 (inclusive education) and Department of Education Order 40, series of 2012 (Child Protection Policy), reflect the institutionalization of *Kalinga* in school systems. These guidelines advocate for inclusive and caring learning environments that protect children from harm while fostering their academic and emotional development.

Leaders who practice *Kalinga* implement wellness programs, peer support groups, and mental health initiatives, especially in light of rising youth mental health concerns post-COVID (UNICEF Philippines, 2021; Schussler et al., 2022). Schools in the province of Sorsogon initiated peer counseling (PIA, 2024).

Cipriano et al. (2020) investigate the critical function of schools in promoting students' mental health through a culture of care. They recommend a comprehensive, school-wide strategy that incorporates psychosocial support, trauma-informed methods, and strong connections between students, educators, and families. By integrating mental health resources into everyday activities and lessons, schools can proactively respond to students' emotional needs, enhancing their overall well-being and academic performance. The authors highlight that caring environments not only alleviate the negative impacts of stress and trauma but also foster resilience and a sense of belonging among learners. Ultimately, the study highlights the importance of prioritizing mental health in educational contexts as a compassionate and strategic effort to facilitate the holistic development of every student (Cipriano et al., 2020).

Leadership with Malasakit: Compassionate Decision-Making.

Malasakit embodies a profound concern for others, along with the readiness to take action for their well-being, even if it requires personal sacrifice. In educational settings, this concept manifests as leadership centered on empathy, openness, and collective responsibility. Instead of rigidly following hierarchical structures, *malasakit*-driven leadership emphasizes authentic care and attentiveness to the needs of stakeholders (Jocano, 2000; del Valle, 2022).

This approach is evident in actions like waiving fees for families facing crises or adjusting school policies to meet challenges. Leaders show *malasakit* by actively listening, communicating with empathy, and engaging both staff and students in finding solutions (Cipriano et al., 2020). They exemplify ethical leadership and foster a shared dedication to a nurturing school environment (Smylie et al., 2016).

Empathy is a central aspect of this leadership style (Goleman et al., 2002), who emphasize that compassionate decision-making builds trust, collaboration, and community resilience. School leaders who exhibit *malasakit* balance organizational needs with human concerns, fostering environments where every individual feels valued and heard.

Leadership with Bayanihan: Strength in Solidarity

Bayanihan, traditionally associated with communal cooperation, remains a powerful value in modern educational leadership. Historically, bayanihan was done when families need to relocate. Relocation becomes a community endeavor to help carry the house being relocated. The whole community helps carries the house made of nipa, a palm used to make traditional Filipino houses. In modern times, bayanihan underscores the belief that collective effort among stakeholders such as teachers, parents, barangay officials, NGOs, and local businesses can significantly uplift schools (Lim. 2024).

Programs like *Brigada Eskwela* (school brigade) exemplify *Bayanihan* in action. Community volunteers help clean, repair, and beautify schools, showing how collective ownership strengthens learning environments and stakeholder engagement (Serafico-Reyes, 2020). Such efforts enhance not just physical spaces but also school-community bonds.

Leaders who exemplify *Bayanihan* foster shared governance through engaged PTAs, volunteer groups, and planning committees. This spirit of collaboration is vital in schools with limited resources, as community involvement helps to fill gaps in funding, infrastructure, and services (Brillantes & Fernandez, 2011). In the end, leadership rooted in Bayanihan links schools to the larger social network, emphasizing that education is a collective societal obligation.

BARRIERS AND CHALLENGES

One of the foremost barriers to transforming school climate through Filipino values is the resistance to change and the cultural inertia embedded in school systems.

Many schools operate within deeply entrenched bureaucratic routines that favor standardization and compliance over innovation and reflection. Some teachers and administrators may feel hesitant to embrace culturally grounded leadership practices for fear of being perceived as unprofessional or "unscientific." Moreover, reforms that attempt to localize educational frameworks are often met with skepticism (Gaston, 2023), especially when they challenge long-standing norms or disrupt hierarchical power dynamics (McMillian, 2023). This inertia is not merely institutional; it is psychological, rooted in comfort with the familiar and fear of the unknown.

A significant challenge lies in merging traditional Filipino values with modern leadership practices. Values such as *pakikisama* (getting along with others) and *utang na loob* (debt of gratitude) promote harmony and enrich relationships; however, they can be misconstrued or misapplied, possibly hindering accountability and essential discussions. For example, a school leader might avoid necessary conflicts to maintain peace or tolerate inadequate performance to preserve relationships. Thus, integrating Filipino values into the school setting demands careful consideration to ensure these values support rather than restrict (Liboon et al., 2024). The emphasis should be on critical reflection, identifying which traditional aspects enhance leadership and which should be modified based on ethical, inclusive, and professional standards.

Systemic challenges like insufficient resources, inadequate infrastructure, and overcrowded classrooms significantly hinder efforts to transform school climate (Zaman et al., 2023). In numerous underfunded public schools in the Philippines, both teachers and school leaders are overwhelmed with administrative duties and strained by high student-to-teacher ratios. With little time and support available, it becomes challenging for them to prioritize building positive relationships, ensuring emotional safety, or participating in community-building initiatives (Kalonga & Mwila, 2025). Additionally, the lack of mental health services, guidance counselors, or opportunities for professional development can further impair a school's capacity to maintain a caring and supportive environment, irrespective of the values promoted by its leaders (Karaman et al., 2021). Further, the lack of feeding programs in high-poverty incidence schools can hamper learning among students who come to school without even eating breakfast.

Political pressures at local and national levels can hinder school leaders' ability to lead with integrity and cultural focus (Marrero, 2024). In certain areas, patronage politics influence school appointments and leadership choices, undermining merit-based and values-driven leadership principles. Additionally, fluctuating national education policies, inconsistent funding, and the politicization of school initiatives can disrupt long-term planning and steady reform implementation (Ullah & Khan, 2023). In such environments, even well-meaning leaders might feel stuck, caught between the need to comply with external demands and their commitment to the fundamental values they want to uphold in their school communities.

Finally, the prevalence of Western educational paradigms can hinder the Filipinization of the school environment. Training for leaders, curriculum standards, and school evaluation frameworks frequently embody Western values of efficiency, individual achievement, and linear progression, offering limited opportunities for culturally relevant methods. Filipino concepts such as *bayanihan* and *kalinga* might be viewed as minor or secondary rather than essential components of educational success. Without intentional efforts to acknowledge and validate indigenous values as legitimate in both pedagogical and organizational contexts, school leaders may perpetuate imported models that fail to resonate with the experiences of Filipino students and educators.

RECOMMENDATIONS FOR SCHOOL LEADERS

Embedding core Filipino values in school leadership starts with purposeful and reflective practices that integrate cultural principles into daily decisions. School leaders can infuse *malasakit* into student welfare policies by exceeding mere compliance and displaying authentic care in developing support structures for at-risk learners (del Valle, 2022). *Kalinga* is exemplified through initiatives that prioritize staff well-being and enhance academic performance. By embodying these values, leaders establish the tone and cultivate compassionate, relational leadership throughout the institution.

Leaders can view school routines and rituals such as flag ceremonies, faculty meetings, and recognition days as chances to strengthen shared Filipino values. Rather than considering these activities as mere procedures, they can be transformed to emphasize *kapwa* and *bayanihan*. This could involve engaging students and parents in collaborative endeavors or acknowledging community service contributions. Such symbolic affirmations embed values into the institutional culture in visible, collective, and emotionally impactful ways.

Educators can leverage mentoring programs and peer learning groups to convey essential values. When experienced teachers or administrators demonstrate *pakikipagkapwa* or *pagkakaisa*, it enhances the chances that newer staff will embrace these behaviors. Similarly, these values can be emphasized in students during their Values Education classes. Creating formal spaces for reflective dialogue such as leadership learning circles or community discussions, encourages inclusive and participatory governance, nurturing a strong sense of shared accountability that embodies Filipino communal values.

Professional development programs must also shift to emphasize Filipino values-based leadership. Rather than focusing solely on managerial efficiency or academic metrics, leadership trainings can include modules on empathy, relationality, and ethical leadership grounded in Filipino cultural narratives. Workshops and seminars can be co-facilitated by educators, cultural practitioners, and community elders, ensuring authenticity and depth in cultural understanding. These sessions can offer case studies of Filipino schools that successfully align leadership practices with core values, allowing others to learn and adapt.

Leadership development should also include immersion experiences and local learning exchanges. For instance, school heads in urban settings could visit rural or indigenous schools to observe how *bayanihan* is operationalized in low-resource environments. Such exposure promotes humility, broadens perspectives, and enhances the cultural literacy of educational leaders, helping them apply Filipino values meaningfully within their own contexts. More research on how indigenous communities apply *bayanihan*, *kalinga*, and *malasakit* can be further examined.

In addition to personal practices, reforms in institutional policies are crucial to maintain the incorporation of Filipino values in school governance. School improvement plans, codes of conduct, and teacher evaluation systems can be updated to emphasize relational indicators and value-driven behaviors and going beyond mere academic performance or adherence to rules. Policies ought to incentivize expressions of care, community engagement, and collaborative leadership. This approach validates Filipino values as essential elements of educational excellence rather than merely as soft skills.

Advocating for values-based educational leadership in national and regional policy dialogues is essential. Educational summits, principal congresses, and regional education meetings should create opportunities for discussions around cultural responsiveness and Filipino leadership philosophies. By enhancing the visibility and legitimacy of these approaches, stakeholders can collaboratively transform the conversation. It is important to shift the perception of Filipino values from simply heritage themes to vital pillars for school reform, system resilience, and nation-building.

FUTURE DIRECTIONS

As the movement toward Filipinizing school leadership gains traction, there is a growing need to deepen our empirical understanding of how Filipino core values—such as *kapwa*, *malasakit*, *kalinga*, and *bayanihan*—influence leadership styles and outcomes in schools. Research can explore how these values are enacted in decision-making, conflict resolution, teacher support, and student engagement. Qualitative studies, especially case studies and ethnographies, can provide rich insights into the everyday realities of school leaders who integrate these values into their practice.

There is also great potential for comparative research. Studies can compare schools with values-informed leadership models to those that follow more traditional, managerial, or Western frameworks. This would allow us to investigate if schools that embody Filipino values have stronger school climates, lower teacher attrition, higher student morale, or more cohesive communities. These metrics could help validate the long-assumed benefits of Filipino relational values in education.

Future research might investigate how Filipino values interact with factors such as gender, class, region, and indigenous identity in educational leadership. For example, how do female school leaders express *kalinga* in ways that differ from their male counterparts? Are there distinct meanings of *bayanihan* in indigenous schools that influence their management of school programs? These questions encourage a deeper understanding of values, highlighting their evolving interpretations through different layers of identity and context.

Action research stands out as an effective approach, especially for empowering school leaders to act as researchers. By thoughtfully examining and refining their leadership methods through the lens of Filipino values, principals and department heads can create relevant knowledge that contributes to school enhancement. This form of inquiry, driven by practitioners, effectively connects theory and practice while also making educational research more accessible.

Alongside research, future strategies should involve broadening the existing values-based leadership framework to effectively accommodate diverse educational environments throughout the Philippines. Urban public schools encounter distinct challenges compared to rural multigrade schools or those affected by conflict in Mindanao. This framework needs to be flexible and adaptable, enabling localized interpretations of values while still being rooted in the overarching cultural narrative of *kapwa* and collective care.

A modular and scalable approach can also help. Rather than imposing a uniform framework, school communities can be encouraged to co-create their own values integration plans, drawing from a national template while customizing it to reflect their linguistic, cultural, and geographical contexts. This bottom-up approach fosters ownership and relevance, increasing the likelihood of sustainable transformation in school climate. Further, future applications of the framework must account for learners with disabilities, indigenous learners, and those in alternative learning systems (ALS). Such students have a different curriculum and can meet in non-school set-ups. Some essential questions asked would be: How can *kalinga* guide inclusive practices? How might *malasakit* be embedded in leadership decisions about curriculum adaptation or learner accommodations? Expanding the framework to address these special considerations ensures that Filipinized leadership must not exclude marginalized learners.

This advocacy can be advanced through coalitions of educators, scholars, school leaders, and the rest of the community working collaboratively to lobby for reform. Policy briefs, research roundtables, and regional forums can be platforms for amplifying the voices of those who are already practicing Filipinized leadership but have yet to be recognized or supported systemically. These efforts can also influence curriculum reforms in teacher education programs, ensuring future school leaders are equipped with a deep appreciation for Filipino cultural values.

Ultimately, the future of school leadership in the Philippines lies in reclaiming and re-centering Filipino identity in how we lead, manage, and envision educational spaces. Values-based leadership rooted in *kapwa*, *malasakit*, *kalinga*, and *bayanihan* is not merely a cultural return, it is a transformative strategy for rebuilding trust, fostering resilience, and ensuring every Filipino learner thrives in a school climate that truly feels like home.

CONCLUSION

Reimagining the school climate through Filipino core values extends beyond cultural relevance. It signifies an urgent need to transform our leadership, nurturing, and sustainability of educational environments in the Philippines. Today's educational challenges that span from student disengagement to systemic inequities, call for more than policies and programs; they necessitate a recommitment to the values that shape our identity as a people and a learning community.

By anchoring school climate in values such as *kapwa. kalinga*, *malasakit*, and *bayanihan*, we move away from transactional leadership models and toward a deeper, relational, and more human approach. These values reflect the best of Filipino culture, the nation's innate ability to care, to collaborate, and to lift each other up. These values provide a powerful compass for building safe, inclusive, and joyful schools.

The integration of Filipino values into leadership practice has the power to transform not just individual schools but the entire education landscape. When school leaders embody *malasakit*, for example, they go beyond compliance and actively seek the well-being of every student and teacher. When *bayanihan* becomes part of how school activities are planned, communities unite and take shared responsibility for student success.

A values-driven school climate fosters a sense of belonging for all members of the school community. Students feel safe, seen, and supported; teachers feel respected and empowered; and families become active partners rather than passive observers.

These outcomes are essential in cultivating schools that do not merely perform well on paper, but truly flourish in spirit and substance.

Moreover, values-based leadership establishes the conditions for resilience. In a VUCA world, school communities require more than just resources; they need relationships, empathy, and shared meaning. Filipino values offer that emotional and cultural grounding, assisting school leaders in navigating crises with grace and strength.

This approach also promotes inclusivity. Filipino values, when viewed through the lens of social justice and equity, can encourage schools to be more responsive to the needs of learners with disabilities, indigenous students, those in conflict-affected areas, and other marginalized groups. *Kalinga* and *kapwa* become guiding lights for inclusive education policies and practices.

The time is ripe for a paradigm shift. Current education reforms should not overlook the cultural wealth embedded in our local traditions and ways of being. Values are not abstract ideals; they are lived principles that shape the tone, texture, and trust within our schools. Thus, reimagining school climate through Filipino values is not an add-on; it is foundational.

Therefore, this chapter urges all school leaders, educators, teacher educators, and policymakers to reflect on their leadership models and ask these essential questions: *Are we leading in a way that reflects who we are as Filipinos?* Are our schools places where *malasakit* is practiced, where *kapwa* is honored, where *bayanihan* thrives not only during disasters but in daily learning life?

Educational policymakers must support structural reforms that embed these values into standards for school leadership, teacher training, and curriculum design. Professional development programs must equip leaders to practice values-based governance, and education departments must celebrate, not just tolerate, Filipino cultural frameworks in school operations.

Reimagining school climate through Filipino core values is more than a nostalgic return to culture. It is a forward-looking strategy for educational transformation. It envisions schools as places of connection, compassion, and community, where every learner can thrive, every teacher can lead with heart, and every leader can be an agent of meaningful change rooted in the richness of the Filipino soul.

RECOMMENDATIONS FOR FUTURE RESEARCH

Malasakit and bayanihan are deeply rooted Filipino values that hold profound significance for school leadership in the Global South, particularly in shaping a positive, nurturing school climate amid persistent structural and resource constraints. As an Asian country that is a part of the Global South, the Philippines has a history of colonialism, resource insufficiency, and a high rate of teacher migration. Given the context of *malasakit*, understood as genuine concern, empathy, and moral responsibility for others, this reframes leadership as an ethic of care that directly shapes how safe, valued, and supported members of the school community feel. With *malasakit* in leadership, teachers, students, and support staff can work hand in hand to create a better school climate to improve learning. More Filipino researchers can explore how malasakit is operationalized in their respective school settings and check on its effect on teacher retention, learning outcomes, and staff satisfaction. On the other hand, other countries in the Global South can start focusing their research on indigenous frameworks of leadership.

Bayanihan strengthens school climate by promoting collective ownership, shared leadership, and collaborative problem-solving. In many Global South contexts, school improvement depends heavily on strong relationships among administrators, teachers, parents, and community partners. Through bayanihan, school leaders cultivate inclusive decision-making processes and encourage stakeholders to work together toward common goals, thereby reducing conflict and enhancing a sense of belonging. This collective engagement not only lightens administrators' leadership burden but also builds collective efficacy, a key driver of positive school climate and sustained organizational improvement. Given this, further research on Bayanihan is recommended, as this is an age-old Filipino value. Perhaps, future research can come up with frameworks supporting this.

Together, malasakit and bayanihan offer a culturally grounded framework for empowering school administrators while intentionally improving school climate in the Global South. These values challenge technocratic and hierarchical leadership models by positioning care, solidarity, and relational trust as central to effective school governance. When embedded in leadership practice, professional development, and policy, malasakit and bayanihan enable administrators to create school environments that are emotionally supportive, socially cohesive, and resilient. In doing so, they empower school leaders to navigate global educational demands while cultivating humane, inclusive, and context-responsive schools that enable both leaders and learners to thrive.

REFERENCES

Bradshaw, C. P., Cohen, J., Espelage, D. L., & Nation, M. (2021). Addressing school safety through comprehensive school climate approaches. *School Psychology Review, 50*(2-3), 221–236. DOI: 10.1080/2372966X.2021.1926321

Brillantes, A. B., & Fernandez, M. T. (2011). Restoring trust and building integrity in government: Issues and concerns in the Philippines and areas for reform. *International Public Management Review, 12*(2), 55–80.

Charlton, C. T., Moulton, S., Sabey, C. V., & West, R. (2021). A systematic review of the effects of schoolwide intervention programs on student and teacher perceptions of school climate. *Journal of Positive Behavior Interventions, 23*(3), 185–200. DOI: 10.1177/1098300720940168

Chua Reyes, V., Hamid, O., & Hardy, I. (2022). When reforms make things worse: School leadership responses to poverty, disasters, and cultures of crises in the Philippine education system. *International Journal of Leadership in Education, 25*(2), 331–344. DOI: 10.1080/13603124.2021.2009038

Cipriano, C., Barnes, T., Jones, S. M., & Osher, D. (2020). Educating the whole child: Improving school climate to support student success. *Phi Delta Kappan, 101*(8), 8–14. DOI: 10.1177/0031721720926230

Creasman, B. K. (2022). *Prioritizing health and well-being: Self-care as a leadership strategy for school leaders*. Rowman & Littlefield. https://rowman.com/ISBN/9781475867367/Prioritizing-Health-and-Well-Being-Self-Care-as-a-Leadership-Strategy-for-School-Leaders

Cruz, R. D. D., & Ormilla, R. C. G. (2022). Disaster risk reduction management implementation in the public elementary schools of the department of education, Philippines. *International Journal of Disaster Risk Management, 4*(2), 1–15. DOI: 10.18485/ijdrm.2022.4.2.1

Darling-Hammond, L., & DePaoli, J. (2020). Why school climate matters and what can be done to improve it. *State Education Standard, 20*(2), 7.

Dayson, C., Ayo, J., Equipado, B., Polon, J., Buban, R., & Digo, G. (2024). A Systematic Review and Meta-synthesis of Indigenous Filipino Values in Educational Leadership. [Humanities And Social Sciences]. *Journal Of Community Development Research, 17*(4), 1–22. DOI: 10.14456/jcdr-hs.2024.24

del Valle, J. L. (2022). Tough teachers actually care: An ethnographic look into the'problematic'role of teachers as figures of authority under learner-centered education. *The Australian Journal of Teacher Education, 47*(6), 19–36. DOI: 10.14221/ajte.2022v47n6.2

Delaliarte, M., Linaugo, J. D., & Madrigal, D. V. (2024). Fire Safety Awareness and Practices of Science, Technology, Engineering, and Mathematics Students in a Philippine Public Secondary School. *Formatif: Jurnal Ilmiah Pendidikan MIPA, 14*(1).

Delos Reyes, Z. M. A. (2019). Assessment of School Critical Safety Concerns. *Delos Reyes, ZMA, 1*, 22–32.

Department of Education. (2009). *DepEd Order No. 72, s. 2009 – Inclusive Education as Strategy for Increasing Participation Rate of Children.* https://www.deped.gov.ph/2009/07/06/do-72-s-2009-inclusive-education-as-strategy-for-increasing-participation-rate-of-children/

Department of Education. (2012). *DepEd Order No. 40, s. 2012 – DepEd Child Protection Policy.* https://www.deped.gov.ph/2012/05/14/do-40-s-2012-deped-child-protection-policy/

Department of Education. (2022, December 7). *DepEd highlights Karapatan, Kalidad, Kalinga at 4th National Summit on Child's Rights in Education.* https://www.deped.gov.ph/2022/12/07/deped-highlights-karapatan-kalidad-kalinga-at-4th-national-summit-on-childs-rights-in-education/

Enriquez, V. G. (1992). *From colonial to liberation psychology: The Philippine experience.* University of the Philippines Press.

Fernandez, M. G. H., Lopez, G. D., Mesa, M. L. R. A., Fernandez, K. T. G., Madrid, B. J., Dans, L. F., & Lina, S. G. A. (2022). An outcomes-based evaluation of the mindfulness for safe schools program. *Acta Medica Philippina, 56*(15). Advance online publication. DOI: 10.47895/amp.v56i15.6157

Gaston, P. L. (2023). *Higher education accreditation: How it's changing, why it must.* Taylor & Francis. DOI: 10.4324/9781003445074

Goleman, D., Boyatzis, R., & McKee, A. (2002). *Primal leadership: Realizing the power of emotional intelligence.* Harvard Business Press.

Grazia, V., & Molinari, L. (2021). School climate multidimensionality and measurement: A systematic literature review. *Research Papers in Education, 36*(5), 561–587. DOI: 10.1080/02671522.2019.1697735

Hammar Chiriac, E., Forsberg, C., & Thornberg, R. (2024). Teacher teams: A safe place to work on creating and maintaining a positive school climate. *Social Psychology of Education, 27*(4), 1775–1795. DOI: 10.1007/s11218-023-09880-1

Jabar, M., Kasilag, R., Collado, Z., & Jamoral, R. (2023). Family capital and parental involvement among parents in public elementary and secondary schools in the Philippines: Perspectives of parents and children. *Asia Pacific Journal of Education, 43*(2), 555–571. DOI: 10.1080/02188791.2021.1944841

Jocano, F. L. (2000). *Filipino value system: A cultural definition*. Punlad Research House.

Kalonga, R., & Mwila, P. M. (2025). Effect of Class Size on Quality of Education in Public Secondary Schools in Temeke Municipality, Tanzania. [ARJI]. *Action Research Journal Indonesia, 7*(2), 251–277.

Karaman, M. A., Eşici, H., Tomar, İ. H., & Aliyev, R. (2021). COVID-19: Are school counseling services ready? Students' psychological symptoms, school counselors' views, and solutions. *Frontiers in Psychology, 12*, 647740. DOI: 10.3389/fpsyg.2021.647740 PMID: 33868121

Konishi, C., Wong, T. K., Persram, R. J., Vargas-Madriz, L. F., & Liu, X. (2022). Reconstructing the concept of school climate. *Educational Research, 64*(2), 159–175. DOI: 10.1080/00131881.2022.2056495

Liboon, C. A., Gutierrez, R. A. E., & Dimagiba, A. G. (2024). Reciprocity in Research and Evaluation: Conceptualizing Utang Na Loob, Pakikipagkapwa, and Alalay as Filipina American Educational Researchers. In *Theories Bridging Ethnography and Evaluation: Making Transformative, Intersectional, and Comparative Connections* (pp. 105-132). Emerald Group Publishing Limited.

Lim, R. P. S. (2024). Filipino values of secondary school heads on school-based management: Basis for values enhancement program. *International Journal of Management* [IJMTS]. *Technology and Social Sciences, 9*(1), 115–132.

Madrid, B. J., Lopez, G. D., Dans, L. F., Fry, D. A., Duka-Pante, F. G. H., & Muyot, A. T. (2020). Safe schools for teens: Preventing sexual abuse of urban poor teens, proof-of-concept study-Improving teachers' and students' knowledge, skills and attitudes. *Heliyon, 6*(6), e04080. DOI: 10.1016/j.heliyon.2020.e04080 PMID: 32566775

Malbas, M., Kilag, O. K., Diano, F., Jr., Tiongzon, B., Catacutan, A., & Abendan, C. F. (2023). In Retrospect and Prospect: An Analysis of the Philippine Educational System and the Impact of K-12 Implementation. *Excellencia: International Multi-disciplinary Journal of Education (2994-9521), 1*(4), 283-294.

Manla, V. H. (2021). School climate: Its impact on teachers' commitment and school performance. *Journal of World Englishes and Educational Practices*, *3*(2), 21–35. DOI: 10.32996/jweep.2021.3.2.3

Marrero, A. (2024). In the Crosshairs: Navigating Political Pressures as a New Leader. In *Pursuing Equity and Success for Marginalized Educational Leaders* (pp. 101-117). IGI Global.

McMillian, R. (2023). Using critical discourse analysis to operationalize discursive violence in school closure education reform policy. *International Journal of Qualitative Studies in Education : QSE*, •••, 1–24. DOI: 10.1080/09518398.2023.2258099

Monsillion, J., Zebdi, R., & Romo-Desprez, L. (2023). School mindfulness-based interventions for youth, and considerations for anxiety, depression, and a positive school climate—A systematic literature review. *Children (Basel, Switzerland)*, *10*(5), 861. DOI: 10.3390/children10050861 PMID: 37238409

Philippine Information Agency. (2024, February 26). *Sorsogon schools train peer counselors to aid students facing mental stress*. https://mirror.pia.gov.ph/news/2024/02/26/sorsogon-schools-train-peer-counselors-to-aid-students-facing-mental-stress

Philippines, U. N. I. C. E. F. (2021). *The state of Filipino children's mental health*. https://www.unicef.org/philippines/reports/state-filipino-childrens-mental-health

Reyes, C. M. (2015). Improving the quality of basic education in the Philippines: Challenges and solutions. *Philippine Institute for Development Studies Discussion Paper Series*, No. 2015-55.

Reyteran, R. (2021). Teaching in IP Schools: An Analysis of Teachers' Experiences as Input to Pre-Service Teacher Education. *International Journal of Language and Literary Studies*, *3*(4), 131–142. DOI: 10.36892/ijlls.v3i4.758

Rizzotto, J. S., & França, M. T. A. (2022). Indiscipline: The school climate of Brazilian schools and the impact on student performance. *International Journal of Educational Development, 94*, 102657. Roldán, S. M., Marauri, J., Aubert, A., & Flecha, R. How Inclusive Interactive Learning Environments Benefit Students Without Special Needs. *Frontiers in Psychology, 12*, 661427. DOI: 10.3389/fpsyg.2021.661427

Schussler, D. L., Jennings, P. A., Sharp, J. E., & Frank, J. L. (2022). Factors affecting educational leader wellbeing: Sources of stress and self-care strategies. *Educational Management Administration & Leadership*, *50*(3), 456–473. DOI: 10.1177/1741143220931260

Second Congressional Commission on Education. (2024). Miseducation: The Failed System of Philippine Education. EDCOMII Year 1 Report. Second Congressional Commission on Education. https://edcom2.gov.ph/media/2024/02/EDCOM-II-Year-One-Report-PDF-022924.pdf

Serafico-Reyes, N. M. A. (2020). Brigada Eskwela: Bayanihan through school-community partnerships (2003–2020). *Philippine Normal University Research Journal, 12*(1), 45–60.

Shields, C. M. (2010). Transformative leadership: Working for equity in diverse contexts. *Educational Administration Quarterly, 46*(4), 558–589. DOI: 10.1177/0013161X10375609

Smylie, M. A., Murphy, J., & Louis, K. S. (2016). Caring school leadership: A multidisciplinary, cross-occupational model. *American Journal of Education, 123*(1), 1–35. DOI: 10.1086/688166

Solitario, J. (2022). Revisiting kapwa: Filipino ethics, subjectivity, and self-formation. *Philippine Studies: Historical and Ethnographic Viewpoints, 70*(4), 539–564. DOI: 10.13185/PS2022.70404

Sruthi, S., & Seema Menon, K.P. (2024). Cultural Adaptation in Educational Systems: Strategies for Integrating Diverse Cultural Values. *ISHAL PAITHRKAM*, (39).

Ullah, A., & Khan, M. (2023). Political instability and its implications on public policy: A comprehensive analysis of governance transitions and their effect on educational policies in Pakistan. *The Lighthouse Journal of Social Sciences, 2*(2), 81–96.

Zaman, M., Sajjad, S., & Gharsheen, L. U. (2023). Exploring The Dynamics of Teacher-Student Relationships in Overcrowded Secondary School Classrooms. *International Research Journal of Management and Social Sciences, 4*(4), 245–261.

Chapter 9
Adapting to Evolution:
Best Practices for Educators Facing Change

R. Velmurugan
http://orcid.org/0000-0002-7925-9757
Karpagam Academy of Higher Education, Coimbatore, India

J. Sudarvel
http://orcid.org/0000-0001-6656-1992
Karpagam Academy of Higher Education, Coimbatore, India

R. Bhuvaneswari
http://orcid.org/0009-0003-9396-6129
Dr. Mahalingam College of Engineering and Technology, Pollachi, India

Ravi Thirumalaisamy
http://orcid.org/0000-0002-5954-5940
Modern College of Business and Science, Oman

ABSTRACT

With rapid changes in educational ideology in the metrics of technology, inclusivity and world uncertainties, educators should adapt with resiliency, working in collaboration within an innovative environment. Today we need to be flexible, digitally integrated and practices that are culturally responsive and so on. Teachers use professional learning to engage in problem solving, solution co construction, policy and systemic change navigation, and through their actions, professional learning becomes an opportunity to turn challenges into opportunities. Emotionally well, inclusive leader and global citizens education are essential. Educators learn

DOI: 10.4018/979-8-3373-1350-4.ch009

with storytelling and reflection, and change through their storytelling. Institutional support provides them the opportunity to flourish as agents of equity, creation, and life long learning in an uncertain world.

INTRODUCTION

Education is experiencing a period of profound transformation shaped by technological advancement, changing learner expectations, and evolving socio-political and economic conditions, requiring educators to take on expanded roles as adaptive leaders, reflective practitioners, and collaborative contributors to institutional change (OECD, 2019; Darling-Hammond et al., 2020; Fullan et al., 2021) . The traditional model of teaching has shifted toward more dynamic and inclusive pedagogies that respond to diverse student needs, promote equity, and encourage personalized learning pathways. In this context, educators are expected not only to deliver subject knowledge but also to navigate complex environments that demand innovation, critical thinking, and responsiveness to broader societal challenges.

While technological developments such as artificial intelligence, machine learning, and digital learning platforms have significantly influenced educational practices, the transformation of education cannot be understood solely through a technological perspective. Broader influencing forces, including institutional culture, policy frameworks, financial constraints, and socio-political dynamics, play a crucial role in shaping how change is implemented and sustained (OECD, 2021; Selwyn, 2020; Weiner et al., 2022). Institutional culture, in particular, serves as a foundational element that determines whether innovation is supported, resisted, or meaningfully integrated. Educational institutions that foster collaboration, shared leadership, and continuous professional development create environments where educators feel empowered to experiment, adapt, and contribute to collective progress (Sahlberg, 2015; Timperley et al., 2019).

At the individual level, mindset development is equally significant in enabling educators to respond effectively to change. A transformational mindset characterized by resilience, flexibility, empathy, and a commitment to lifelong learning supports educators in embracing new practices and navigating uncertainty (Dweck, 2006). However, developing such a mindset is not sufficient in isolation; it must be supported by institutional structures and cultures that encourage reflection, dialogue, and professional growth. Without such support, even highly motivated educators may struggle to sustain adaptive practices in the face of ongoing change.

A critical yet often underexplored dimension of educational transformation is the emotional and psychological experience of educators during periods of transition. Change processes frequently generate stress, resistance, uncertainty, and identity-related challenges, as educators are required to adjust their roles, responsibilities, and professional identities (Collie et al., 2020; Howard & Johnson, 2018; Jennings & Greenberg, 2009). Increased workload, heightened expectations, and the pressure to integrate new technologies or pedagogical approaches can contribute to burnout and reduced well-being. These emotional dynamics are not peripheral concerns but central factors that influence educators' capacity to engage with and sustain change. Recognizing and addressing these challenges through supportive leadership, professional development, and well-being initiatives is essential for fostering resilience and long-term adaptation.

The changing nature of learners further adds complexity to the educational landscape. Students today come from diverse cultural, linguistic, and socio-economic backgrounds, with varying learning preferences and needs. The growing recognition of neurodiversity and the increasing importance of mental health awareness have prompted a shift toward more inclusive and flexible approaches to teaching and learning. Frameworks such as Universal Design for Learning and Social Emotional Learning have gained prominence in addressing these diverse needs, promoting accessibility, engagement, and holistic development (CAST, 2021; Meyer et al., 2019). Educators must therefore design learning environments that accommodate variability and support all learners in meaningful ways.

Digital transformation continues to play a significant role in reshaping education, with tools such as online learning platforms, virtual classrooms, and data-driven instructional systems becoming integral to teaching practices. However, the effectiveness of these technologies depends on how they are integrated into pedagogical strategies rather than their mere adoption. The use of digital tools should enhance student engagement, collaboration, and autonomy while supporting informed instructional decisions (Lim et al., 2021; Ng & Goh, 2023). At the same time, reliance on technology raises concerns related to equity, access, and digital divides, highlighting the need for thoughtful and inclusive implementation.

Educational change is also influenced by policy environments and broader societal debates, including curriculum reforms, assessment practices, and issues related to equity and inclusion. Educators are increasingly required to engage with policy processes, advocate for meaningful reforms, and contribute to shaping educational practices at multiple levels (Darling-Hammond et al., 2020). This expanded role underscores the importance of policy literacy and professional agency in navigating complex educational systems.

Institutional support remains a critical factor in enabling effective adaptation. Organizations that prioritize collaboration, innovation, and continuous learning are better positioned to respond to change and support their educators. Professional learning communities, shared leadership structures, and opportunities for collective problem-solving contribute to building organizational resilience and fostering a culture of ongoing improvement (Larmer & Mergendoller, 2019; Barron et al., 2021). Such environments not only enhance teacher effectiveness but also positively influence student engagement and outcomes.

Thus, educational transformation is a multifaceted process shaped by the interaction of technological, institutional, and human factors. A comprehensive understanding of this transformation requires attention to institutional culture, emotional and psychological dimensions, and broader socio-political influences, alongside technological developments. By integrating these elements in a coherent and concise manner, the discussion achieves greater clarity and continuity, contributing to a more cohesive and academically rigorous understanding of how educators adapt to and lead change in contemporary education (Darling-Hammond et al., 2020; Fullan et al., 2021; Zhao et al., 2023).

FOSTERING AN ADAPTIVE MINDSET

To sustain teaching and learning in a rapidly evolving educational environment, the adoption of innovative approaches and adaptive thinking has become essential, as change is now an inherent aspect of education. This transformation is influenced not only by technological advancements but also by institutional, socio-political, and economic forces, requiring educators to respond with flexibility, resilience, and a commitment to continuous improvement. An adaptive mindset enables teachers to engage positively with uncertainty and refine their practices in alignment with changing learner needs. This perspective is grounded in the concept of a growth mindset, where abilities can be developed through effort and persistence (Carol Dweck, 2006; Johnson & Mims, 2018; Lee et al., 2021). Developing such a mindset involves strengthening emotional awareness, self-reflection, and professional identity. Reflective practices, such as journaling, support educators in examining their experiences, understanding classroom dynamics, and responding to challenges. Larrivee (2000) highlights that reflective journaling enhances awareness and supports identity development during periods of change, while continued reflection helps educators refine instructional approaches (Brinko, 2019; Chang & Lin, 2023). These processes are particularly important in addressing the emotional and psychological challenges associated with transitions, including stress, uncertainty, and evolving professional roles.

Institutional culture plays a central role in enabling effective adaptation. Environments that promote collaboration, shared leadership, and continuous professional learning create conditions that support meaningful change. Peer mentoring and instructional coaching provide opportunities for dialogue, feedback, and problem-solving, as seen in the New Teacher Induction Program in Ontario, which supports early career teachers in managing challenges and integrating into professional communities (Cherubini, 2009). Such initiatives demonstrate that adaptation is not an individual effort but a collective process shaped by institutional support. Professional Learning Communities further strengthen this culture by encouraging shared inquiry and reflective dialogue. DuFour, DuFour, and Eaker (2008) emphasize that PLCs enable educators to collaboratively analyze student learning and refine teaching practices. Structured implementation of such communities, as observed in Singapore, supports continuous improvement by facilitating knowledge exchange among teachers (Tan & Lim, 2021; Wong & Chia, 2023). These collaborative environments also enhance emotional resilience by reducing isolation and promoting shared responsibility.

The emotional and psychological dimensions of change are critical to understanding adaptation. Transition periods often involve increased workload, shifting expectations, and uncertainty, which can affect educators' well-being and professional identity. Supportive institutional practices, including mentoring and collaboration, help mitigate these challenges and enable educators to adapt more effectively. Ultimately, educational transformation is shaped by the interaction of mindset, institutional culture, and broader contextual influences rather than technology alone. While digital tools contribute to new possibilities, their effectiveness depends on educator readiness and institutional support. As noted by Fullan (2020), successful change relies on individuals with the appropriate mindset, skills, and motivation (Fullan, 2020; Fullan & Quinn, 2018; Robinson et al., 2019). A concise and integrated approach that connects these dimensions ensures a more cohesive and academically rigorous understanding of how educators adapt to and lead change.

INTEGRATION OF TECHNOLOGY THAT SERVES A PURPOSE

Digital tools such as e learning platforms, artificial intelligence, virtual and augmented reality, gaming, and learning analytics have the potential to transform education when they are meaningfully integrated with pedagogy rather than used in isolation. These technologies support access, engagement, personalization, and continuous feedback, enabling more responsive learning environments. However, their impact depends on how effectively they align with instructional goals and broader educational contexts rather than on their mere presence (Kimmons & Hall, 2020; Selwyn, 2019). Educational change, therefore, must be understood not only through technological advancement but also through institutional, social, and policy influences that shape how these tools are adopted and sustained. Effective integration of technology requires collaboration between educators and technologists to ensure that digital tools are relevant, user friendly, and pedagogically meaningful. The TPACK framework emphasizes the need for teachers to develop a balanced understanding of technology, pedagogy, and content to design effective learning experiences. This calls for continuous professional development that goes beyond technical skills to include instructional design and contextual adaptation. Institutional culture plays a critical role in supporting such development, as environments that encourage collaboration, experimentation, and shared learning enable educators to adopt technology more confidently and effectively (Sahlberg, 2015; Voogt et al., 2018).

Technological applications demonstrate their value when they enhance learning processes. Adaptive systems and gamified platforms can support engagement and individualized learning, while immersive technologies such as virtual and augmented reality provide new ways of understanding complex concepts (Yarnall et al., 2016; Chen et al., 2021; Byun & Joung, 2020; Radianti et al., 2020; Fowler, 2019). Yet, these innovations must be supported by institutional practices that ensure equitable access, thoughtful implementation, and alignment with educational objectives. Without such support, technological initiatives risk becoming fragmented or ineffective. The integration of technology also introduces significant challenges that extend beyond technical concerns. Issues of digital equity, access to infrastructure, and inclusivity can widen existing disparities if not carefully addressed. Ensuring equal opportunities for all learners requires institutional commitment through initiatives such as access programs and inclusive policies (Van Dijk, 2020; Warschauer & Matuchniak, 2019; DiMaggio & Bonikowski, 2020). At the same time, ethical considerations related to data privacy and governance must be managed responsibly to protect student information and maintain trust (Slade & Prinsloo, 2019; Regan & Jesse, 2019). Equally important are the emotional and psychological implications of technology integration for educators and learners. Increased reliance on digital tools can contribute to cognitive overload, fatigue, and stress, particularly during periods of transition when expectations and practices are rapidly changing. Educators may experience pressure to adapt quickly while managing new responsibilities, which can affect their well being and professional identity. Addressing these challenges requires supportive institutional cultures that prioritize balance, well being, and professional support systems, enabling educators to engage with change in a sustainable manner (Montag et al., 2021; Sweller, 2020).

In conclusion, technology should enhance meaningful learning rather than dominate it. Its successful implementation depends on the interaction of pedagogical design, institutional culture, and broader contextual factors. When supported by collaborative environments, ethical practices, and attention to human dimensions, technological integration contributes to a more cohesive and academically rigorous educational transformation (Kirkwood & Price, 2020; Trust & Whalen, 2021).

PEDAGOGICAL FLEXIBILITY AND INCLUSIVITY

As classrooms become increasingly diverse, educators must adopt flexible and inclusive approaches that respond to differences in culture, language, cognition, and socio economic background. Moving beyond a one size fits all model, frameworks such as Universal Design for Learning, culturally responsive teaching, and differentiated instruction support equitable learning by addressing varied student needs (CAST, 2018; Ladson Billings, 2021; Tomlinson & Imbeau, 2019). However, the success of these approaches depends not only on pedagogical strategies but also on institutional culture, which shapes how inclusivity, collaboration, and innovation are supported within educational settings. Universal Design for Learning promotes multiple means of representation, engagement, and expression, allowing students to access content in diverse ways and demonstrate understanding through varied formats. This flexibility enhances participation among students with disabilities, English language learners, and those previously disengaged, fostering a sense of belonging and inclusion (Meyer et al., 2019; Rao et al., 2020). Similarly, culturally responsive teaching emphasizes the importance of integrating students' cultural identities into the learning process, making education more relevant and meaningful. When such approaches are supported by institutional environments that value diversity and inclusivity, they contribute to stronger relationships and improved learning outcomes (Bishop et al., 2009; Lee & Green, 2023).

Differentiated instruction further supports diverse classrooms by adapting content, process, and assessment based on students' readiness, interests, and learning preferences. Practices such as varying levels of challenge and using formative assessment strategies enable educators to address individual differences while maintaining shared learning goals (Tomlinson, 2014; Tomlinson & Imbeau, 2019; Headden & Cowan, 2019). These approaches are most effective when embedded within institutional cultures that encourage reflective practice, collaboration, and continuous professional learning. Educational change in this context extends beyond instructional methods and involves shifts in professional roles, expectations, and identities. Teachers are increasingly required to act as facilitators, mentors, and co learners, which can create emotional and psychological challenges during transitions. Increased workload, uncertainty, and the need to adapt to new pedagogies may affect well being and confidence. Supportive institutional cultures that promote collaboration, shared responsibility, and professional support systems play a critical role in helping educators manage these challenges and sustain adaptive practices.

Student centered approaches that emphasize voice and choice further strengthen engagement and ownership of learning. When learners participate in shaping goals and activities, they develop deeper involvement and critical thinking skills. Such practices reflect broader educational changes influenced by social, cultural, and institutional factors rather than technology alone. Thus, flexible teaching approaches are essential for addressing diversity and promoting equity, but their effectiveness depends on the interaction of pedagogy, institutional culture, and human factors. A concise and integrated focus on these dimensions enhances clarity and continuity, contributing to a more cohesive and academically rigorous understanding of inclusive education (CAST, 2020; Ladson Billings, 2021).

POLICY ENGAGEMENT AND INSTITUTIONAL ALIGNMENT

Amid continuous systemic change, educators must actively engage with curriculum reforms, assessment redesign, and inclusivity mandates to shape the future of education. Teachers are not merely implementers but key contributors whose experiences and insights inform effective reform. Their involvement enables a shift from passive compliance to leadership and co creation, strengthening the relevance and impact of educational change (Fullan, 2020; Lieberman & Miller, 2019). However, such engagement is influenced by institutional culture, which determines whether educators are empowered to participate meaningfully in decision making processes. Curriculum reforms often respond to broader societal shifts, including global competencies and changing knowledge demands. Yet, without teacher involvement, these reforms risk misalignment with classroom realities. Sustainable change requires educators to be included from the outset, ensuring that reforms are contextually grounded and practically viable. The implementation of Scotland's Curriculum for Excellence demonstrates how involving educators enhances coherence, interdisciplinary learning, and relevance (Priestley & Minty, 2013; Priestley et al., 2020). Institutional cultures that support collaboration and shared responsibility enable such alignment and promote long term success.

Assessment redesign represents another critical dimension where teacher participation is essential. Moving beyond standardized testing, approaches such as formative and performance based assessments better reflect complex skills and learning outcomes. Teacher led initiatives, such as those within the New York Performance Standards Consortium, illustrate how innovative assessment practices can enhance engagement and achievement (Hsieh, 2016; Darling Hammond et al., 2020). These developments require supportive institutional environments that value professional judgment and encourage innovation. Inclusivity mandates further highlight the importance of teacher agency in shaping equitable educational practices. Educators play a central role in addressing diverse learning needs and ensuring access for all students. Collaborative approaches, such as co planning individualized education plans in British Columbia, demonstrate how teacher involvement strengthens inclusive practices and aligns them with student needs (Rodriguez, 2020; Florian &

Black Hawkins, 2018). Such efforts are most effective within institutional cultures that prioritize equity, trust, and shared learning.

At the same time, educational transitions create emotional and psychological challenges for educators, including increased workload, uncertainty, and evolving professional identities. These factors influence their capacity to engage with reform and sustain innovation. Supportive institutional cultures that promote adaptive leadership, transparency, and distributed decision making help mitigate these challenges and foster resilience (Heifetz et al., 2009). Collaborative structures such as leadership teams and innovation hubs further enable educators to experiment, reflect, and share practices (Anderson et al., 2019; Sutherland & Howard, 2021). Professional development, recognition of innovation, and a shared vision are essential pillars of institutional resilience. Programs such as the Teacher Learning and Leadership Program support teacher led inquiry, reflection, and collaboration, contributing to stronger professional cultures and improved retention (Campbell et al., 2016; Zhao & Lei, 2022). Overall, when institutional culture supports educator participation and well being, teachers become active agents of change, enhancing both policy relevance and classroom impact (Meyers & Lieberman, 2021; Ball & Forzani, 2020).

PRIORITIZING EDUCATOR WELL-BEING

Educational change involves a critical yet often overlooked dimension of emotional labor in teaching. As educators adapt to new pedagogical models, curricular demands, inclusivity expectations, and broader systemic pressures, they experience significant emotional and psychological strain. This includes burnout, identity shifts, career uncertainty, and ongoing stress. Adaptive change requires educators to reassess their professional values, teaching philosophy, and sense of purpose, making the process deeply personal and complex (Heifetz et al., 2009). These challenges highlight that educational transformation extends beyond technology to include human and institutional dimensions. Burnout among educators has reached concerning levels. Maslach and Leiter (2016) describe burnout as emotional exhaustion, depersonalization, and reduced personal accomplishment. During the COVID 19 pandemic, rapid transitions to remote and hybrid learning intensified these pressures. Evidence indicates that many educators considered leaving the profession due to stress and lack of support, affecting teacher retention, student outcomes, and overall school climate (Steiner & Woo, 2021). Such patterns demonstrate that emotional well being is central to sustaining educational systems.

Institutional culture plays a decisive role in addressing these challenges. Supportive environments that prioritize well being, collaboration, and professional trust enable educators to manage stress and adapt more effectively. Structured initiatives such as counseling services, peer support groups, and resilience training contribute to emotional stability. The Compassionate Systems Framework developed in Massachusetts integrates mindfulness and systemic awareness into school culture, leading to reduced stress and stronger collaboration among educators (Senge et al., 2019; Jennings et al., 2020). These approaches show how institutional culture shapes adaptation processes. Work life balance is another essential component. Policies that regulate workload, encourage planning time, and protect personal boundaries help educators sustain their professional roles. In Finland, manageable working hours, professional autonomy, and supportive structures contribute to reduced burnout and higher job satisfaction (Sahlberg, 2015; Tirri & Nokelainen, 2021). Such practices reflect how institutional design influences well being and long term engagement.

Trauma informed practices further extend the scope of support by addressing the emotional impact of working in challenging environments. Educators exposed to student trauma may experience secondary stress, requiring targeted professional development and emotional support. Initiatives in Spokane Public Schools demonstrate that training in stress recognition and empathetic communication can reduce teacher turnover and improve school climate (Blodgett & Dorado, 2016; Dorado et al., 2019). Supporting educator well being is both an ethical responsibility and a strategic necessity. Emotionally supported teachers are better equipped to build positive relationships and foster effective learning environments (Jennings & Greenberg, 2009). Therefore, institutional frameworks that integrate mental health support, balanced workloads, and collaborative cultures are essential for sustaining resilience and ensuring meaningful educational change (Greenberg et al., 2021; Roffey, 2020).

INTERDISCIPLINARY COLLABORATION AND LIFELONG LEARNING

In a dynamic educational environment, interdisciplinary collaboration has emerged as a key approach to enhancing teaching relevance and student engagement. Educators are increasingly required to move beyond subject specific boundaries and engage with professionals from diverse sectors such as community organizations, industry, and scientific fields. These collaborations enrich learning experiences by connecting academic knowledge with broader societal contexts, fostering creativity, critical thinking, and innovation among both educators and learners (Graham et al., 2019; Sandoval et al., 2020) . However, the effectiveness of such practices depends not only on pedagogical intent but also on institutional culture, which determines the extent to which collaboration is encouraged, supported, and sustained. Interdisciplinary approaches enable the integration of multiple knowledge domains, making learning more meaningful and applicable. Curriculum designs that incorporate authentic, real world connections are more likely to engage students and promote deeper understanding (Beane, 1997). Collaborative initiatives that connect education with environmental, health, and social contexts demonstrate how integrated learning can enhance both academic and life skills (Barab & Luehmann, 2021). Similarly, partnerships with industry contribute to aligning education with evolving workforce demands, enabling students to engage with contemporary knowledge and practices while supporting educators in updating curricula (Friedman, 2014).

Global collaborations further expand learning opportunities by connecting classrooms across geographical boundaries. Programs that facilitate cross cultural exchanges promote digital literacy, global awareness, and intercultural understanding, reflecting the broader social and cultural dimensions of educational change (Voogt et al., 2018; Johnson & Wells, 2021). These developments illustrate that educational transformation extends beyond technology, encompassing social, cultural, and institutional influences that shape learning experiences. Sustaining such innovative practices requires strong institutional support for continuous professional learning. Opportunities such as micro credentialing, academic conferences, and peer learning networks enable educators to remain engaged with evolving pedagogical trends and disciplinary advancements (Henderson et al., 2022; Brallier & Gaudreau, 2020). Micro credentials, in particular, provide flexible pathways for skill development in areas such as digital pedagogy, inclusive education, and leadership, supporting ongoing professional growth (Dede et al., 2019). Professional networks and collaborative platforms further contribute to knowledge sharing and reflective practice (Trust et al., 2021).

At the same time, transitions associated with collaborative and innovative practices can present emotional and psychological challenges for educators. Adapting to new roles, engaging in interdisciplinary work, and meeting evolving expectations may lead to uncertainty, increased workload, and stress. Institutional cultures that prioritize collaboration, trust, and professional support are essential in helping educators navigate these challenges and maintain resilience. Supportive environments enable educators to experiment with new approaches while managing the demands of change. Institutional culture thus serves as a foundational element in enabling sustainable educational transformation. Schools that promote shared vision, collaboration, and continuous learning create conditions where innovation can thrive. As emphasized by Fullan (2020), collaborative cultures function as the core driver of school improvement, reinforcing the importance of collective effort in achieving meaningful change (Fullan et al., 2022; Stoll & Louis, 2023). By integrating interdisciplinary collaboration with sustained professional development and attention to human dimensions, educational systems can build adaptive, resilient, and forward looking learning communities.

GLOBAL CITIZENSHIP AND SUSTAINABILITY EDUCATION

Addressing global challenges such as social injustice, climate change, and economic inequality requires a reorientation of how educators design and deliver learning experiences. In an increasingly interconnected world, education must move beyond content delivery to cultivate critical thinking, empathy, and ethical responsibility among learners. Global competencies are therefore essential, enabling students to develop curiosity, compassion, and a shared sense of responsibility toward society (Reimers & Chung, 2018; Zhao et al., 2021). This shift reflects broader educational change shaped by social, cultural, and institutional forces rather than technology alone. The integration of sustainability education and global citizenship into curricula supports learners in developing systems thinking, ethical awareness, and problem solving abilities. Frameworks such as UNESCO's Global Citizenship Education promote values of inclusion, social justice, environmental stewardship, and peaceful coexistence, guiding educators in fostering responsible and engaged citizens (UNESCO, 2019; Oxley & Morris, 2020). These approaches emphasize collaborative and cross cultural learning, preparing students to address complex global issues (Schulz et al., 2022). Their effectiveness, however, depends on institutional cultures that prioritize inclusivity, shared vision, and interdisciplinary engagement.

Institutional culture plays a central role in enabling educators to implement such transformative practices. Schools that support collaboration, reflective practice, and professional autonomy create conditions where sustainability and global education can be meaningfully integrated. At the same time, educators navigating these shifts may experience emotional and psychological challenges, including uncertainty, increased expectations, and evolving professional roles. These challenges influence their ability to adapt and engage with new pedagogical approaches. Supportive institutional environments that foster trust, collaboration, and well being are therefore essential for sustaining educator resilience and effective practice. Experiential and context based learning further strengthens sustainability education by connecting theory with practice. Initiatives that engage students in environmental and social projects enhance both academic learning and civic responsibility. Such approaches encourage learners to apply knowledge in meaningful contexts, reinforcing the relevance of education to real world challenges (Mwamwenda & Mokoena, 2019). Similarly, programs that promote ethical reflection through the study of historical and contemporary issues support the development of informed and active citizenship (Berkowitz & Bier, 2019).

These educational transformations also require educators to adopt new roles as facilitators of inquiry, ethical reflection, and collaborative learning. This transition can be demanding, as it involves balancing traditional responsibilities with new expectations related to global and sustainability education. Institutional support through professional development, collaborative networks, and reflective spaces helps educators manage these demands and sustain adaptive practices. Thus, sustainability and global citizenship education represent a comprehensive approach to preparing learners for complex global realities. Their successful implementation depends on the interaction of pedagogical strategies, institutional culture, and human factors. By addressing both structural and emotional dimensions of change, educational systems can create more cohesive and academically rigorous learning environments that promote agency, empathy, and responsibility (Nolan & O'Higgins, 2020; Crane & Matten, 2022).

CULTURAL COMPETENCE AND INCLUSIVE LEADERSHIP

Education cannot thrive under exclusive leadership; it requires inclusive practices that recognize diversity, address systemic inequities, and foster equitable participation. Inclusive leadership involves more than representation, extending to the promotion of academic success, cultural affirmation, and critical awareness among learners (Ladson-Billings, 1995). This approach reflects broader educational change shaped by social, cultural, and institutional forces rather than technology alone, emphasizing the need for educators to engage with equity as a central dimension of practice. At the core of inclusive leadership is the development of awareness and action. Educators and school leaders must be equipped to identify implicit biases that influence disciplinary practices, assessment, and expectations. Training initiatives, such as those provided by the Kirwan Institute for the Study of Race and Ethnicity, support educators in recognizing and addressing these biases, enabling more equitable and reflective decision making. Such efforts are most effective within institutional cultures that encourage continuous learning, critical reflection, and shared responsibility for inclusion.

Institutional culture plays a decisive role in embedding inclusive practices within educational systems. Schools that prioritize diversity, equity, and collaboration create environments where inclusive leadership can be sustained. This includes reviewing curriculum content, communication practices, and organizational policies to ensure representation of diverse perspectives and experiences. Policy initiatives, such as Ontario's Equity and Inclusive Education Strategy, demonstrate how institutional frameworks can guide inclusive practices by promoting equitable language, inclusive hiring, and engagement with diverse communities (Ontario Ministry of Education, 2009). These structural efforts reinforce the importance of aligning institutional values with inclusive goals. At the same time, transitions toward inclusive education can present emotional and psychological challenges for educators. Engaging with issues of bias, equity, and systemic inequality may create discomfort, uncertainty, and increased professional demands. Educators may need to reassess long held beliefs and adapt their practices, which can affect their confidence and well being. Supportive institutional environments that provide professional development, collaborative spaces, and emotional support are essential in helping educators navigate these challenges and sustain inclusive practices.

Trauma informed approaches further extend inclusive leadership by addressing the needs of students affected by adversity and systemic discrimination. These practices emphasize psychological safety, empathy, and consistent support, contributing to improved engagement and behavior (Overstreet & Chafouleas, 2016). At the same time, educators working in such contexts may experience emotional strain, highlighting the need for institutional support systems that promote well being and resilience. Inclusive leadership is therefore an ongoing, reflective process that requires continuous engagement with equity and student well being. By integrating inclusive practices with supportive institutional cultures and attention to emotional dimensions, education systems can create environments where all learners are recognized, valued, and supported, contributing to a more cohesive and academically rigorous approach to educational transformation.

NAVIGATING UNCERTAINTY AND EMBRACING COMPLEXITY

In contemporary education, educators operate within conditions of uncertainty shaped by rapid technological change, policy reforms, increasing student diversity, and global disruptions such as pandemics and climate events. This environment requires not only adaptability but also the capacity to make informed decisions under evolving conditions. Educational change, therefore, is not driven by rigid planning but by the ability to respond creatively and contextually to emerging challenges (Fullan, 2011; Fullan & Quinn, 2018; Hargreaves & O'Connor, 2018). Such transformation extends beyond technology to include institutional, social, and environmental influences that shape educational practices. Developing adaptive capacity among educators involves structured approaches that enhance foresight, flexibility, and reflective thinking. Scenario planning, for instance, enables educators to anticipate multiple future possibilities and prepare strategic responses. This approach, applied in leadership initiatives such as those in Victoria, Australia, supports educators in addressing complex issues including enrolment changes, climate resilience, and systemic transformation (Australia Gonski Institute for Education, 2020; Wright, 2022). These practices are most effective within institutional cultures that encourage innovation, shared decision making, and proactive engagement with change.

Reflective practice is another critical component in building resilience. Schön (1983) emphasizes the importance of reflection in action, where educators continuously reassess and adapt their approaches during real time problem solving. Activities such as journaling, peer coaching, and professional dialogue support educators in interpreting their experiences and strengthening their adaptive responses (Edwards et al., 2019; Farrell & Ives, 2021). These reflective processes also help educators navigate the emotional and psychological demands associated with uncertainty, including stress, role ambiguity, and shifting expectations. Simulation based learning further enhances preparedness by allowing educators to rehearse responses to complex situations in controlled settings. Role playing and scenario based exercises enable teachers to develop strategies for managing classroom disruptions, conflict, and crisis situations (Johnson & Yates, 2020; Martinez et al., 2023). Such experiences build confidence and competence while reducing anxiety associated with unpredictable environments.

Institutional culture plays a central role in sustaining these adaptive practices. Schools that promote collaboration, trust, and distributed leadership create supportive environments where educators can experiment, reflect, and co create solutions. During periods of transition, educators may experience emotional strain, including uncertainty and increased workload. Institutional support systems, including professional learning communities and collaborative platforms, help mitigate these challenges and foster resilience. Ultimately, when educators are empowered within supportive institutional cultures, they are more likely to approach uncertainty with curiosity and collaboration rather than fear. Their active involvement in problem solving and innovation strengthens both individual capacity and organizational resilience (Anderson & Krajcik, 2021; Lewis et al., 2024). A cohesive integration of adaptive strategies, institutional support, and attention to emotional dimensions ensures a more rigorous and sustainable approach to educational change.

SHARING NARRATIVES OF ADAPTATION

Storytelling has emerged as a powerful medium for professional learning, enabling educators to humanize change and foster collective growth. By sharing personal experiences through blogs, podcasts, case studies, and social platforms, educators contribute to a culture of reflection, experimentation, and collaboration. These narratives make complex educational changes more accessible by illustrating how educators navigate evolving practices and challenges. As Emdin (2016) suggests, storytelling bridges the gap between personal and professional identities, allowing educators to bring authenticity into their practice and strengthen connections with colleagues and learners. The value of storytelling extends beyond individual reflection to institutional development. When supported by institutional cultures that encourage openness and dialogue, storytelling becomes a mechanism for shared learning and continuous improvement. Platforms such as Edutopia and TeachThought showcase teacher authored accounts that document classroom innovations, challenges, and insights related to pedagogy and inclusivity. These narratives validate the intellectual and emotional dimensions of teaching while offering adaptable strategies for diverse contexts. Such environments highlight that educational change is influenced not only by technology but also by social, cultural, and organizational factors.

Storytelling also plays a critical role in addressing the emotional and psychological challenges educators face during transitions. Teaching involves significant emotional labor, particularly when adapting to new expectations, pedagogies, or institutional reforms. Narratives provide spaces for educators to express uncertainty, resilience, and growth, helping to reduce isolation and build professional solidarity. Listening to or sharing experiences through podcasts such as "Truth for Teachers" or "The Cult of Pedagogy" enables educators to connect with wider communities, reinforcing a sense of belonging and shared purpose. These interactions support well being and contribute to adaptive capacity. Furthermore, storytelling links theory with practice by translating research into relatable experiences. Initiatives such as the Harvard Graduate School of Education's "Usable Knowledge" series demonstrate how narrative accounts can illustrate the application of educational theories in real contexts. These accounts allow educators to learn from both successes and challenges, encouraging reflective practice and informed decision making. When embedded within supportive institutional cultures, storytelling enhances professional learning by promoting dialogue, critical thinking, and collaborative problem solving.

Institutional culture remains central in sustaining the impact of storytelling. Schools that value educator voice, encourage vulnerability, and support knowledge sharing create environments where innovation can flourish. During periods of change, educators may experience stress, uncertainty, and shifting professional identities. Storytelling helps navigate these challenges by providing a structured means of reflection and emotional expression, thereby strengthening resilience. Thus, storytelling contributes to a cohesive and academically rigorous approach to educational transformation by integrating human experiences with professional knowledge. By fostering collaboration, addressing emotional dimensions, and extending beyond technology to broader influences, storytelling supports the development of adaptive and reflective educational communities (Kimmons & Veletsianos, 2018).

CONCLUSION

Educational change requires continuous adaptation through collaboration, innovation, and responsiveness to evolving contexts. Rapid transformations shaped by technological advances, policy reforms, social diversity, and global disruptions demand that educators combine technical competence with emotional resilience and professional judgment (Fullan, 2020; Hargreaves & Fullan, 2012). Change is not uniform or predictable; it is influenced by broader institutional, cultural, and social forces that shape how educators experience and implement reform (Borko, 2018; Trust et al., 2020). A central element in navigating this complexity is the development of an adaptive mindset. Drawing from Dweck (2006), a growth mindset encourages educators to view challenges as opportunities for learning and improvement. Teachers who embrace uncertainty, reflect on practice, and learn from setbacks are better positioned to respond to change effectively (Reeves, 2019). Professional Learning Communities further support this process by fostering collaboration, shared inquiry, and collective problem solving, enabling educators to sustain engagement even during periods of disruption (Trust et al., 2020; Vescio, Ross, & Adams, 2018). Such collaborative practices are strengthened within institutional cultures that promote trust, shared responsibility, and continuous learning.

Technology integration remains an important dimension of educational change, but its effectiveness depends on alignment with pedagogical goals and institutional support. Frameworks such as the SAMR model guide educators in using technology to enhance learning experiences rather than simply replacing traditional methods (Puentedura, 2009; Zhao et al., 2021). Innovative practices involving digital tools can increase student engagement and creativity when implemented thoughtfully (Becker et al., 2020; Koh et al., 2019). However, technology alone cannot drive transformation; its success is shaped by broader contextual factors, including institutional readiness, teacher capacity, and equity considerations. Inclusivity and equity are fundamental to meaningful educational change. Culturally responsive teaching emphasizes the importance of recognizing and valuing students' identities, thereby promoting engagement and academic success (Gay, 2010; Ladson-Billings, 2021). Practices such as restorative approaches strengthen relationships and build trust within diverse learning communities (Gonzalez et al., 2019; Gregory et al., 2018). These initiatives require supportive institutional cultures that prioritize inclusion, collaboration, and shared values.

Educator involvement in policy development further enhances the relevance and effectiveness of reform. When teachers participate in shaping policies, they ensure alignment with classroom realities and contribute to more practical implementation (Hargreaves & Fullan, 2012; Priestley & Minty, 2018). Institutional environments that encourage participation and distributed leadership enable educators to act as active contributors rather than passive implementers. Attention to educator well being is essential in sustaining change. Emotional labor, role ambiguity, and workload pressures can hinder adaptation and innovation. Support systems that address mental and emotional needs, such as collaborative networks and well being initiatives, help educators manage these challenges and maintain resilience (Jennings &

Greenberg, 2009; Collie et al., 2018). Thus, educational transformation depends on the integration of adaptive mindsets, inclusive practices, institutional support, and attention to human dimensions. When educators are supported within collaborative and reflective cultures, they can respond to complexity with confidence and creativity, enabling schools to evolve as resilient and future oriented learning environments (Fullan, 2020; Leithwood et al., 2020).

REFERENCES

Anderson, C., & Krajcik, J. (2021). Co-creation in education: Building institutional resilience through collaborative solutions. *Journal of Educational Change, 22*(3), 341–360.

Anderson, L., Smith, J., & Johnson, K. (2019). Teacher leadership and innovation: Building professional learning communities in schools. *Journal of Educational Change, 20*(4), 475–492.

Australia Gonski Institute for Education. (2020). *Scenario planning in education leadership: Responding to future challenges*. University of New South Wales.

Ball, D. L., & Forzani, F. M. (2020). Teacher professional learning and development: What works? *Educational Researcher, 49*(7), 456–464.

Barab, S., & Luehmann, A. (2021). Curriculum integration in environmental education: Promoting ecological literacy through interdisciplinary teaching. *The Journal of Environmental Education, 52*(2), 123–135.

Barron, B., Schwartz, D., Vye, N., Moore, A., Petrosino, A., Zech, L., & Bransford, J. (2021). Doing with understanding: Lessons from research on problem- and project-based learning. *Journal of the Learning Sciences, 7*(3-4), 271–311. DOI: 10.1080/10508406.1998.9672056

Beane, J. A. (1997). *Curriculum integration: Designing the core of democratic education*. Teachers College Press, PO Box 20, Williston, VT 05495-0020.

Becker, S. A., Cummins, M., Davis, A., Freeman, A., Hall, C., Ananthanarayanan, V., & Koenig, J. (2020). *NMC Horizon Report: 2020 Higher Education Edition*. EDUCAUSE.

Berkowitz, M. W., & Bier, M. C. (2019). What works in character education: A research-driven guide for educators. *Journal of Moral Education, 48*(4), 378–392.

Bishop, R., Berryman, M., Cavanagh, T., & Teddy, L. (2009). Te Kotahitanga: Addressing educational disparities facing Māori students in New Zealand. *Teaching and Teacher Education, 25*(5), 734–742. DOI: 10.1016/j.tate.2009.01.009

Blodgett, C., & Dorado, J. (2016). A trauma-informed approach to systems change: A case study of Spokane public schools. *Children & Schools, 38*(4), 207–215.

Borko, H. (2018). Professional development and teacher learning: Mapping the terrain. *Educational Researcher, 47*(5), 252–258.

Brallier, S., & Gaudreau, P. (2020). Micro-credentials in teacher professional development: A new paradigm for lifelong learning. *Professional Development in Education, 46*(4), 555–570.

Brinko, K. T. (2019). Reflective journaling as a tool for teacher development. *Teacher Education Quarterly, 46*(4), 75–90.

Byun, S., & Joung, S. (2020). The effects of gamification in education: A systematic review. *Educational Technology Research and Development, 68*(4), 1915–1935.

Campbell, C., Malkus, N., & Miller, L. (2016). *Evaluation of the Teacher Learning and Leadership Program (TLLP)*. Ontario Ministry of Education.

CAST. (2018). *Universal Design for Learning Guidelines version 2.2.*

CAST. (2020). *Universal Design for Learning: Theory and practice*. CAST Professional Publishing.

CAST. (2021). *Universal Design for Learning Guidelines version 2.2.*

Chang, C., & Lin, S. (2023). Enhancing teacher self-awareness through reflective writing. *International Journal on Teaching and Learning in Higher Education, 35*(1), 45–59.

Chen, X., Xie, H., & Hwang, G.-J. (2021). Effects of adaptive learning systems on students' learning outcomes: A meta-analysis. *Computers & Education, 168*, 104212.

Cherubini, L. (2009). The New Teacher Induction Program in Ontario: Supporting early career educators. *Canadian Journal of Education, 32*(3), 531–554.

Collie, R. J., Shapka, J. D., & Perry, N. E. (2020). School climate and social–emotional learning: Predicting teacher stress, job satisfaction, and teaching efficacy. *Journal of Educational Psychology, 112*(4), 664–678.

Crane, A., & Matten, D. (2022). *Business ethics: Managing corporate citizenship and sustainability in the age of globalization* (6th ed.). Oxford University Press.

Darling-Hammond, L., Flook, L., Cook-Harvey, C., Barron, B., & Osher, D. (2020). Implications for educational practice of the science of learning and development. *Applied Developmental Science, 24*(2), 97–140. DOI: 10.1080/10888691.2018.1537791

Dede, C., Edwards, L., & Huang, Y. (2019). Micro-credentials for teacher learning: Trends and challenges. *Journal of Digital Learning in Teacher Education, 35*(3), 151–160.

DiMaggio, P., & Bonikowski, B. (2020). Digital divide: The impact of technology on inequalities in education. *Annual Review of Sociology, 46*, 441–463.

Dorado, J., Martinez, M., McArthur, L., & Leibovitz, T. (2019). Secondary traumatic stress and trauma-informed schools. *Psychology in the Schools*, *56*(6), 1005–1019.

Dufour, R., Dufour, R., & Eaker, R. (2008). *Revisiting professional learning communities at work*. Solution Tree, Llc.

Dweck, C. S. (2006). *Mindset: The new psychology of success*. Random House.

Edwards, A., Martin, A., & Knight, P. (2019). Reflective practice and teacher resilience: Expanding adaptive capacity in schools. *Teaching and Teacher Education*, *79*, 96–104.

Farrell, S., & Ives, P. (2021). Professional dialogue and peer coaching: Enhancing teacher reflective practices. *Journal of Teacher Education*, *72*(1), 36–50.

Florian, L., & Black-Hawkins, K. (2018). Exploring inclusive pedagogy. *British Educational Research Journal*, *44*(3), 1–16.

Fowler, C. (2019). Virtual reality and learning: Where is the pedagogy? *British Journal of Educational Technology*, *50*(2), 437–450.

Friedman, T. (2014). *The world is flat: A brief history of the twenty-first century* (Updated and expanded edition). Picador.

Fullan, M. (2020). *The new meaning of educational change* (5th ed.). Teachers College Press.

Fullan, M., & Quinn, J. (2018). *Coherence: The right drivers in action for schools, districts, and systems*. Corwin.

Fullan, M., Quinn, J., & McEachen, J. (2021). *Deep learning: Engage the world change the world*. Corwin.

Gay, G. (2010). *Culturally responsive teaching: Theory, research, and practice* (2nd ed.). Teachers College Press.

Gonzalez, T., Vélez, V., & Núñez, C. (2019). Restorative justice in Oakland Unified School District: Reducing suspensions and fostering equity. *Journal of School Violence*, *18*(1), 23–39.

Graham, C. R., Burgoyne, N., Cantrell, S., Smith, L., & Harris, R. (2019). Using technology-enhanced interdisciplinary collaboration to support STEM education. *Journal of Educational Computing Research*, *57*(5), 1176–1198.

Greenberg, M. T., Brown, J. L., & Abenavoli, R. M. (2021). Teacher stress and health: Effects on teachers, students, and schools. *EdArXiv*.

Gregory, A., Clawson, K., Davis, A., & Gerewitz, J. (2018). The promise of restorative practices to transform teacher-student relationships and achieve equity in school discipline. *Journal of Educational & Psychological Consultation, 28*(4), 344–370.

Hargreaves, A., & Fullan, M. (2012). *Professional capital: Transforming teaching in every school.* Teachers College Press.

Hargreaves, A., & O'Connor, M. T. (2018). *Collaborative professionalism: When teaching together means learning for all.* Corwin. DOI: 10.4135/9781506328133

Headden, S., & Cowan, J. (2019). Formative assessment and differentiated instruction. *Educational Leadership, 76*(3), 48–53.

Heifetz, R., Grashow, A., & Linsky, M. (2009). *The practice of adaptive leadership: Tools and tactics for changing your organization and the world.* Harvard Business Press.

Henderson, M., Alluvia, A., & Glazer, N. (2022). Sustaining teacher professional development through micro-credentials and communities of practice. *Teaching and Teacher Education, 112*, 103615.

Howard, S., & Johnson, B. (2018). Resilient teachers: Resisting stress and burnout. *Social Psychology of Education, 21*(5), 1253–1273.

Hsieh, C. (2016). The impact of performance-based assessments on student engagement and academic achievement. *Educational Assessment Journal, 21*(3), 223–240.

Jennings, P. A., Frank, J. L., Snowberg, K. E., Coccia, M. A., & Greenberg, M. T. (2020). Improving classroom learning environments by cultivating awareness and resilience in education (CARE): Results of a randomized controlled trial. *School Psychology Quarterly, 35*(4), 471–484. PMID: 24015983

Jennings, P. A., & Greenberg, M. T. (2009). The prosocial classroom: Teacher social and emotional competence in relation to student and classroom outcomes. *Review of Educational Research, 79*(1), 491–525. DOI: 10.3102/0034654308325693

Johnson, K., & Mims, C. (2018). Growth mindset in teacher professional development. *Educational Review, 70*(5), 624–638.

Johnson, L., & Wells, J. (2021). Global digital citizenship education: Developing skills for a connected world. *Journal of Global Education and Research, 5*(1), 22–38.

Johnson, L., & Yates, K. (2020). Blended learning simulations: Preparing educators for crisis management. *Educational Technology Research and Development, 68*(4), 2123–2141.

Kimmons, R., & Hall, C. (2020). Technology integration in education: A critical examination of TPACK in teacher preparation. *Journal of Digital Learning in Teacher Education, 36*(2), 78–90.

Kirkwood, A., & Price, L. (2020). Technology-enhanced learning and teaching in higher education: What is 'enhanced' and how do we know? A critical literature review. *Learning, Media and Technology, 45*(2), 145–163.

Koh, J. H. L., Chai, C. S., & Wong, B. (2019). Teachers' 21st century competencies and readiness for educational technologies: A Singapore study. *Educational Technology Research and Development, 67*, 293–310.

Ladson-Billings, G. (1995). Toward a theory of culturally relevant pedagogy. *American Educational Research Journal, 32*(3), 465–491. DOI: 10.3102/00028312032003465

Ladson-Billings, G. (2021). Culturally relevant pedagogy: Asking a different question. *Teachers College Record, 123*(1), 101–120.

Larmer, J., & Mergendoller, J. R. (2019). *Gold standard PBL: Essential project design elements*. Buck Institute for Education.

Larrivee, B. (2000). Transforming teaching practice: Becoming the critically reflective teacher. *Reflective Practice, 1*(3), 293–307. DOI: 10.1080/713693162

Lee, A., Chen, Y., & Wang, H. (2021). Growth mindset and adaptive teaching in the 21st century classroom. *Teaching and Teacher Education, 97*, 103217.

Lee, C., & Green, M. (2023). Culturally responsive teaching in diverse classrooms: New directions. *Journal of Education, 21*(2), 78–94.

Leithwood, K., Harris, A., & Strauss, T. (2020). *Leading school turnaround: How successful leaders transform low-performing schools*. Jossey-Bass.

Lewis, M., Thompson, G., & Rodriguez, A. (2024). Innovation and resilience in uncertain educational landscapes: The role of teacher collaboration. *International Journal of Educational Research, 115*, 102097.

Lieberman, A., & Miller, L. (2019). Teacher leadership as a lever for change: Promoting professional collaboration and student success. *Educational Leadership, 76*(8), 36–42.

Lim, C., Tan, C., & Chee, Y. (2021). Smart classrooms and student learning: A Singapore case study. *International Journal of Educational Technology in Higher Education, 18*(1), 50. PMID: 34778537

Martinez, L., Brown, T., & Kim, S. (2023). Trauma-informed approaches in teacher training: Simulation as a preparation tool. *Journal of School Psychology*, *95*, 1–15.

Maslach, C., & Leiter, M. P. (2016). Burnout. In Fink, G. (Ed.), *Stress: Concepts, cognition, emotion, and behavior* (pp. 351–357). Academic Press.

Meyer, A., Rose, D. H., & Gordon, D. (2019). *Universal Design for Learning: Theory and practice* (2nd ed.). CAST Professional Publishing.

Meyers, E. M., & Lieberman, A. (2021). Teacher voice and agency in education reform: Strategies for empowerment. *International Journal of Leadership in Education*, *24*(5), 605–622.

Montag, C., Duke, É., & Markowetz, A. (2021). Toward psycho-informatics: The study of online behavior as a window to psychological phenomena. *Current Opinion in Behavioral Sciences*, *18*, 1–8.

Mwamwenda, T., & Mokoena, A. (2019). Environmental education and sustainability: Impact of student projects on water conservation in South Africa. *International Journal of Environmental and Science Education*, *14*(9), 567–579.

Ng, C. F., & Goh, H. K. (2023). Artificial intelligence in education: Smart classrooms and pedagogical integration. *Computers & Education*, *181*, 104519.

Nolan, K., & O'Higgins, N. (2020). Educating for global citizenship: Reflections on theory and practice. *Journal of Global Ethics*, *16*(1), 1–17.

OECD. (2019). *Trends shaping education 2019*. OECD Publishing.

OECD. (2021). *The state of school education: One year into the COVID pandemic*. OECD Publishing.

Ontario Ministry of Education. (2009). *Equity and Inclusive Education Strategy*. Government of Ontario.

Overstreet, S., & Chafouleas, S. M. (2016). Trauma-informed schools: Introduction to the special issue. *School Mental Health*, *8*(1), 1–6. DOI: 10.1007/s12310-016-9184-1

Oxley, L., & Morris, P. (2020). Global citizenship: A typology for distinguishing its multiple conceptions. *British Journal of Educational Studies*, *68*(2), 159–182.

Priestley, M., Biesta, G., & Robinson, S. (2020). *Teacher agency: An ecological approach*. Bloomsbury Academic.

Priestley, M., & Minty, S. (2013). Curriculum for excellence and teachers' work: A Scottish perspective. *Journal of Curriculum Studies*, *45*(6), 779–801.

Puentedura, R. R. (2009). Transformation, technology, and education.

Radianti, J., Majchrzak, T. A., Fromm, J., & Wohlgenannt, I. (2020). A systematic review of immersive virtual reality applications for higher education: Design elements, lessons learned, and research agenda. *Computers & Education, 147*, 103778. DOI: 10.1016/j.compedu.2019.103778

Rao, K., Okolo, C., & Thurlow, M. L. (2020). Impact of Universal Design for Learning on students with disabilities: A meta-analysis. *Journal of Special Education Technology, 35*(1), 3–15. PMID: 38911489

Reeves, D. (2019). *Leading change in your school: How to conquer myths, build commitment, and get results* (3rd ed.). ASCD.

Regan, P. M., & Jesse, J. (2019). Ethical challenges of ed-tech, big data and personalized learning: Twenty-first century student sorting and tracking. *Ethics and Information Technology, 21*(2), 85–99.

Reimers, F. M., & Chung, C. K. (2018). *Education for shared prosperity: Principles and practices for educating children and youth to engage with the world.* Harvard Education Press.

Robinson, K., Harris, A., & Rainer, D. (2019). Mindsets in education reform: Changing perspectives. *Educational Leadership, 76*(3), 44–51.

Rodriguez, A. (2020). Inclusive education reforms in British Columbia: Teacher collaboration and student success. *Canadian Journal of Education, 43*(1), 124–142.

Roffey, S. (2020). *Positive relationships: Evidence-based practice across the world.* Springer.

Sahlberg, P. (2015). *Finnish lessons 2.0: What can the world learn from educational change in Finland?* Teachers College Press.

Sandoval, W., Vigotsky, A., & Callahan, J. (2020). Interdisciplinary collaboration in education: Fostering innovation through shared inquiry. *Educational Researcher, 49*(7), 486–495.

Schön, D. A. (1983). *The reflective practitioner: How professionals think in action.* Basic Books.

Schulz, W., Ainley, J., Fraillon, J., Kerr, D., & Losito, B. (2022). *International civic and citizenship education study 2022: Assessment framework.* International Association for the Evaluation of Educational Achievement (IEA).

Selwyn, N. (2019). *Should robots replace teachers? AI and the future of education.* Polity Press.

Selwyn, N. (2020). What should digital education be? *Learning, Media and Technology, 45*(3), 225–237.

Senge, P. M., Hamilton, H., & Kania, J. (2019). The dawn of system leadership. *Stanford Social Innovation Review, 11*(1), 27–33.

Slade, S., & Prinsloo, P. (2019). Learning analytics ethical issues and dilemmas. *The American Behavioral Scientist, 63*(10), 1430–1440.

Steiner, E. D., & Woo, A. (2021). Job-related stress threatens the teacher supply: Key findings from the 2021 state of the U.S. teacher survey. RAND Corporation.

Stoll, L., & Louis, K. S. (2023). *Professional learning communities: Divergence, depth, and dilemmas*. Routledge.

Sutherland, R., & Howard, S. (2021). Distributed leadership and teacher collaboration: Building adaptive professional cultures. *Journal of Educational Administration, 59*(2), 141–158.

Sweller, J. (2020). Cognitive load theory and educational technology. *Educational Technology Research and Development, 68*(1), 1–16. DOI: 10.1007/s11423-019-09701-3

Tan, J., & Lim, M. (2021). Professional learning communities in Singapore schools: Policy and practice. *Asia Pacific Journal of Education, 41*(4), 492–507.

Timperley, H., Wilson, A., Barrar, H., & Fung, I. (2019). *Teacher professional learning and development: Best evidence synthesis iteration.* Ministry of Education.

Tirri, K., & Nokelainen, P. (2021). Teacher well-being and burnout in Finnish schools. *Educational Psychology, 41*(3), 253–266.

Tomlinson, C. A. (2014). *The differentiated classroom: Responding to the needs of all learners* (2nd ed.). ASCD.

Tomlinson, C. A., & Imbeau, M. B. (2019). *Leading and managing a differentiated classroom.* ASCD.

Trust, T., Carpenter, J. P., & Krutka, D. G. (2020). Leading schools in a digital world: Teacher learning communities and the use of digital tools to support students. *Educational Media International, 57*(2), 104–120.

Trust, T., Krutka, D. G., & Carpenter, J. P. (2021). "Together we are better": Professional learning networks for teachers. *Computers & Education, 163*, 104099.

Trust, T., & Whalen, J. (2021). K-12 teachers' experiences and challenges with using technology for emergency remote teaching during COVID-19. *Journal of Technology and Teacher Education, 29*(2), 189–199. DOI: 10.70725/307718pkpjuu

UNESCO. (2019). *Global citizenship education: Topics and learning objectives.* UNESCO Publishing.

Van Dijk, J. (2020). *The digital divide.* Polity Press.

Vescio, V., Ross, D., & Adams, A. (2018). A review of research on the impact of professional learning communities on teaching practice and student learning. *Teaching and Teacher Education, 24*(1), 80–91. DOI: 10.1016/j.tate.2007.01.004

Voogt, J., McKenney, S., & Yang, Y. (2018). Designing international collaborative projects for digital literacy and global competence. *Computers & Education, 123*, 131–143.

Warschauer, M., & Matuchniak, T. (2019). New technology and digital worlds: Analyzing evidence of equity in access, use, and outcomes. *Review of Research in Education, 34*(1), 179–225. DOI: 10.3102/0091732X09349791

Weiner, M., Wilcox, D., & Lieberman, R. (2022). Teachers as advocates: A framework for teacher leadership. *Education Policy Analysis Archives, 30*(14), 1–24.

Wong, H. K., & Chia, S. (2023). Building collaborative teacher communities. *International Journal of Educational Research, 115*, 102102.

Wright, S. (2022). Foresight in education leadership: Implementing scenario planning in schools. *Leadership and Policy in Schools, 21*(2), 203–220.

Yarnall, L., Kozak, S., & Rueda, R. (2016). Adaptive learning and its impact on student success in mathematics courses. *Journal of Learning Analytics, 3*(2), 45–61.

Zhao, Y., Cao, Y., & Liu, X. (2023). Transformational leadership in education: Supporting professional growth and institutional change. *Educational Management Administration & Leadership, 51*(2), 234–252.

Zhao, Y., & Lei, J. (2022). Professional development for teacher leadership: A meta-analysis. *Review of Educational Research, 92*(1), 54–82.

Zhao, Y., Lei, J., Yan, B., Tan, H. S., & Lai, C. (2021). What makes the difference? A practical analysis of research on the effectiveness of technology integration in education. *Computers & Education, 69*, 403–415.

Chapter 10
How to Weather a Social Media Storm

Samuel D. Duncan
Southeast Missouri State University, USA

Emilie R. Torres
http://orcid.org/0009-0000-3160-3056
Southeast Missouri State University, USA

Michelle A. Kilburn
http://orcid.org/0000-0002-9959-1378
Southeast Missouri State University, USA

ABSTRACT

Amid a sharp increase in threats of school violence, the intersection of social media and school safety has captured the nation's attention. Recent events highlight the challenge of identifying and responding to online threats. While some of these events may turn out to be hoaxes, they can cause significant disruptions, place added strain on law enforcement, and instigate fear in a community. The impact might be school closures, canceled events, increased security, or heightened anxiety. Schools and law enforcement agencies have adapted by implementing threat assessment strategies, increasing School Resource Officer's (SROs) presence, and incorporating technology to mitigate dangers. However, concerns regarding over-policing, privacy, and disproportionate disciplinary actions remain. Legal frameworks struggle to keep up and to strike a balance between First Amendment rights and criminal actions. This chapter explores the impact of social media threats, examines the role of SROs, and proposes strategies for effective prevention, response, and community engagement.

DOI: 10.4018/979-8-3373-1350-4.ch010

INTRODUCTION

On January 22, 2025, an 18-year-old female at Coral Springs High School in Miami was arrested for making threats about her school on Snapchat. She was charged with one count of intimidation threats to kill or injure others. An additional part of her sentence included denying access to the internet, ordering her to stay away from the school, and wearing a GPS ankle monitor (Beyer et al., 2025).

A similar case occurred at Somerset Academy in Florida on November 9, 2024, when two male students, ages 14 and 15, along with one of the student's mothers, were arrested for making threats against the school and deleting evidence pertaining to the investigation. The students made multiple threats against the school over the phone, with a promise to create violence and chaos. These threats continued for a week before both students were charged with threatening violence with a dangerous device, harassment, and disrupting an academic institution. The mother was also arrested and charged for deleting evidence from her son's phone (Beyer, et al., 2025).

These examples are important because they demonstrate not only the legal repercussions that may occur, but the consequences of social media threats on schools, regardless of their credibility. False threats can cause closures in schools, waste the resources of law enforcement, and greatly impact the mental health of the community. However, credible threats reinforce the need for first responder action, proper threat assessment, and efficient communication (see Ellen, 2000). Throughout this chapter various topics such as: social media threats of violence, threat assessments, first line response, collaboration, legal implications, and challenges will be discussed to provide a deeper understanding of the many elements present when a school threat is made online.

This chapter draws on two theoretical perspectives to understand social media threats and school responses. First, Routine Activity Theory suggests that crime or harmful behavior occurs when a motivated offender, a suitable target, and the absence of capable guardians converge (Cohen & Felson, 1979). In the context of social media threats, social media may increase opportunities for individuals to issue threats while reducing traditional forms of social guardianship. Second, Crisis Communication Theory provides insight into how organizations respond to emerging crises and public fear. Research suggests that timely, transparent communication from trusted institutional sources can reduce misinformation and restore public confidence during crises (Coombs, 2007; Austin et al., 2012). Together, these frameworks help explain both the conditions under which social media threats emerge and the strategies schools use to respond to them.

The Intersection of Social Media, Schools, & Threats

Threats to the safety of children in schools are not a new phenomenon. In decades past, a threat may have come in the form of a phone call from a pay phone to a secretary in a principal's office. Common practices during that era included evacuating a school because of an anonymous bomb threat (Trump, 1998). Present day social media saturation makes it possible for a threat to be public within a matter of seconds.

Social media threats, whether hoaxes or credible plans of violence, have real world implications for educational communities. Students represent one of the most valued and vulnerable segments of a community, and social media platforms offer unique avenues for individuals seeking to incite significant disruptions within school environments. The exact percentage of social media school threats that materialize into actual violence annually is not clearly quantified in available data. However, many of such threats are hoaxes. In one study found in the *Journal of Threat Assessment and Management,* out of more than 1,000 shooting threats made at K-12 public schools over a four-year academic period, 40% of cases were categorized as unclear if the threat was real, 29% were genuine, and 31% were identified as a joke/hoax (Peterson et al., 2024).

Regardless of how a social media threat appears, the Cyber Security and Infrastructure Security Agency (CISA, 2023) stresses that all threats should be treated as credible until law enforcement and a threat assessment team determine otherwise. Over time and with each occurrence, schools commonly develop protocols to engage local authorities to catch the perpetrators, and deal with the impact of their actions. Some school districts have hired public relations specialists who, as part of their assignments, assist schools in crafting releases with appropriate talking points to combat fear (Moore et al., 2024).

Social media often shapes public opinion, effectively becoming a court of public opinion. School administrators frequently find themselves helplessly watching social media twist and bend as it weaves public perceptions. As the saying goes, "perception is reality," meaning public perception can often overshadow facts during a social media frenzy. Public perception by social media narratives can quickly become more influential than facts (Dizikes, 2018).

Regarding social media and school threats, an analysis of a recent social media storm can inform practice. A social media storm is a rapid escalation of public comments and reactions that can demonstrate fear/anxiety in all stakeholders. The event tends to generate intense scrutiny toward the leaders of the organization. On September 13, 2024, at 5:30 AM KFVS-12, a local news station in Southeast Missouri, aired that the R-2 School District in Jackson, Missouri, would be closed. The news reporter mentioned they were reading "*directly from the district Facebook page*" and noting that "*parents were getting messages and alerts.*" The reporter went on to say, "*The school district notes that this is out of an abundance of caution. They say that they will be closed today. This is to investigate a potential threat that was received late last night and is circulating through social media that involves school violence.*" Later that day, a second news report aired on KFVS-12 that Jackson R-2 had "*canceled all sporting contests and practices for that night, including their first home football game of the season.*" In this second news report, it was stated that "*this threat follows a string of similar threats, nationwide and elsewhere in the Heartland this week.*" (Heller, 2024)

This appears to have marked the beginning of a region-wide social media "storm", as it was soon accompanied by multiple other school threats in numerous communities throughout Southeast Missouri. As a direct result, additional schools in the region closed for a day, while others proceeded to cancel evening and week-end co-curricular activities. In many cases, local law enforcement was dispatched to schools to help patrons and students feel safer. That day, and over the course of the next several days, social media threats to schools were made at an alarming and unusual rate throughout Southeast Missouri (Brown, 2024).

One of the authors of this chapter recently retired as superintendent of schools in one of the districts that was impacted. Two of the researchers had family members in Jackson schools. As the scenario unfolded, a discussion began among the authors. With the second researcher being a criminal justice professor, the conversation encompassed both school and law-enforcement challenges in dealing with social media threats, as well as resulting social media fallouts. The decision was made to research the nature of the threats, the responses of school leaders, and the reactions from the community. The goal was to identify themes and develop recommendations based on a content analysis of Facebook posts and a review of current literature.

Although many of the details were never released by area school districts, several posted messages on their social media feeds regarding the threats, either real or perceived. In all cases studied, school officials quickly, and sometimes preemptively, posted messages on social media which effectively curtailed the spread of misinformation. Messages were generally characterized as demonstrating awareness of the threats and assuring action to address them.

The authors of this chapter's analysis of seven school districts' posts in the Southeast region of Missouri on or shortly after September 13, 2024, revealed commonalities as well as differences. The content analysis involved organizing raw qualitative data into categories based on recurring themes or patterns. Researchers systematically coded and interpreted the data to uncover underlying messages, ideas, or trends, including:

1. Districts adopted a range of social media approaches when responding to threats.

2. Reactions varied among school districts in terms of the nature and quantity of comments; as well as "likes, dislikes, caring" emojis etc.
3. Several other characteristics appeared to contribute to the varied responses, including follow-up communication, empathetic tone, content diversification, messaging consistency, posting frequency, and the method or source of communication.

Months later, December 4, 2024, KFVS 6 o'clock news reported that "*a Heartland student suspended in connection with a social media threat reported earlier this year, is suing the school district that punished her.*" According to the news report the seventh-grade student bringing the lawsuit argued that the circumstances leading up to the suspension "*were just a misunderstanding.*" The report tied this lawsuit directly back to the threat which prompted the Jackson School closures in September of that same year. The reporter stated the 12-year-old was suspended for 180 days as a result. Lawyers essentially argued lack of intent in the lawsuit implying that the student "*was concerned about a threat she saw online so she re-created it and shared it with another student only to see if they knew anything about it. The second student then circulated a screenshot of the message, which made it look like the 12-year-old was the source of the threat*" (Johnson, 2024).

While analyzing social media engagement provides insight into public perception and school communication strategies, it is also important to recognize how schools coordinate with law enforcement to respond effectively to such threats. This directs the focus on the crucial collaboration between schools and law-enforcement and managing threats.

Collaboration Between Schools and Law Enforcement

Educators understand student behavior in a school setting, while law-enforcement has investigative tools and training to assess threats. Working collaboratively has often become the norm and appears to be facilitated by the increased presence of specially trained law enforcement officers in schools, commonly referred to as school resource officers (SROs) (Counts et al., 2018).

Effective school and law enforcement collaboration often centers on the School Resource Officer (SRO) role. SRO programs have evolved over decades – beginning in 1953 in Flint, Michigan and expanding significantly in the late 1990s after incidents such as Columbine. This historical growth underpins why today collaboration is often personified by the SRO.

The concept of school resource officers appears to have gained traction across the nation in recent years. (New York State School Boards Association, 2019). Research has shown that school resource officers can reduce physical fights, threats, and attacks by about 30% and improve safety planning such as emergency protocols. However, due to higher detection rates, firearm related offenses often increase. Armed officers have not been shown to prevent school shootings. A study by Peterson et al. (2021) examined 133 cases of school shootings and attempted shootings in the United States from 1980 to 2019. The findings revealed that the presence of an armed officer on the scene was associated with 2.83 times increase in the rate of deaths. The authors suggest that this correlation might be due to the "weapons effect," where the presence of a weapon can increase aggression, or because shooters with suicidal intent may anticipate and even desire armed confrontation.

SROs can deter social media threats through swift identification and intervention. SROs have successfully identified suspects instigating social media threats, presenting descriptions, or identifying potential violence. The officers illustrate the importance of educating students about the consequences of online threats and encourage reporting suspicious behavior. This approach helps mitigate risk (Farlow, 2022).

Joint responses to social media threats (investigations, arrests, public safety updates) appear to build trust and transparency within the community. With the primary responsibility of school resource officers being to protect schools, these officers may become well-versed in recognizing and quickly addressing threats. (Devlin & Fisher, 2021). It is not uncommon for school districts to counter social media threats with increased law enforcement presence from city and county forces. Although concerns sometimes exist among school administrators that an over-abundant show of force may make students and parents feel uncomfortable. Right or wrong such efforts appear normalized and seem to be widely accepted by educational communities (Tocci et al., 2024). Social media threats can be dealt with in a way that eases tensions. For instance, one large, public regional school in the authors' content analysis posted "*no credible threat, extra security on site*." In this instance, it appears the combination of a quick and concise response on social media combined with increased law enforcement presence received strong engagement from the community.

Another example involved a joint response in which the department of public safety (DPS) shared a message about the situation on its website. The school district then shared the DPS post. The post was received well overall, though some community members expressed frustration over school absences due to the threats. To fully understand and manage these responses, it is essential to understand the broader legal and procedural context that helps shape how threats are interpreted and addressed. The importance of the criminal justice influence and perspective should not be understated.

Importance of the Criminal Justice Perspective

Social media created new opportunities and dangers for threats against schools, often causing disruption, anxiety, fear, and legal consequences. Assessing these threats with a criminal justice perspective helps create order, clarity, standards, and legal protocols for protecting the safety of the schools. Through this perspective, legal frameworks are created, law enforcement agencies' strategies are cultivated, and threats are assessed. This perspective helps mitigate threats, prevent violence, and hold individuals accountable for online threats (Moore et al., 2024).

Law enforcement officers are often the first to respond and analyze threats made against schools. As a threat is reported, specific protocols are followed to determine the legitimacy of the threat, analyze the contents of the threats, and identify where the threat has been made. If the threat is found to be legitimate, response plans are created, school security increases, and arrests are potentially made (Alathari et al., 2018).

While school resource officers often play a visible role in responding to threats, school safety increasingly relies on multidisciplinary collaboration among educators, mental health professionals, families, and community partners. Research on comprehensive school safety models emphasizes that school counselors, psychologists, and teachers are often best positioned to identify early warning signs such as social withdrawal, emotional distress, or changes in student behavior (Cornell 2020; Thapa et al., 2013).

School counselors and psychologists provide critical mental health support and can intervene when online threats reflect emotional distress rather than credible intent to harm others. Teachers also play an essential role in recognizing behavioral warning signs because they interact with students daily and are often the first to notice concerning changes in student behavior (Astor et al., 2005).

Parents and caregivers represent another important component of prevention. Studies of school violence prevention consistently highlight the importance of family engagement and communication between schools and parents when concerning behaviors emerge (Cornell, 2020). Multidisciplinary threat assessment teams therefore rely on collaboration among educators, mental health professionals, parents, and law enforcement to ensure both safety and student well-being are addressed. With an examination through a criminal justice lens, it is important that there is a balance between Constitutional rights and social media threats. While the First Amendment (free speech protection) ensures the right of free speech it does not protect individuals from threatening violence or disrupting public safety. Courts have established that making a threat as a joke is not protected under the First Amendment if they create fear and disruption to the public. While it is true that many social media threats are made by minors who may be unaware of the legal consequences, there are a multitude of rehabilitation programs and educational opportunities to help prevent any future incidents (Golob, 2022).

Regarding threat assessment, it is essential that law enforcement agencies and schools collaborate for effective perpetrator identification and prevention. These assessment and prevention plans consist of monitoring social media to identify early threats, educating minors and parents about the legal consequences that follow making an online threat, ensuring students and teachers feel comfortable reporting any threats they see, along with offering mental health services to help prevent any students from feeling the need to post threats online (Ellington, 2019).

Social Media threats can impact the mental health of students, educators, parents, and communities. Whether threats are legitimate or a hoax, they may enable fear and anxiety all the same. If a school system consistently experiences threats, academic success, community trust, and the school environment may be negatively impacted (Astor et al., 2005). These collaborations produce a safe learning environment by mitigating potential risks and holding perpetrators accountable. As social media continues, law enforcement strategies must evolve with it.

THE ROLE OF SOCIAL MEDIA

Social media can play a crucial role in connecting communities instantly, but it can also be a conduit to spread harmful rumors and threats like wildfire. In recent years, law enforcement has reported a surge in threats via platforms such as X (aka Twitter), Instagram, TikTok, and others (Moore et al., 2024). Online challenges can quickly become viral, and result in tragic outcomes and hoaxes alone can have a traumatizing effect on communities, parents, administrators, teachers, and students. Compounding the complexity of social media threats are the legal accountability of online communities as well as the balance of First Amendment rights and criminal law. To adequately address these issues, a discussion of the following topics is warranted: (1) how social media is used for school threats, (2) the influence of viral trends and online challenges, (3) legal implications, and (4) solutions and policy recommendations.

How Social Media is Used for School Threats

In one study of 973 K–12 schoolteachers, almost 35 percent reported a disruption in school events due to social media threats from students. More than half of the teachers said there was more than one event over the 2021- 2022 school year. Teachers at the middle and high school levels were more likely to report classroom disruptions compared to elementary teachers (Jackson et al., 2023). In the authors' home state of Missouri, a tip, posted to the Missouri State Highway Patrol's Courage2Report hotline, about an online threat to a school was reported and the armed suspect was apprehended within half an hour. According to the Highway Patrol, its hotline received 102 tips for school shooting threats and 14 tips for planned attacks between July 1 and Sept. 17, 2024 (Rock, 2024).

Threats can come in the form of direct public posts about bringing a weapon, private messages from texts to peers, and anonymous posts on apps or public forums. These threats can be explicit (e.g., "I will shoot up the school tomorrow") or vague warnings and ominous images (*FBI Threat Intimidation*, n.d.).

The data collected and content analyzed from the regional Missouri schools by the authors revealed that social media plays an important role in school threats as those types of posts received higher engagement compared to routine school updates.

Influence of Viral Trends and Online Challenges

After highly publicized school incidents there is often an influx of "copycat" threats on social media. Some threats originate from viral social media trends rather than explicit threats with an intent to harm. For example, on December 17, 2021, a viral TikTok hoax, "*National Shoot Up Your School Day*," warned of a nationwide school attack. This event caused numerous schools across the U.S. to close out of caution, despite authorities finding no credible threat (Hahn, 2021). The September 13, 2024, threat in Southeast Missouri the authors analyzed highlights this concern for students' safety across the U.S. with one community member posting, "*It's all over the country...Someone is scaring our children for no reason.*"

Current research on the reasoning behind why students may choose to participate in these challenges suggests a variety of motivations, but consistent findings suggest peer influence and the structure of social media to reward high engagement are driving factors. The social rewards of gaining "likes", followers, or a sense of belonging to a group (trend) may sufficiently motivate students to participate despite the potential harm and risk (Metzler et al., 2023).

In a 2022 study analyzing this phenomenon through the lens of "Uses and Gamification Theory", interviews with teens suggest that entertainment and social status were the top motivators and some explicitly mentioned aiming for "fame or virality" as their reward (Falgoust et al, 2022). Research also suggests that the "Fear of Missing Out" (FOMO) may be a factor as teens may join a risky challenge because they see their peers participating and they do not want to be left out. Developmentally, the adolescent brain is still maturing in areas related to impulse control, which may limit teens' ability to fully consider the consequences of their actions. Anonymity in the online environment might also provide an illusion regarding a teen's sense of confidence or toughness. In this way social media may appear as an environment where teens feel there are fewer "real-life" consequences. They may feel comfortable saying or doing one thing online that they would never do in a face-to-face setting (Lahti et al., 2024; Metzler et al., 2023; Gupta & Sharma, 2021).

The results of these online threats can be (1) school lockdowns and evacuations, (2) law enforcement resources drained, (3) psychological impacts on students and staff, and (4) difficulty differentiating hoaxes from real threats. In a perfect storm of high social media usage (e.g., as high as 95% of teens have smartphones according to Pew Research Center), strong peer influence, and high reward structures, we have a potentially volatile environment where viral trends can be normalized, even those that are dangerous and unlawful. From a Routine Activity Theory perspective, social media platforms may increase opportunities for threats by expanding access to potential targets while decreasing direct adult supervision (Cohen & Felson, 1979). Understanding the legal implications for online threats is crucial for the community, schools, law enforcement, parents, and students.

Reflections from Practice: Student Perspective

As a student, the rapid increase in school threats and advancements in the level of violence that is being threatened has created high anxiety in schools all over the nation. While safety measures are put into place, such as an increased number of SROs and lockdown drills, these methods themselves often contribute to feelings of unease. School threats, whether a hoax or real, create disruptions not only in our education but also mental health, which can lead to a decrease in academic performance. Social media can rapidly spread threats, and disinformation, however, when the threats that are being posted are written by fellow peers it can make reporting them even more difficult.

In regard to lockdown drills there have been multiple situations where students haven't been told that they were drills and teachers have been shot by pellet guns (execution style in Indiana, leaving multiple injured) to simulate what could potentially happen (KRON4, n.d.). These drills have been considered to have gone too far and have even led to lawsuits. In 2015 in Oregon an elementary teacher filed a lawsuit because during an unannounced active shooter drill, a masked man pointed a fake gun in her face, pulled the trigger causing smoke to fill the room and he "announced" her dead. This gave her PTSD while other students/teachers have been physically harmed in other drills (Hammond, 2015).

Research suggests that repeated exposure to school violence threats can negatively affect students' psychological well-being and academic engagement. Studies examining school climate indicate that students who perceive their schools as unsafe are more likely to have trouble concentrating, increased absenteeism, and declines in academic performance (Astor et al., 2005; Thapa et al., 2013). Exposure to crisis events or repeated safety threats can also contribute to anxiety and heightened vigilance, which may interfere with learning and classroom participation. Even when threats ultimately prove to be hoaxes, the resulting lockdowns, school closures, and law enforcement responses may disrupt instructional time and contribute to emotional stress for students. These findings highlight the importance of addressing both the safety and mental health dimensions of social media threats within school environments.

Legal Implications of Online Threats

A social media threat, even if determined a hoax, is not exempt from legal consequences. These types of events can trigger not only school discipline, but criminal prosecution as well. The Federal Bureau of Investigations warns that making threats is not a prank, it is a federal crime. Convicted individuals could spend up to five years in federal prison (*FBI Threat Intimidation*, n.d.).

In the September 13, 2024, event the authors analyzed, several school districts posted the following image or like their Facebook pages (see Figure 1). The image specifies that posting threats online is “not a joke” with the hashtag:thinkbeforeyoupost.

Figure 1. FBI social media post graphic

Source: www.fbi.gov

However, with the rise in online pranks we have also witnessed a rise in challenges to legal frameworks including First Amendment rights and the difficulty in proving intent (mens rea). In the U.S., the First Amendment does provide protection for free speech, but that freedom does not extend to "true threats". A "true threat" is a statement with the intent to communicate or commit violence. A common example is shouting "fire" in a crowded movie theatre. The Supreme Court has long-standing precedent that some types of speech are so dangerous that they are not protected by the First Amendment (*Chaplinsky v. New Hampshire,* 1942). However, determining what is a "true threat" in the online environment has proven complicated. In the landmark case, *Elonis v. United States* (2015), the Supreme Court overturned a conviction, not on the grounds of the First Amendment, but on the premise that the prosecution needed to prove intent to threaten, rather than just how a reasonable person would perceive the posts. This case was in reference to Elonis publishing threatening posts on Facebook against his wife and others. The Court determined that it was not enough that Elonis's wife and others *felt* threatened, the prosecution must establish that Elonis *had the intent (mens rea) to commit violence*. Defendants sometimes argue their provocative online posts are satire, art (as Elonis claimed his were rap lyrics), or jokes taken out of context. Courts are continuing to grapple with this issue (see *Counterman v. Colorado*, 2023 examining whether a stalking victim's perspective or the poster's intent defines true threat). To summarize, even though in person and outright threats of violence are generally considered unprotected speech, the social media environment has raised debates on how to assess intent.

The *Tinker v. Des Moines Independent Community School District* (1969) decision established that students retain their First Amendment rights in school, if their speech does not cause a "material and substantial disruption" to the educational environment. This ruling, known as the *Tinker Standard*, has served as a landmark test for evaluating the limits of student expression. However, applying this standard to digital speech is challenging. Unlike traditional, on-campus expressions of free speech, online communication—especially when generated off-campus—can spread rapidly, reach large audiences, and impact the school climate. Courts have struggled to consistently determine when schools have the authority to discipline students for social media posts made outside school hours, particularly when the speech is offensive, threatening, or critical of school personnel. As a result, while Tinker remains influential, the rise of digital platforms has complicated its application, leading to ongoing legal and policy debates.

While we have established that the FBI has repeatedly held that even hoax threats online are treated as serious crimes, not juvenile pranks, many states have their own statutes addressing online threats. In response to school shooting tragedies some states have strengthened these laws. Tennessee enacted a law, for example, making threats of violence on school grounds an automatic felony regardless of whether the threat is discredited and posted by a minor (*Tennessee Mass Violence Threats Act;* Swaby, 2024). Supporters of these strict laws argue they deter potential threats, but critics warn of the dangers of criminalizing, labeling, and causing trauma to young juveniles whose intent might be a joke or attention-seeking behavior (Baumgartner et al., 2021). An additional layer of difficulty develops when posts are made across state lines, or even internationally. In summary, the legal environment is struggling to keep up with the technology and balance protection of free speech while holding individuals accountable when online threats reach the threshold of unlawful and "true threats."

The U.S. Secret Service's NTAC (National Threat Assessment Center) study of school attackers found most perpetrators leaked their violent intentions online or in messages before the event. This study underscores the concern that law enforcement must treat online threats with the utmost urgency (Department of Homeland Security, 2024). This recognized need for heightened awareness of online threats underscores the critical role of School Resource Officers (SROs) in identifying potential dangers within the school community.

It should also be noted that although this chapter focuses primarily on the United States, international policy approaches provide additional context for understanding school safety governance. In the United Kingdom, the Prevent Duty under the Counter-Terrorism and Security Act (2015) requires schools and public institutions to identify individuals who may be vulnerable to radicalization or violence and refer them to multidisciplinary intervention programs (Home Office, 2015). In Canada, school safety initiatives such as Ontario's Safe and Accepting Schools strategy emphasize bullying prevention, mental health support, and positive school climate rather than relying primarily on law enforcement responses (Ontario Ministry of Education, 2018). These approaches illustrate how international models often integrate education, social services, and community engagement to address the underlying factors that contribute to threats and violence in schools.

THE ROLE OF SCHOOL RESOURCE OFFICERS (SROS)

School Resource Officers (SROs) often play a central role in school threat response strategies. When used effectively, SROs can help safer school climates and act as first responders in emergency situations. A recent study found that 85% of middle and high school teachers felt safer with an SRO present at the school (Jackson et al., 2023). Positive relationships with students and SROs have been associated with an increased likelihood of students reporting potential threats (Sorensen et al., 2023).

SROs in Threat Prevention and Response

The nature and frequency of threats vary. However, school personnel can recognize that deadly school violence often occurs without warning; and sometimes threats are made ahead of time and are carried out. The very basic need of feeling safe is fragile in such a climate. The presence of School Resource Officers (SROs) is primarily intended to enhance perceptions of safety within the school community and serve as a deterrent to potential threats or acts of violence (Sorensen et al., 2023).

School Resource Officers can quickly assess and respond to threats in ways non-commissioned citizens, including building principals, simply cannot. Educators are primarily trained in caring for students and educating them. SROs are expected to be equipped to deal effectively with individuals who pose a threat to schools (Schweit & Mancik, 2017).

Another benefit of having SROs is training of personnel. Many school resource officers are tasked with conducting drills prior to the start of school and throughout the school year focused on keeping students safe in the event of an active shooter. Other training may involve bringing additional law enforcement agencies into the school during after school hours to coordinate efforts in the event of a school shooting. It is important to note that drills should be conducted responsibly. Excessively realistic simulations can cause injury or trauma. The best outcomes occur when drills are well communicated and designed to educate without instigating unnecessary fear. The execution of these drills requires careful consideration. Research by Zhe and Nickerson (2007) shows that while these drills can improve understanding of emergency procedures, they may also heighten anxiety, particularly when the scenarios are very realistic. To be effective, drills should inform and prepare without inducing unnecessary fear or distress.

The National Child Traumatic Stress Network (2019) also stresses the need for a trauma-informed approach to active shooter drills. Their recommendations include giving advance notice, adapting drills to the developmental stages of students, and avoiding graphic or intense simulations that could trigger psychological harm. This strategy helps ensure safety training does not come at the expense of emotional well-being. Furthermore, the National Child Traumatic Stress Network (2019) emphasizes the importance of trauma-informed approaches when conducting active shooter drills. Their guidelines suggest that drills should be announced in advance, tailored to the developmental levels of students, and should avoid the use of simulations that can cause psychological distress. This approach aims to balance preparedness with the mental well-being of students and staff.

It is important to note a growing body of research raises concerns about the unintended consequences of school-based policing. Studies suggest the presence of SROs correlates with increased arrests, suspensions, and referrals to the juvenile justice system - especially for Black, Latinx, and disabled students (Riddle & Sinclair, 2019; Fisher & Hennessy, 2016). This over-policing has been linked to the school-to-prison pipeline, which suggests that minor infractions traditionally managed by educators are now criminalized (Fisher et al., 2023). However, when clearly defined roles, accountability structures, and collaborative relationships with school staff are in place, SROs are more likely to support rather than hinder a positive school environment (Reynolds & Astor, 2024).

Reflections from Practice: School Resource Officer Perspective

My SRO role has shifted from strict law enforcement to a more holistic, student-centered approach. After the Sandy Hook Elementary shooting (2012) and the Parkland shooting (2018), many schools increased SRO presence to give more emphasis on active shooter response training and emergency preparedness. It has led us to a greater use of security technology, including surveillance systems and metal detectors. I focus on building relationships with students and try to act as a trusted figure rather than just an authority figure. The SRO training I participate in now has expanded over time to include Mental health awareness, de-escalation techniques, trauma-informed policing, and cultural competency. I feel I have grown to be better equipped to handle students in crisis rather than defaulting to arrests. I believe SROs should focus on relationship-building rather than intimidation.

Something else I do is ensure my actions comply with FERPA (Family Educational Rights and Privacy Act) when sharing student information. I am part of all school safety planning and emergency drills. I also work collaboratively on threat assessment teams rather than taking a punitive approach. Balancing security measures with maintaining a welcoming learning environment is something I believe is important.

I'm not involved in routine discipline (e.g., dress code violations, minor disruptions). I appreciate that the school administrators I work with make sure teachers do not over-rely on me for behavior management. Another example of my role is developing and maintaining clear protocols for when law enforcement intervention is necessary. My administrators are aware of school district policy about student rights regarding searches, questioning, and arrests. We make a great team!

THE ROLE OF ARTIFICIAL INTELLIGENCE (AI) IN THREAT DETECTION

A study published in *Social Media + Society* examined social media use among public mass shooters in the United States and found many perpetrators exhibited concerning behaviors online prior to their attacks, including posting threats or violent content (Peterson et al., 2023). An intentional focus on prevention has led some schools to utilize Artificial Intelligence (AI) in detecting such events before they escalate into real-world violence. Schools are turning to machine learning tools, such as AI, and natural language processing algorithms to monitor student communications. Sources might include social media, email, and cloud-based content for language that indicate potential harm (Burke & Bloss, 2020). While promising, evidence suggests these tools may have mixed effectiveness. For example, the U.S. Secret Service analysis (2021) found social media monitoring was one indicator in threat discovery in a minority of averted attack cases. However, research has also indicated that AI monitoring tools can generate false positives and may disproportionately identify marginalized populations based on linguistic or cultural expressions (Xu et al., 2021). In some cases, content related to LGBTQIA+ identities or specific mental health topics has been incorrectly identified as threatening (Brennan Center for Justice & Center for Democracy and Technology, n.d.). This highlights the continued need for human observation and reporting despite advances in AI.

In Pinellas County, Florida, students used an anonymous reporting system to disclose threats made on Snapchat. Such tools allow students to submit tips about potential dangers, including violent threats, directly to law enforcement and school officials. For example, FortifyFL enables users to include descriptions, photos, and videos of threats while keeping their identity optional. In one instance, a Snapchat threat involving racist language and a claim of a planned school shooting caused widespread concern among students in the county. The threat was reported by a student through these channels and investigated by law enforcement, who determined it was not credible and originated outside the state (Solochek, 2021).

In one pilot study, AI systems reached 91% accuracy in identifying high risk students. However, this was in a controlled setting; in real world use, accuracy may vary, and such tools must be continuously validated (Barzman et al., 2017). Some school districts have already begun using AI driven monitoring tools, such as scanning social media posts or school emails for keywords to get an early warning of threats. However, these systems have faced scrutiny. Critics argue they sacrifice student privacy for improving security gains and there is evidence they disproportionately flag minority students or those with certain keywords, unrelated to violence (Brennan Center for Justice & Center for Democracy and Technology, n.d.). Unlike static, traditional tools, artificial intelligence models may continue to improve by adding new data allowing for better identification of risk factors overtime (Barzman, 2018).

Law enforcement officials themselves use a multitude of digital tools and methods to locate the geographical location of the perpetrators making these threats. In social media threats, these digital tools allow investigators to view perpetrators' internet activity, internet protocol address, analyze their metadata activity, and recover any deleted posts or information that could be critical to the threat (Burley, 2025). Through a process called cyber-tracking, artificial intelligence monitors online activity and specific keywords (Rigano, 2018). Throughout this process law enforcement agencies collaborate with experts in cybersecurity to second encrypted messages and ensure that social media accounts are not only credible but that the proper individuals are being held accountable for their actions (Miller, 2024).

The use of AI-based monitoring systems also raises ethical concerns related to surveillance, consent, and student rights. Scholars have noted that digital monitoring technologies in schools may create tensions between safety efforts and student privacy protections, particularly when automated systems collect large amounts of student data or flag language without contextual review (Regan & Jesse, 2019; Williamson & Eynon, 2020). Schools implementing these systems must therefore balance safety objectives with transparency, oversight, and clear data governance policies.

Due to potential ethical concerns, schools may prefer using techniques already in place. One technique is to systematically and consistently encourage students, parents, and school staff to report suspicious behaviors or threats (Hsieh et al., 2022). Another is to provide specific training to staff to recognize where a troubled individual's intent surfaces through their statements and actions. Resource officers also assist school personnel in implementing systems such as online forms or apps, which will allow for anonymous reporting (Stohlman et al., 2020).

CHALLENGES AND CONCERNS

Educators, law enforcement, and communities struggle to navigate a landscape marked by false alarms, misinformation, and rapidly evolving technology. These challenges highlight the need for transparent messaging, proactive engagement, and protocols that can help build trust and resilience. This portion of the chapter explores concerns surrounding school threat management, community partnerships, as well as collaboration between schools and law enforcement. Together these efforts can help create a sense of shelter in the face of uncertainty.

Managing False Alarms and Misinformation

Because a relatively high percentage of threats are not carried out, the ability to manage false alarms, and to successfully respond to associated misinformation is vital to maintaining order throughout the educational community. As learned through the study of the September 2024 social media scare in Southeast Missouri, school administrators have adopted a variety of approaches:

Example 1: A rural school in Southeast Missouri used a mass notification product to simultaneously post an immediate response across multiple social media platforms. This saturation strategy yielded high engagement with 137 likes and 33 comments on Facebook applauding safety updates and post-incident communication.

Utilizing mass notification systems to disseminate information simultaneously across various social media platforms ensures timely and widespread communication during crises. This approach leverages the immediacy and broad reach of social media, facilitating rapid information sharing and community engagement. A systematic review of crisis communication literature emphasizes the importance of using social media's potential to create dialogue and choosing the right message, source, and timing during crises (Austin et al., 2012).

Example 2: In the same incident, another district coordinated its message with the town's Department of Public Safety. The DPS first shared a post, emphasizing the seriousness of the threat, which the school district re-shared. This demonstrated a unified front and established an authoritative tone to the message and was met with strong support.

Collaborating with outside agencies, such as the Department of Public Safety, to deliver unified messages enhances the credibility and authority of school communications during emergencies. Coordinated efforts ensure consistency, reduce misinformation, and demonstrate a cohesive response strategy. Research highlights the significance of performing pre-crisis work and developing an understanding of social media logic to effectively manage crisis communication.

Example 3: A third school superintendent took a personal approach by posting an in-person video on the school district's Facebook page with a clear message that such threats will not be taken lightly, and perpetrators will be held accountable. This direct, face-to-face style also received broad community support.

These examples illustrate the application of research-backed strategies in crisis communication, emphasizing timely dissemination, authoritative coordination, and transparent engagement to effectively manage school emergencies (Lachlan et al., 2016).

Potential for Over-Policing and Disproportionate Impacts

Due to the highly publicized nature of many of these threats, Boards of Education may find themselves under pressure from community members and educators to pursue long-term suspensions and expulsions. This can become problematic, because the Board of Education will most likely review information the public will not have access to. In situations where leniency may be appropriate, community pressure or heightened public attention can unduly influence disciplinary outcomes, potentially resulting in disproportionately harsh penalties for students. The principle of "better safe than sorry" may serve as a protective measure for School Boards; however, its application can also result in disproportionately adverse consequences for students, potentially affecting their future opportunities (Skiba & Losen, 2016).

Another consideration is that while no clear evidence appears to exist indicating that students from one demographic are more frequently found guilty of posting social media threats, studies do suggest that punitive measures for social media threats disproportionately impact students of Color (Welch & Payne, 2010). This disparity may result from systemic biases in how schools and law enforcement assess and respond to such incidents, as well as broader inequities in disciplinary practices (Quereshi, n.d).

Regarding over policing, an obvious indicator would be increases in school resource officers. In the 2015–16 school year, approximately 52,000 school resource officers (SROs) were reported to work in schools at least once per week. As illustrated in Figure 2 by 2019–20, 51.4% of public schools had an armed law enforcement officer present on campus, reflecting a steady increase from 45% in 2017–18 and 32% in 2005–06 (Ghavami et al., 2021).

Figure 2. Percentage of armed law enforcement officers present in schools

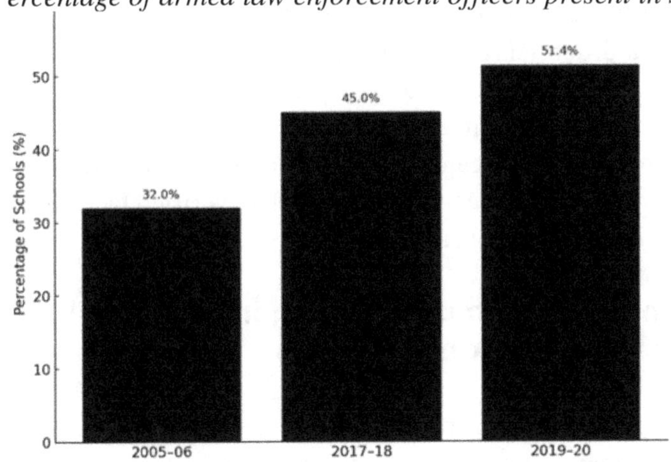

Another indicator of the level of policing may be the use of metal detectors in schools. (Kremser, n.d) As with school resource officers, the use of metal detectors in public schools has increased over the past decade, though growth has been gradual and varies by region and school level. In 2013-2014, 8.7% of public high schools used metal detectors, up from 5% in 2010. By the 2022-2023 school year, estimates show 10% of high schools, 7% of middle schools, and 2% of elementary schools had implemented them (Statista, n.d.). This trend may reflect heightened concerns about school safety following incidents of gun violence and other security threats. However, the overall adoption rate remains relatively low, with some schools using them only temporarily or for random checks (Price & Khubchandani, 2019).

Some methods to curtail social media threats may raise ethical concerns. The significant implications associated with school safety may lead to overextension in accessing and utilizing student data, raising concerns about privacy and ethical boundaries. Balancing privacy against threats may at some point in the future warrant further emphasis.

Communication & Transparency

Schools that handle social media threats effectively have a clear, consistent communication plan. Timely and transparent messaging helps reduce panic, address misinformation, and build trust with families and the community.

Effective communication during a crisis begins long before the threat appears. Districts should pre-design templates for alerts, identify who will take the lead regarding communication, and coordinate with local law enforcement and emergency services. During an event, real-time updates using multiple channels, including social media, encourage the community to stay informed (Moore et al., 2024; Mazer et al., 2015). Afterward, debriefs and follow-up communication reinforce transparency and accountability (Su et al., 2022; O'Malley et al.,2009).

The best responses are grounded in empathy. Messaging acknowledging public concern while providing facts can help prevent confusion and restore confidence. In a time when online rumors can spread faster than verified facts, trusted communication directly impacts school safety.

The Importance of Community Engagement and Proactive Intervention Strategies

Community engagement can play a pivotal role in enhancing school safety by fostering an environment where parents, educators, students, and local organizations work together to identify and address potential threats or challenges. Research indicates that schools with high community involvement experience fewer incidents of violence and higher level of trust among the stakeholders (Cueller & Theriot, 2017). In addition, high community participation can aid accurate information dissemination. A positive outcome is often increasing public confidence and support

(Wilson, 2015). Keep in mind a majority of community members may support school decisions, even if they are less vocal than the critics.

To garner and maintain community support, it is important for school leaders to implement strategies that are perceived as reasonable and practical by most in the community. This might include clear communication of safety plans, regular engagement with the community, and an openness to be responsive to community feedback (Sullivan, 2022). These approaches can not only address immediate safety concerns but also contribute to the cultivation of a supportive community and school relationship.

Reflections from Practice: School Superintendent

In my experience as a superintendent of schools, our School Resource Officers became adept at detecting, assessing, and addressing social media threats before they materialized. Lessons are learned by perpetrators of threats when tech-savvy SROs flush out such mischief and show up at a student's home in evening hours to alert parents and work with them to verify whether weapons or explosives are present in the home.

Of course, this can be an emotional roller coaster for students and staff, especially the accused. Balancing these types of scenarios with thorough investigating coupled to counseling and support for families is vital and necessary. As bad as a scenario may appear on the front end, even when kids make bad decisions, school personnel have a responsibility to care for them and look out for their well-being. However, if a student is adjudicated, school policy, which is most likely based on state statute, will serve as the guide for efforts of school leaders in working with the student going forward.

As superintendent, one experience I recall was an early morning threat that spread quickly on social media resulting in a high school student being detained, questioned, and taken into custody when they arrived at school and discretely escorted away in handcuffs before the school day began. Even in instances where things flow this efficiently, the academic process loses traction, and significant emotional stress may be evident among faculty, students, and parents. After that incident, our school counselors and principal sent emails to all parents to let them know the threat had passed, and held meetings with staff to hear their concerns and help them decompress. The community, though shaken, appeared grateful that decisive action was taken.

The Missouri Information Analysis Center (MIAC) monitors threats on all social media platforms and notifies SROs if something threatening appears. Administrators of MIAC and similar systems can run a threat through their database to determine if it has been seen elsewhere. They can also provide IP addresses of where a threatening thread originated. What follows is police work by going to the person that shared it until officers can confirm who originated the message.

Strengthening School & Law Enforcement Collaboration

It has been established throughout this chapter that schools and law enforcement agencies are often expected to collaborate. Some fundamentals of this collaboration include integrated communications, shared protocols, as well as training and drills. A highly publicized example of a deadly school shooting where warning signs were missed, and collaboration broke down, was the Uvalde, Texas, school shooting on May 24, 2022. Among several other warning signs, the shooter was known for violent tendencies, including threatening peers and showing an obsession with violence. He shared disturbing content online, including threats and images of weapons, but these were not reported to law enforcement or acted upon (El-Bawab, 2022).

Social media activity surrounding the Uvalde shooting incident revealed significant warning signs and sparked widespread public reaction. The gunman sent private Facebook messages shortly before the attack, stating his plans to shoot his grandmother and later target an elementary school. These were discovered post-tragedy and were not public posts. However, the gunman had previously shared concerning content on social platforms, including images of firearms and cryptic warnings like "*Kids be scared*" on TikTok (El-Bawab, 2022).

The tragedy garnered massive attention online, with engagement peaking in the immediate aftermath. However, conflicting reports from law enforcement created an information vacuum that fueled public discourse. Comparisons were drawn to previous shootings, such as Parkland, where survivors sustained activism for gun control. In contrast, Uvalde's focus shifted more quickly due to the increase of mass shootings in the U.S. The incident highlighted both the role of social media in detecting threats and its impact on public awareness and debate around gun control.

Schools are primarily tasked with educating students. However, examples of school violence remind us that the safety of students must remain a primary goal. Investigating and addressing every issue can be a daunting, but necessary task. Threats that appear on social media, and the storms that follow, may seem unrevivable when they become intense. However, remaining vigilant in all security efforts, and implementing thoughtful communications strategically can ease tensions (Macias, n.d.). Looking forward to recommendations and areas of future research, the value of systematically analyzing best practices is paramount to success.

RECOMMENDATIONS AND AREAS OF FUTURE RESEARCH

The authors' content analysis along with the research discussed throughout this chapter emphasizes the importance of a proactive and balanced approach in responding to social media threats. These strategies align with Routine Activity Theory by increasing guardianship through community collaboration and communication strategies consistent with Crisis Communication Theory. As stated by one parent on a social media post from an area Missouri school regarding a potential threat, "*Thank you, Dr.——, for keeping our kids safe*." There is an implied social contract among the community, parents, schools, and students that the educational environment should be a safe environment. The themes that developed from the content analysis are illustrated in Figure 3: Timely updates, trust, community, communication, and engagement.

Figure 3. Themes from Southeast Missouri threat event analysis

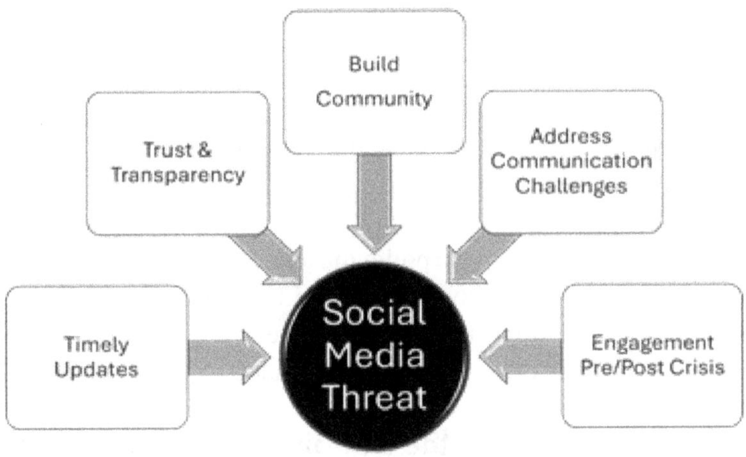

Within the school system it will be important for schools to develop strong relationships with the community as well as law enforcement. Building a school culture where students feel connected and heard can help reduce the likelihood of attacks as some threats are a cry for help. Bolstering mental health resources and anti-bullying programs, as well as incorporating social workers into the system, are preventative strategies that might help address root causes. Table 1 *Implications for Practice* integrates the identified themes, provides an overview of the authors' observations, and suggests implications for practice that will assist school administrators weathering the social media storm that may "brew" from posted online threats.

Table 1. How to weather the storm: Observations & implications

How to Weather the Storm	Observations	Implications
Issue Storm Warnings. *Regular updates reduce panic and misinformation.*	Timely updates are crucial. *Communities value rapid updates during crises, which boosts online engagement and trust.*	Enhance follow-up communication. *Develop crisis templates and assign response leads.*
Maintain Visibility. *Clear skies follow clear communication.*	**Build Trust and Transparency.** *Honest messaging (e.g., 'No credible threat, extra security on-site') was positively received.*	**Focus on clarity.** *Directly address negative or frustrated comments to demonstrate understanding.*
Construct a Shelter. *Cultivate a strong support base before the storm hits.*	**Build Community.** *Safety-related posts often generate strong community support.*	**Encourage two-way dialogue.** *Frame safety updates as shared community responsibility.*
Forecast. *Anticipate potential miscommunications and prepare messaging strategies in advance.*	**Communication Challenges.** *Empathetic responses and follow-up updates reduced community frustration.*	**Focus on engagement.** *Continue using compassionate language and direct engagement.*
Ride Out Storms with the Community. *Show up consistently before and after the event to demonstrate accountability and care.*	**Consistent Engagement.** *Regular posting prior to a crisis improved response and community support during a threat.*	**Diversify content.** *Share success stories, school events, and achievements to maintain engagement outside of emergencies.*

Having critically explored the intersection between school safety, social media threats and law enforcement response, a careful review of the case examples, content analysis, and existing research results in several actionable recommendations for school leaders, law enforcement, and policymakers. Policymakers may also consider national standards for threat reporting and guidance for responsible monitoring in schools, which would support consistent prevention strategies across districts. Based on these assessments the authors of this chapter recommend the following:

Clarify and Operationalize the Definition of "True Threat"

It is important districts work with law enforcement to establish clear, legally sound criteria to help distinguish between hoaxes, vague threats, and credible acts of violence. These recommendations might include adopting standardized threat assessment protocols that address both the content and intent of the message. The authors suggest avoiding disproportionate consequences and an intentional balance between "in loco parentis" (acting in the best interests of the student) and public safety. Protocols should be examined for implicit bias to ensure fair treatment across racial, ethnic, socioeconomic, etc., lines.

Support Proactive, Not Only Reactive, Law Enforcement Presence

SROs should be seen as not only first responders, but also as community partners. SROs should continue to receive professional development, particularly in digital threat detection, adolescent mental health, and de-escalation strategies. Proactive visibility, positive relationships, and early intervention, consistent with community policing principles should remain the cornerstone of effective school policing.

Leverage Artificial Intelligence Ethically

AI-assisted monitoring and predictive threat assessment tools can certainly aid in early detection of potential threats, but schools should be mindful of the ethical implications. False positives can unjustly target students, particularly students with behavioral challenges or from marginalized populations. Districts adopting AI should include clear policies, transparency, and an appeals mechanism to avoid overreach or misuse.

Enhance the Role of Non-Law Enforcement Mental Health Professionals

The presence of guidance counselors, social workers, and school psychologists is essential. These professionals are often best positioned to identify students in distress, provide interventions, and offer alternative solutions for students who may post threats as a form of expression or cries for help. These services can function as an integrated part of a district's safety infrastructure.

Foster Community Engagement and Shared Responsibility

Building community trust before a crisis is essential. Communities must be a part of the conversation surrounding school safety. From anonymous tip reporting systems to parent education programs, schools can increase safety through partnerships.

Prepare for Social Media Storms Before They Begin

Every school should develop a comprehensive media response plan. This plan should include templates, a crisis communication team, and protocols with local law enforcement and public health officials. In addition, the authors suggest the following areas for future research:

Areas for Future Research

- Long-term psychological effects of repeated hoax threats on student learning.
- Ethics of AI-based threat detection programs in K-12 settings.
- Disparities in discipline and law enforcement referrals resulting from online threat assessment protocols.
- Best practices for interdisciplinary training between school personnel and law enforcement.
- The role of non-law enforcement mental health professions in early detection, assessment, and intervention.

SUMMARY

While this chapter has established there is not a single solution or "quick fix," the authors have established that a combination of education, technology, collaboration, and policy can reduce the risks. It requires all stakeholders (e.g., technology corporations, students, educators, parents, law enforcement, and lawmakers) to have a vested interest, play a role, and create a safer school environment. This chapter has explored the issues and challenges of responding to social media threats, particularly the disruption of school climates, strain on law enforcement resources, and impact on public trust.

The impact of social media threats, whether credible or hoaxes, extends beyond immediate security concerns. This chapter provides case examples, including regional and real-time responses, demonstrating that rapid, transparent, and empathic communication is essential.

Central to this discussion is the role of School Resource Officers (SROs), whose responsibilities have quickly expanded beyond traditional campus safety. In recognition of the complexities of the motivations to make online threats, mental health professionals such as social workers and counselors play a critical role in threat prevention and emotional recovery for everyone involved.

The chapter also established that the use of Artificial Intelligence (AI) in threat detection holds promise but also warrants caution as it introduces ethical challenges around student privacy, misidentification, and accountability. School districts must strike a balance between ensuring safety and avoiding punitive overreach, especially with minors, through contextually grounded threat assessments and equitable discipline practices.

Finally, community engagement and education remain foundational to school safety. Schools must not only respond to social media storms but prepare for them by fostering a culture of trust and shared responsibility. It is only through proactive communication, collaborative planning, and strategic integration of technology, law enforcement, and support services that schools can truly weather the storm and maintain their crucial central mission: providing safe, supportive, and focused learning environments.

REFERENCES

Alathari, L., Drysdale, D., Blair, A., McGarry, J., Canilletti, C., Snook, A., & Driscoll, S. (2018). *Enhancing school safety using a threat assessment model: An operational guide to preventing target school violence*. U.S. Department of Homeland Security. https://www.secretservice.gov/sites/default/files/reports/2020-10/USSS_NTAC_Enhancing_School_Safety_Guide.pdf

Astor, R., Meyer, H., Benbenishty, R., Marachi, R., & Rosemond, M. (2005). School safety interventions: Best practices and programs. *Children & Schools*, *27*(1), 17–32. DOI: 10.1093/cs/27.1.17

Austin, L., Fisher Liu, B., & Jin, Y. (2012). How audiences seek out crisis information: Exploring the social-mediated crisis communication model. *Journal of Applied Communication Research*, *40*(2), 188–207. DOI: 10.1080/00909882.2012.654498

Barzman, D. (2018, May 1). *Pilot study validates AI to help predict school violence*. Cincinnati Children's. https://scienceblog.cincinnatichildrens.org/pilot-study-validates-ai-to-help-predict-school-violence/

Barzman, D., Ni, Y., Griffey, M., Patel, B., Warren, A., Latessa, E., & Sorter, M. (2017). A pilot study on developing a standardized and sensitive school violence risk assessment with manual annotation. *The Psychiatric Quarterly*, *88*(3), 447–457. DOI: 10.1007/s11126-016-9458-7 PMID: 27528455

Baumgartner, F. R., Daniely, T., Huang, K., Johnson, S., Love, A., May, L., Mcgloin, P., Swagert, A., Vattikonda, N., & Washington, K. (2021). Throwing away the key: The unintended consequences of "tough-on-crime" laws. *Perspectives on Politics*, *19*(4), 1233–1246. DOI: 10.1017/S153759272100164X

Beyer, B., Levy, G., Boulandier, K., & Senior, T. (2025). Student arrested for making online threats to Coral Springs Charter School faces judge. *WSVN 7News*. https://wsvn.com/news/local/broward/student-arrested-for-making-online-threats-to-coral-springs-charter-school-faces-judge/

Brennan Center for Justice & Center for Democracy and Technology. (n.d.). *Social media monitoring in K–12 schools: Civil and human rights concerns*. Brennan Center for Justice. https://www.brennancenter.org/our-work/research-reports/social-media-monitoring-k-12-schools-civil-and-human-rights-concerns

Brown, S. (2024). Some area schools are closed for the day in response to school threats. *Cape Girardeau Fox23 Now*. https://www.kbsi23.com/news/some-area-schools-are-closed-for-the-day-in-response-to-school-threats/

Burke, C., & Bloss, C. (2020). Social media surveillance in schools: Rethinking public health interventions in the digital age. *Journal of Medical Internet Research*, *22*(11), e22612. DOI: 10.2196/22612 PMID: 33179599

Burley, B. (2025). Social media as an investigative tool: OSINT strategies for law enforcement. *Police1*. https://www.police1.com/investigations/social-media-as-an-investigative-tool-osint-strategies-for-law-enforcement

Chaplinsky v. New Hampshire, 315 U.S. 568 (1942).

Cohen, L. E., & Felson, M. (1979). Social change and crime rate trends: A routine activity approach. *American Sociological Review*, *44*(4), 588–608. DOI: 10.2307/2094589

Coombs, W. T. (2007). Protecting organization reputations during a crisis: The development and application of situational crisis communication theory. *Corporate Reputation Review*, *10*(3), 163–176. DOI: 10.1057/palgrave.crr.1550049

Cornell, D. (2020). Threat assessment as a school violence prevention strategy. *Criminology & Public Policy*, *19*(1), 235–252. DOI: 10.1111/1745-9133.12471

Counterman v. Colorado, 600 U.S. 66 (2023).

Counts, J., Randall, K. N., Ryan, J. B., & Katsiyannis, A. (2018). School resource officers in public schools: A national review. *Education & Treatment of Children*, *41*(4), 405–430. DOI: 10.1353/etc.2018.0023

Cueller, M. J., & Theriot, M. T. (2017). Exploring the effects of parental and community involvement on the implementation of school safety strategies in United States high schools. *Journal of School Public Relations*, *36*(3), 320–345. DOI: 10.3138/jspr.36.3.320

Cybersecurity and Infrastructure Security Agency. (2023). *Social media threats guidance for school staff and authorities (infographic)*. U.S. Department of Homeland Security. https://www.cisa.gov/sites/default/files/2023-12/Social%20Media%20Threat%20Guidance%20for%20School%20Staff%20and%20Authorities%20Infographic_508.pdf

Devlin, D. N., & Fisher, B. W. (2021). An examination of school resource officers as an approach to reduce social disturbances in schools: Evidence from a national longitudinal study. *Journal of School Violence*, *20*(2), 228–240. DOI: 10.1080/15388220.2021.1875843

Dizikes, P. (2018, March 8). Study: On Twitter, false news travels faster than true stories. *MIT News*. https://news.mit.edu

El-Bawab, N. (2022). New evidence alleges Ethan Crumbley exhibited more warning signs ahead of school shooting. *ABC News*. https://abcnews.go.com/US/evidence-alleges-ethan-crumbley-exhibited-warning-signs-ahead/story?id=90321148

Ellen, M. (2000). *The school shooter: A threat assessment perspective*. Federal Bureau of Investigation. https://www.ojp.gov/ncjrs/virtual-library/abstracts/school-shooter-threat-assessment-perspective

Ellington, B. L. (2019). *Preventing targeted school violence using a threat assessment model: An exploration of school counselors' prevention and intervention practices* (Order No. 27731899). Available from ProQuest Central. (2380641460). Retrieved from https://ezproxy.semo.edu:2443/login?url=https://www.proquest.com/dissertations-theses/preventing-targeted-school-violence-using-threat/docview/2380641460/se-2

Falgoust, G., Winterlind, E., Moon, P., Parker, A., Zinzow, H., & Madathil, K. (2022). Applying the uses and gratifications theory to identify motivational factors behind young adults' participation in viral social media challenges on TikTok. *Human Factors in Healthcare*, *2*, 100014. DOI: 10.1016/j.hfh.2022.100014

Farlow, J. (2022, May 19). SRO success story: School resource officer's fast suspect identification avoids disruption from social media threat. *National Association of School Resource Officers*. https://www.nasro.org/news/2022/05/19/news-releases/sro-success-story-school-resource-officer-s-fast-suspect-identification-avoids-disruption-from-social-media-threat/

Federal Bureau of Investigation. (n.d.). *Threat intimidation guide*. https://www.fbi.gov/file-repository/threat-intimidation-guide-english-022322.pdf/view

Fisher, B., & Hennessy, E. (2016). School resource officers and exclusionary discipline in U.S. high schools: A systematic review and meta-analysis. *Adolescent Research Review*, *1*(3), 217–233. DOI: 10.1007/s40894-015-0006-8

Fisher, B., Petrosino, A., Sutherland, H., Guckenburg, S., Fronius, T., Benitez, I., & Earl, K. (2023). School-based law enforcement strategies to reduce crime, increase perceptions of safety, and improve learning outcomes in primary and secondary schools: A systematic review. *Campbell Systematic Reviews*, *19*(4), e1360. DOI: 10.1002/cl2.1360 PMID: 38024781

Ghavami, N., Thornton, B. E., & Graham, S. (2021). School police officers' roles: The influence of social, developmental and historical contexts. *Journal of Criminal Justice*, *72*, 101724. DOI: 10.1016/j.jcrimjus.2020.101724 PMID: 32958967

Golob, B. (2022). Finding me in social me-dia: Teaching students to use social networking platforms legally and mindfully. *Communication Teacher, 37*(2), 151–158. DOI: 10.1080/17404622.2022.2118340

Gupta, M., & Sharma, A. (2021). Fear of missing out: A brief overview of origin, theoretical underpinnings and relationship with mental health. *World Journal of Clinical Cases, 9*(19), 4881–4889. DOI: 10.12998/wjcc.v9.i19.4881 PMID: 34307542

Hahn, J. (2021, December 16). Schools in the U.S. taking precautions over disturbing TikTok threat for Dec. 17: 'I'm scared'. *People Magazine*. https://people.com/human-interest/schools-taking-precautions-over-tiktok-violence-warning-despite-no-credible-threats/

Hammond, B. (2015, April 28). Teacher 'terrified' by surprise active shooter drill at Oregon school. *The Oregonian*. https://www.oregonlive.com/pacific-northwest-news/2015/04/teacher_terrified_by_surprise.html

Heller, M. (2024, September 13). Jackson R-2 School District closed Friday due to threat investigation. *KFVS12*. https://www.kfvs12.com/2024/09/13/jackson-r-2-school-district-closed-friday-threat-investigation/

Home Office. (2015). *Prevent duty guidance for England and Wales: Guidance for specified authorities in England and Wales on the duty of schools and other bodies to have due regard to the need to prevent people from being drawn into terrorism*. UK Government. https://www.gov.uk/government/publications/prevent-duty-guidance

Hsieh, H. F., Lee, D. B., Zimmerman, M. A., Pomerantz, N., Cunningham, M. C., Messman, E., Stoddard, S. A., Grodzinski, A. R., & Heinze, J. E. (2022). The effectiveness of the Say Something Anonymous Reporting System in preventing school violence: A cluster randomized control trial in 19 middle schools. *Journal of School Violence, 21*(4), 413–428. DOI: 10.1080/15388220.2022.2105858

Jackson, B., Diliberti, M., Moore, P., & Schwartz, H. (2023). *Teachers' views on school safety: Consensus on many security measures, but stark division about arming teachers*. RAND Corporation. https://www.rand.org/pubs/research_reports/RRA2641-1.html

Johnson, M. (2024, December 4). Jackson, MO school district sued over year-long student suspension. *KFVS12*. https://www.kfvs12.com/2024/12/04/jackson-mo-school-district-sued-over-year-long-student-suspension/

Kremser, J. (n.d.). *Do metal detectors improve school security?* Forensic Security & Protection. https://forensicsp.com/2023/07/27/do-metal-detectors-improve-school-security/

KRON4. (n.d.). Teachers shot execution-style with rubber bullets during active shooter drill. *KRON4*. https://www.kron4.com/news/teachers-shot-execution-style-with-rubber-bullets-during-active-shooter-drill/

Lachlan, K., Spence, P., Lin, X., Najarian, K., & Del Greco, M. (2016). Social media and crisis management: CERC, search strategies, and Twitter content. *Computers in Human Behavior*, *54*, 647–652. DOI: 10.1016/j.chb.2015.05.027

Lahti, H., Kokkonen, M., Hietajärvi, L., Lyyra, N., & Paakkari, L. (2024). Social media threats and health among adolescents: Evidence from the Health Behaviour in School-aged Children study. *Child and Adolescent Psychiatry and Mental Health*, *18*(1), 62. DOI: 10.1186/s13034-024-00754-8 PMID: 38812043

Macias, O. (n.d.). *Combating social media threat hoaxes: A unified approach*. Association of California School Administrators. https://content.acsa.org/combating-social-media-threat-hoaxes-a-unified-approach/

Mazer, J., Thompson, B., Cherry, J., Russell, M., Payne, H., Kirby, E., & Pfohl, W. (2015). Communication in the face of a school crisis: Examining the volume and content of social media mentions during active shooter incidents. *Computers in Human Behavior*, *53*, 238–248. DOI: 10.1016/j.chb.2015.06.040

Metzler, L., Lazaros, E., Davison, C., & Truell, A. (2023). Social media challenges are causing disruptions in schools. *Issues in Information Systems*, *24*(1), 84–92. DOI: 10.48009/1_iis_2023_107

Miller, J. (2024). Perspective: Bridging the digital divide: The crucial partnership between law enforcement and cybersecurity professionals. *Homeland Security Today.US*. https://www.hstoday.us/featured/perspective-we-need-a-new-information-sharing-model-for-cyber-defense/

Moore, P., Jackson, B., Leschitz, J., Wolters, N., Goode, T., Diliberti, M., & Pham, P. (2024). *Developing practical responses to social media threats against K–12 schools: An overview of trends, challenges, and current approaches*. RAND Corporation. https://www.rand.org/pubs/research_reports/RRA1077-5.html

National Child Traumatic Stress Network. (2019). *Creating school active shooter/intruder drills: A trauma-informed approach*. https://www.nctsn.org/resources/creating-school-active-shooter-intruder-drills

New York State School Boards Association. (2019, December 3). *Armed security in schools: A report on current trends and policy recommendations*. https://www.nyssba.org/clientuploads/nyssba_pdf/Reports/armed-security-report-12032019.pdf

O'Malley, P., Rainford, J., & Thompson, A. (2009). Transparency during public health emergencies: From rhetoric to reality. *Bulletin of the World Health Organization*, *87*(8), 614–618. DOI: 10.2471/BLT.08.056689 PMID: 19705012

Ontario Ministry of Education. (2018). *Safe and accepting schools: Policy and program requirements.* Government of Ontario.http://www.edu.gov.on.ca/eng/safeschools/

Peterson, J., Densley, J., & Erickson, G. (2021). Presence of armed school officials and fatal and nonfatal gunshot injuries during mass school shootings, United States, 1980–2019. *JAMA Network Open*, *4*(2), e2037394. DOI: 10.1001/jamanetworkopen.2020.37394 PMID: 33591364

Peterson, J., Densley, J., Reidman, D., Spaulding, J., & Malicky, H. (2024). An exploration of K-12 school shooting threats in the United States. *Journal of Threat Assessment and Management*, *11*(2), 106–120. DOI: 10.1037/tam0000215

Peterson, J., Densley, J., Spaulding, J., & Higgins, S. (2023). How mass public shooters use social media: Exploring themes and future directions. *Social Media + Society*, *9*(1), 20563051231155101. Advance online publication. DOI: 10.1177/20563051231155101

Price, H., & Khubchandani, J. (2019). School firearm violence prevention practices and policies: Functional or folly? *Violence and Gender*, *6*(3), 154–167. DOI: 10.1089/vio.2018.0044

Quereshi, A. (n.d.). *Locked out of the classroom: How implicit bias contributes to disparities in school discipline*. LDF Thurgood Marshall Institute. https://dignityinschools.org/wp-content/uploads/2017/12/Bias_Reportv2017_30_11_FINAL.pdf

Regan, P. M., & Jesse, J. (2019). Ethical challenges of edtech, big data and personalized learning: Twenty-first century student sorting and tracking. *Ethics and Information Technology*, *21*(3), 167–179. DOI: 10.1007/s10676-018-9492-2

Reynolds, H., & Astor, R. (2024). School policing, climate, and safety: The challenges for local policy makers in urban districts. *Education and Urban Society*, *56*(3), 286–308. DOI: 10.1177/00131245221121798 PMID: 39991466

Riddle, T., & Sinclair, S. (2019). Racial disparities in school-based disciplinary actions are associated with county-level rates of racial bias. *Proceedings of the National Academy of Sciences of the United States of America*, *116*(17), 8255–8260. https://www.jstor.org/stable/26703499. DOI: 10.1073/pnas.1808307116 PMID: 30940747

Rigano, C. (2018). Using Artificial Intelligence to Address Criminal Justice Needs. *National Institute of Justice*. https://nij.ojp.gov/topics/articles/using-artificial-intelligence-address-criminal-justice-needs

Rock, A. (2024). School shooting threats continue to skyrocket on social media after Apalachee shooting. *Campus Safety Magazine*. https://www.campussafetymagazine.com/news/school-shooting-threats-skyrocket-on-social-media-after-apalachee-shooting/

Schweit, K. W., & Mancik, A. M. (2017, May 10). School resource officers and violence prevention: Best practices (part two). *FBI Law Enforcement Bulletin*. https://leb.fbi.gov/articles/featured-articles/school-resource-officers-and-violence-prevention-best-practices-part-two

Security, H. Department of. (2024). *Improving school safety through bystander reporting*. U.S. Department of Homeland Security. https://www.cisa.gov/sites/default/files/2023-05/CISA-USSS%20K-12%20Bystander%20Reporting%20Toolkit_508.pdf

Skiba, R. J., & Losen, D. J. (2016). From reaction to prevention: Turning the page on school discipline. *American Educator*, *39*(4), 4.

Solochek, S. J. (2021, January 11). Pinellas parents take precautions against non-credible school threat. *Tampa Bay Times*. https://www.tampabay.com/news/education/2021/01/11/pinellas-parents-take-precautions-against-non-credible-school-threat/

Sorensen, L. C., Avila-Acosta, M., Engberg, J. B., & Bushway, S. D. (2023). The thin blue line in schools: New evidence on school-based policing across the U.S. *Journal of Policy Analysis and Management*, *42*(4), 941–970. DOI: 10.1002/pam.22498

Statista. (n.d.). *Security measures taken by public schools in the United States from 2001 to 2022*. https://www.statista.com/statistics/183770/security-measures-at-school/

Stohlman, S., Konold, T., & Cornell, D. (2020). Evaluation of threat assessment training for school personnel. *Journal of Threat Assessment and Management*, *7*(2), 105–119. DOI: 10.1037/tam0000142

Su, Z., Zhang, H., McDonnell, D., Ahmad, J., Cheshmehzangi, A., & Yuan, C. (2022). Crisis communication strategies for health officials. *Frontiers in Public Health*, *10*, 796572. DOI: 10.3389/fpubh.2022.796572 PMID: 36483240

Sullivan, C. (2022, July 19). How superintendents can restore public trust in schools. *Yale Insights*. Edited by T. O'Callahan. https://insights.som.yale.edu/insights/how-superintendents-can-restore-public-trust-in-schools

Sullivan, P. (2022). *Building community support for school safety initiatives*. Education Strategies Press.

Swaby, A. (2024, October 30). An 11-year-old denied making a threat and was allowed to return to school. Tennessee police arrested him anyway. *ProPublica*. https://www.propublica.org/article/tennessee-school-threat-law-kids-arrested

Tennessee Mass Violence Threats Act, TN Pub. Act No. 72-2022.

Thapa, A., Cohen, J., Guffey, S., & Higgins-D'Alessandro, A. (2013). A review of school climate research. *Review of Educational Research*, *83*(3), 357–385. DOI: 10.3102/0034654313483907

Tocci, C., Stacy, S. T., Siegal, R., Renick, J., LoCurto, J., Lakind, D., Gruber, J., & Fisher, B. W. (2024). Statement on the effects of law enforcement in school settings. *American Journal of Community Psychology*, *73*(3-4), 360–377. DOI: 10.1002/ajcp.12723 PMID: 37994201

Trump, K. S. (1998). *Practical school security: Basic guidelines for safe and secure schools*. Corwin Press.

U.S. Secret Service, National Threat Assessment Center. (2021). *Averting targeted school violence: A U.S. Secret Service analysis of plots against schools*. https://www.secretservice.gov/

Welch, K., & Payne, A. (2010). Racial threat and punitive school discipline. *Social Problems*, *57*(1), 25–48. DOI: 10.1525/sp.2010.57.1.25

Williamson, B., & Eynon, R. (2020). Historical threads, missing links, and future directions in AI in education. *Learning, Media and Technology*, *45*(3), 223–235. DOI: 10.1080/17439884.2020.1798995

Wilson, A. D. (2015). The effect of community involvement programs on school violence. *Journal of Multidisciplinary Graduate Research*, *3*(1), 1–19. https://www.shsu.edu/academics/education/journal-of-multidisciplinary-graduate-research/documents/2015/Wilson-3.pdf

Xu, A., Pathak, E., Wallace, E., Gururangan, S., Klein, D., & Sap, M. (2021). Detoxifying language models risks marginalizing minority voices. *arXiv Preprint arXiv:2104.06390*. https://arxiv.org/abs/2104.06390 DOI: 10.18653/v1/2021.naacl-main.190

Zhe, E. J., & Nickerson, A. B. (2007). Effects of an intruder crisis drill on children's knowledge, anxiety, and perceptions of school safety. *School Psychology Review, 36*(3), 501–508. https://www.researchgate.net/publication/340666939. DOI: 10.1080/02796015.2007.12087936

KEY TERMS AND DEFINITIONS

First Amendment Rights and Accountability: The legal balance between protecting free speech and prosecuting individuals for threats.

Hoax Threats: False alarms or deceptive messages intended to cause panic, disruption, or notoriety.

Social Media: Digital platforms that allow users to create, share, and interact real-time and allow for communication, networking, and distributing information. Examples include Facebook, X (formerly Twitter), Instagram, TikTok, and Snapchat.

Social Media Storm: A rapid escalation of public comments and reactions that can demonstrate fear/anxiety, in all stakeholders. The event tends to generate intense scrutiny toward the leaders of the organization.

Social Media Threat: Online communications that imply harm or violence toward individuals, groups, or institutions.

School Resource Officer: A trained and certified law enforcement officer assigned to a school to provide security, prevent crime, and build a positive relationship among schools and law enforcement.

Threat Assessment: A systematic approach to evaluate the credibility, intent, and potential danger of threats.

Viral Threats: Also known as "Copycat Threats" are online challenges or social media trends that encourage students to participate in harmful or disruptive behavior.

Chapter 11
The Impact of War on Higher Education:
Disruptions and Resilience

Patricia Fidalgo
http://orcid.org/0000-0002-2620-9983
Emirates College for Advanced Education, UAE

Joan Thormann
http://orcid.org/0000-0002-5043-8479
Lesley University, USA

Oleksandr Kulyk
http://orcid.org/0000-0001-5581-7040
Oles Honchar Dnipro National University, Ukraine

ABSTRACT

This chapter explores how the conflict in Ukraine has affected higher education, emphasising the real-life experiences of instructors and students dealing with unprecedented difficulties. Responses to an online survey show that distance learning is driven by need rather than preference. The participants mentioned significant challenges, such as frequent power outages, erratic Internet, increased anxiety, decreased motivation, and worries about personal safety. Faculty and students reported that increased workloads, poor communication, and irregular attendance also hampered effective teaching and learning during wartime. The chapter provides practical suggestions for overcoming these problems during and after conflict based on recent evidence. It suggests online mental health services, international partnerships, and infrastructure support to improve educational resilience. Although

DOI: 10.4018/979-8-3373-1350-4.ch011

focused on Ukraine, the knowledge and lessons presented here can be applied to other armed conflict-affected contexts and offer helpful advice for maintaining education during emergencies.

INTRODUCTION

Higher education underwent a global transformation in 2020 due to the COVID-19 pandemic, which accelerated universities' shift to digital platforms. This decision exposed several problems in the educational system and underscored the importance of flexibility in the face of unforeseen difficulties. In the meantime, war-torn areas, like Ukraine, suffered twice as much since their educational systems had to deal with the pandemic and the conflict.

Traditionally, universities have played a significant role in economic development, social cohesion, and political stability. During conflicts, universities face many problems, such as student and faculty displacement, infrastructure damage, and loss of intellectual capital. For instance, in the Libyan civil war, university operations were severely undermined, with some universities closing down entirely, thereby hindering the nation's capacity to pursue post-conflict reconstruction (Heleta, 2015).

The pandemic and war crisis have questioned the need for new mechanisms to maintain learning continuity. One of the key strategies that have arisen to help institutions overcome geographical barriers and access remote or displaced learners is digital transformation. The Ministry of Higher Education and Higher Education and UNESCO initiated the Education Resilience Initiative in Lebanon during prolonged instability. This program has trained over 2,500 teachers in ICT, enabling them to provide uninterrupted teaching amid adverse circumstances (UNESCO, 2024).

This chapter attempts to explain various effects of the 2022 war on the higher education sector in Ukraine, including coping mechanisms used and temporary closure of the sector. It also aims to add valuable insights regarding how to ensure the resilience of higher education institutions against global crises by learning from the setbacks and responding quickly.

BACKGROUND

Recent Ukrainian Higher Education Background

Higher education in Ukraine has undergone several changes since its independence from Soviet Union. In addition, due to the pandemic, Ukraine experienced another period of changes in its educational system. Also, the HEIs have had to adapt to education provision because of the war. To understand the disruption caused by the ongoing war in Ukraine, it is essential to be familiar with the Ukrainian higher education system before 2022.

From 1922 to 1991, Ukraine was part of the Soviet Union and governed by the Communist Party, characterized by Soviet ideology and a planned economy. Significant challenges marked this era. Ukraine became independent from the Soviet Union, but its geopolitical position between Russia and NATO member states has resulted in further problems. Transitioning from a command-planned economy to a market economy was challenging, as Ukraine was closely connected to the Soviet Union. Creating a national identity while dealing with territorial, ethnic, and linguistic issues was turbulent (Zakharchuk, 2020). According to Kutsyuruba (2011), the interpretation of Soviet history by different governments and the hierarchical system prevailing in academia hindered the shift towards a more active role for individuals and institutional changes in Ukrainian universities.

Despite these challenges, recent changes in social structure, along with a desire to move away from the Soviet model, have prompted Ukraine to align its education trajectory with European standards. In the early 2000s, Ukraine started changing its higher education system, quality assurance system, and state-university relationship to align with European standards (Zakharchuk, 2020). The legal framework was adjusted to encourage the integration of Ukrainian education into the European and global community. Ukraine introduced distinctive degrees, a three-degree system, and transferable credits per the principles outlined in the Bologna Declaration. The introduction of these changes started before Ukraine formally committed to the Bologna Process in 2005. The process involved legislative adjustments, ministerial orders, and decrees to align with European requirements (Zakharchuk, 2020).

Quality assurance transformations have also been implemented, introducing external and internal quality control measures (Svyrydenko, 2017). The Ministry of Education and Science of Ukraine, the State Accreditation Commission, and the State Inspectorate have been responsible for licensing, accreditation, and inspections (Kutsyuruba, 2011). Reforms in the state-university relationship were the subject of debate. Decentralization, autonomy, and university self-governance, as advocated by the Bologna Process, were gradually implemented through changes in the Law of Ukraine "On Higher Education" (Shevchenko, 2019). However, there was still a reliance on state financing, leading to discussions about the extent of university autonomy (Friedman & Trines, 2019).

An examination of Ukraine's education system shows deviations from the European model. However, Ukraine has made an effort to be competitive in the global education sector by following pro-European policies (Svyrydenko, 2017). Public debates about the effectiveness of an internationalization strategy are not consensual. Some views are supportive of European integration, while others are resistant (Zakharchuk, 2020).

The intense democratic activity in 2013-2014 led to a political crisis with Russia, resulting in the annexation of Crimea (Shevchenko, 2019). As a result of the ongoing military conflict, large-scale migration has occurred, with millions of Ukrainians fleeing areas directly affected by the conflict (Friedman & Trines, 2019). Academic and scientific institutions had to relocate and faced challenges retaining staff (Svyrydenko, 2017). The Ukrainian government provided support and resources for those institutions, aiming to preserve their potential (Shevchenko, 2019).

The relocation of universities has raised questions and contradictions, with some institutions establishing affiliated branches in safer regions while others remain in uncontrolled territories (Svyrydenko, 2017). The Ukrainian government reoriented itself towards European integration, aiming to reduce energy dependence on Russia and implement social reforms (Zakharchuk, 2020). While progress was made in law enforcement, reforms in areas such as education were still in the early stages (Friedman & Trines, 2019). The visa-free regime with the EU and reduced trade duties reflect Europe's response to Ukraine's aspirations and reforms (Shevchenko, 2019). However, internal factors also drove the need for reforms, such as reallocating state budget funds from the social sphere to defence, changes in education and healthcare management, demographic crises, and migration processes (Svyrydenko, 2017). Many Ukrainian students were forced to continue their studies in the occupied territories, Russia, or other countries (Friedman & Trines, 2019). At the same time, some have had to stop studying and start working to support their families (Shevchenko, 2019). The economic challenges, including a decline in GDP, have led to decreased spending on the education system in Ukraine (Shevchenko, 2019; Friedman & Trines, 2019; Svyrydenko, 2017).

Higher Education in Ukraine During the Pandemic

In 2020, a study by Berezhna and Prokopenko about the impact of the pandemic on DE showed that not all HEIs were ready for DE. Students missed live communication and had a more significant workload and family duties after the closure of universities. Faculty felt complications in providing individual guidance and spent more time corresponding with students. Despite these challenges, Berezhna and Prokopenko (2020) pointed out the following positive features of DE: discipline, self-organization, flexibility, and equal opportunities for getting an education. Adaptation varied in different subject areas, but specialities which required physical space, specific conditions, and professional equipment faced more challenges than others. New problems emerged during the pandemic, yet new opportunities appeared for Ukrainian HEIs. Students needed self-discipline and self-education; faculty had to master DE pedagogies. Also, universities' technical capabilities were tested during the pandemic (Stukalo & Simakhova, 2020; Sultanova, 2020; Yelinska, 2020).

The Impact of War on Education

Literature discussing the impact of war on education is most relevant and directly applicable to the case in Ukraine. The synthesis of such studies highlights prioritizing and safeguarding education amidst conflict as it can minimize the long-term ill effects for both students and schools.

Diwakar (2015) cited the case of war in Iraq, which ignored education while it was in conflict. Now that the conflict is gone, Iraq faces increasingly congested schools, with generally lowered quality and returns on educational investment. Liu (2022) also looks at the many years of setbacks in educational attainment in Liberia, focusing on balancing security interests with the academic needs of that nation. According to Liu, remedial education will contribute to the potential economic consequences of lengthy-time spent away from school.

Yeter et al. (2022) examined the cognitive and linguistic disadvantages of students displaced because of war. They noted that attention to trauma's potential cognitive consequences may help inform future interventions better. Milton (2013) has underlined that in conditions of post-conflict rehabilitation, higher education plays a significant role. In this context, understanding the problems of students and academic staff will help establish stability and rebuilding.

Bernardo and Baranovich (2014) have identified student affairs as promoting peacebuilding in higher education. Poirier (2012) also speaks to harmful conflicts that hinder educational performance in Sub-Saharan Africa by underlining the need for educational resource allocation. A study by Shemyakina (2011) found that in Tajikistan, girls in conflict-affected regions were less likely to complete their compulsory schooling compared with those who had a chance to complete it before the conflict. The conflict created regional and generational disparities in women's education. In Bosnia and Herzegovina, the war's long-lasting consequences include dividing young people according to nationality, religion, or language, unequal position given to education, and discrimination in school textbooks (Kreso, 2008). On a more positive note, innovative technologies have been able to bridge the war-imposed obstacles in pursuing higher learning in countries like Nigeria by offering remote education delivery, interactive experiences, and practical training sessions that ensure accessible and quality learning for students (Kichuk et al., 2024).

According to Torlone and Ishchenko (2024) research, students living under war conditions rarely have an academically appropriate environment. There are many complaints about social exclusion. This is the most inhibiting factor out of all the factors that cause damage to learners and their psychic condition. Many consider the learning situation unsatisfactory, including scant study materials or disturbing factors that impede academic engagement. Emotional difficulties and psychosocial problems concerning family separation were among the issues mentioned by the participants in Torlone and Ishchenko's study. Besides these, worsening material conditions in their lives, including adaptation to a new home and the unavailability of medical support, added to their sufferings, yet these are much less strong compared to social isolation and studying conditions. The decline in students' mental health was significant, which makes it very pressing to address their psychological needs. These challenges may indicate that social-emotional and material challenges have profound implications for students' concentration and academic goals during wartime.

Elhag & Baleela (2025) studied the effects of displacement among Sudanese academicians, researchers, administration officials, and students, and found a high displacement rate, a preference to cross borders into Egypt, problems with furthering education, and dramatic changes in monthly expenses due to war-related inflation. Such results may be linked to broader issues such as food insecurity, healthcare destruction, and environmental challenges aggravated by climate change. Therefore, this calls for an acute need for international support, fellowships, and collaboration in rebuilding Sudan's academic sector and its recovery and development processes in the post-conflict era.

Research by Herrera and Byndas (2023) provides a narrative of 14 Ukrainian refugee students who fled to Poland after the Russian invasion. Based on a qualitative case study approach, the research forwards some of the main implications of interrupted education, such as linguistic problems that have affected students' emotional stability and identity. Participants shared a sense of uncertainty and instability, often describing their situation metaphorically as adrift. Students highly valued the supportive environment in Poland, but substantial linguistic barriers obstructed their academic participation, especially in Polish, whereas those continuing their education in English felt more included. It thus puts into the limelight the dire need for special integration policies, access to educational opportunities in their mother tongue, and increased availability of mental health resources to help these refugees cope with their new academic environment. Generally, continuous study is needed to deal with the ever-evolving needs of this vulnerable population.

The various studies reviewed carry several essential lessons that can help Ukraine deal with the challenges evident in the ongoing conflict. The most crucial lessons are maintaining access to quality learning, including remedial programs, and ensuring the mental health support of the displaced students. HEIs should be included in rebuilding during stabilization and recovery. Resources should be carefully allocated so that higher education is made functional during and after the conflict. Special attention is recommended for the most vulnerable groups, especially girls, in order not to disrupt their academic cycle. Lessons learned from other countries affected by active conflict help adapt policies and strategies in Ukraine to mitigate the war's impact on the nation's educational system for immediate and long-term recovery. It is thus essential that the context of pre-war changes is identified, as this will provide a starting point in understanding the current impact brought about by the war on education and planning the post-war recovery.

Current Reports on the Impact of the War on Ukrainian Higher Education

Early in 2022, Ukraine adopted the Higher Education Development Strategy for 2022-2032 to align the educational system with the Bologna process (Cabinet of Ministers of Ukraine, 2022a, 2022b). However, the invasion on February 24, 2022, disrupted education; DE was implemented, and a two-week vacation was recommended to take measures regarding the safety of students, including evacuations and special conditions for students involved with military actions. By March 2022, the Ministry of Education and Science had already provided recommendations for the completion of the academic year in a flexible manner, including remote assessments, presenting the changes to martial law for relocated students and employees by providing internal academic mobility for institutions' benefit that have suffered from the conflict (Cabinet of Ministers of Ukraine, 2022a, 2022b). The war made it obligatory to establish a National Council for Recovery concerning education (Skokova, 2022), creating an action plan to restore education in newly liberated territories. The war has necessitated a rapid shift to online learning modalities. While this transition has enabled the continuation of education, it has also exposed significant challenges, including inadequate digital infrastructure and limited access to necessary technologies. A 2023 study highlighted that both local and international students faced substantial disruptions, with many experiencing psychological trauma and educational stress due to the abrupt changes in their learning environments (Al Gharaibeh et al., 2023).

According to the World Bank (2022), 16 per cent of students and six per cent of educators have fled Ukraine, while high levels of displacement and destruction of facilities affected education. The Kyiv School of Economics (2022) estimates the war has caused over $8 billion in damages to Ukraine's educational infrastructure. The loss of educated individuals seriously threatens the country's human capital and economic stability and could hinder the post-war reconstruction efforts (Ivanenko, 2024).

For Ukrainian HEIs, the war has worsened their financial problems. After the invasion, the government cut back on public funding for higher education by 10%. Financial issues have also forced international aid organisations to halt their support and education programs. As an illustration of the precarious financial situation of many organisations, GoGlobal, an educational charity, was forced to cease operations due to a halt in US development support (Harmash, 2025).

Since the beginning of the war, the Ukrainian Ministry of Education and Science has been trying to keep the education system functioning (Kahanec et al., 2022). As an example of educational support, the All-Ukrainian Online School was created. It is an educational website developed for pupils' extra-mural education and teachers' methodological support. It reflects the goals of UNICEF on equal, accessible, and safe access to quality content. This website is a model example of cooperation between national and international organizations in pursuing global objectives while aiding the systemic transformation of Ukraine to European standards (Akgün et al., 2024).

The International Science Council (2022) indicated a need to support faculty and students who stay behind in Ukraine. European countries provided help to displaced scholars and students (European University Association, 2022; Nikolaiev et al., 2023; Maiev, 2023). For instance, the University of Sheffield donated £20,000 to Kyiv Polytechnic Institute to build a bomb shelter on campus and gave access to lab equipment and Internet resources (University of Sheffield, 2022). These partnerships aim to increase Ukrainian universities' ability to deal with the conflict.

Nikolaiev et al. (2023) noted that the centre of attention shifted from ensuring personal safety to addressing the challenges of disrupted remote learning. The war has also shown its psychological impacts on faculty, reflected in the quality of teaching and more psychological assistance requests (Lavrysh et al., 2022).

The Ukrainian government set up over 12,000 secondary education institutions catering to four million pupils. Some of those students were evacuated from occupied territories. Thousands of students received full-time education, with over two million studying remotely. The heads of institutions made their choices regarding the educational delivery mode depending on factors such as distance from the war zone and proximity to military facilities. Since the beginning of the war in February 2022, approximately 20% of Ukrainian universities have suffered damage or destruction, particularly in the eastern regions where hostilities are most intense. Numerous institutions have tried to carry out their educational missions despite these obstacles. For example, to continue operating after significant damage to its campus, Mariupol State University moved to Kyiv. The state framework for the Quality of Education in Ukraine recommended creating a safe educational environment, training on safety measures, and allowing institutions to change their educational processes rapidly (Zaplatynskyi, 2022; Yuzyk et al., 2022).

Education becomes one of the rays of hope for Ukrainian refugees who face barriers in their unsettled immigration status and financial constraints (Morrice, 2022). Equally, fear and stress have resulted from this war, leading to an academic decline since the study hours have been reduced, and the quality of remote learning has decreased (Kurapov et al., 2023). Strategic interventions to address the challenges experienced should be informed by lessons identified from global experiences.

The following table provides approaches to support HEIs after the war. Several organizations' recommendations are similar, emphasizing the expected challenges the rebuilding process will face.

Table 1. Recommendations to support Ukrainian HEIs after the war

Organization	Recommendation
Ministry of Education and Science of Ukraine (2022b)	Manage higher education, align with European standards, rebuild trust, and restore personnel.
Institute of Educational Analytics of the Ministry of Education and Science of Ukraine (2022a)	Restore infrastructure, rehabilitate personnel, and foster public-private partnerships.
(UNESCO, 2023)	Ensure uninterrupted learning, professional development, and renovate infrastructure.
European University Association (2022)	Promote online education, provide technical support, and coordinate with governments.
United Nations (2023a)	Provide mental health support, financial aid, repair infrastructure, and engage in peacebuilding.

Building a resilient and adaptive education framework in Ukraine has been increasingly demanded, striving to cope with the ongoing aftermath of armed conflict interwoven with a disruption period because of recent transformations of its educational system. HEIs can play a significant role in the trajectory of post-conflict recovery. Rosenberg and Miquelino (2002) noted that the development and reconstruction of these institutions coincide with the broader national recovery goals, as this puts HEIs in a position to serve as engines of stability and development. The war in Ukraine is likely to cause significant future earning losses, potentially placing affected students among the lowest earners in Europe (Sheather, 2022). Drawing from Poirier's (2012) research, conflicts have adversely affected educational performance and efficient use of resources; hence, prioritizing social expenditures in education is vital to Ukraine. In supporting HEIs after the war, a multi-layer approach will be essential in tackling immediate and long-term needs (Marston & Tsolakis, 2022; Galynska & Bilous, 2022).

THEORETICAL FRAMEWORK

Educational Disruption

While "Educational Disruption Theory" is not defined or recognized in much of the literature, the concept of educational disruption can be found in numerous instances, especially regarding dramatic interruptions of mainstream education (Christensen, 2006). These can be caused by one or many other variables, including technology advancements, crises, pandemics, economic recessions, and wars (Carter, 2013; Reimers & Schleicher, 2020). In these contexts, "educational disruption" refers to forces that are either external or internal, fundamentally changing traditional models or practices in education.

Some even describe the COVID-19 pandemic as a significant disruptor in education since it compelled many teachers to give up old teaching practices; simultaneously, remote learning technologies accelerated significantly (UNESCO, 2020; Hodges et al., 2020). Similarly, wars and conflicts have devastated the educational systems by closing schools and going online or using alternative learning formats, which has consequences for students' and teachers' academic and psychological well-being (Shemyakina, 2011; Kreso, 2008).

Different theories also support the idea of disruption in education, some including:

- Disruptive Innovation Theory: Mainly propounded by Clayton Christensen, the theory describes the dynamics whereby technologies or new ways begin to challenge old educational methods to overtake them eventually. This is partially shown in how education introduces online learning, AI-powered learning, and, more generally, how traditional approaches of schools move to flexible, personalized education. Christensen et al. (2011) and Horn & Staker (2014) discussed that this theory has suggested that educational institutions must adapt to disruptive innovations, which can subsequently result in long-term systemic changes.
- Education in Crisis and Disaster: Many research works have focused on how education systems respond to emergencies created by wars, natural disasters, and pandemics. Along the way, other works have referred to other education provisions, e.g., digital education, mobile learning, and hybrid instructions, as ways of providing educational continuity (Shemyakina, 2011; Kreso, 2008; Davies, 2005). Research on post-disaster education has shown that though immediate disruptions result in loss of learning, strategic interventions could minimize long-term damage and even inspire innovation in pedagogical modes of teaching (Reimers & Schleicher, 2020; UNESCO, 2023b).

These frameworks span both the short-run and long-run impacts due to disruption for students and faculty. In the case of Ukraine, the interruption of education began with the COVID-19 pandemic and continued with war. The change in traditional face-to-face education in Ukraine to a digital and distance format because of war presents a classic case of crisis-forced disruptions that must be subjected to rapid transformation. The features of this change are:

- Disruption of Curricula: The war has caused school closure, disruption of curricula delivery, and displacement of teachers and students. Several institutions had to change syllabi and exam procedures to cater to displaced students and conflict areas (European University Association, 2022).
- Shifting of Pedagogies: The limitations accelerated the diffusion of online and asynchronous learning environments and blended methods. Besides, teaching and learning digital literacy for schooling rose significantly (UNESCO, 2022; Reimers & Schleicher, 2020).
- Learning Gaps: The disruption has widened the gap in learning achievements, especially for children from low-income families who may be deprived of access to digital devices and stable Internet connectivity. Learning gaps resulting from war are estimated to have long-term effects on labour preparedness and national development (Shemyakina, 2011; Kreso, 2008).

This study explores how conflict reshapes various aspects of higher education and aligns with theoretical frameworks of educational disruption. The phenomenon is investigated based on emerging literature concerned with the need to preserve continuity in education during periods of crisis (European University Association, 2022; UNESCO, 2022; Reimers & Schleicher, 2020).

EXAMINING EDUCATION AND LEARNING IN THE CONTEXT OF WAR

Higher education faculty and students have experienced daily disruptions to their academic lives due to the ongoing conflict in Ukraine; these disruptions are both logistical and profoundly emotional and psychological. This chapter documents those lived realities and provides a voice to the adaptations, challenges, and strengths of teachers and students negotiating an uncertain environment.

The chapter was guided by two questions: How has the war affected students' learning and faculty's teaching? How have these individuals adjusted, and what difficulties do they perceive in one another's situations?

To answer these questions, we polled teachers of higher educational institutions and students from different universities of Ukraine who studied and worked actively in 2022/2023. The realities of war that they shared were rich and textured.

Faculty members were candid about the practical and emotional aspects of teaching through power failures and air raid sirens, and students discussed their barriers to learning. Both groups were asked to think about each other's experiences and their own, forming a common story that cut across the educational gap. Important themes naturally surfaced from the surveys. Frequently used terms like "disruption," "connection," "resilience," and "fatigue" vividly depicted education under siege. While both groups discussed emotional strain, for instance, students highlighted feelings of loneliness. At the same time, faculty members were more likely to express concern for the motivation and well-being of their students. Faculty and student quotes are shared to illustrate the effect of the war concretely. It is an effort to pay tribute to the tales of people who decided to keep teaching and learning even in the face of conflict. Their experiences demonstrate both the extraordinary adaptability of those within a system and its vulnerabilities when under stress. This chapter can help guide re-establishing direction once peace returns.

Teaching and Learning in Wartime: Stories from Ukrainian Classrooms

In the classrooms of Ukraine, learning has not stopped—but it has changed profoundly. Students and teachers continue to show up, and their stories reveal the logistical challenges and the emotional, psychological, and institutional toll of war.

The shift to virtual instruction grew more intense as the conflict did. Online instruction swiftly took over as the most popular approach. This change was an institutional survival tactic. Significant changes to the curriculum and pedagogy were reported by both students and faculty. To adapt to the realities of air raid sirens, power outages, and mental exhaustion, academic calendars were changed, teaching strategies were reconsidered, and lessons were reorganised. Teachers reported unpredictable participation, erratic attendance, and fewer students.

Faculty and students experienced many technical issues. Access to classes was erratic due to power outages and unreliable Internet connections. These challenges persisted, particularly for students. One described the disruption: "The biggest difficulties are caused by problems of a technical nature—failures in the Internet." Another shared, "It undoubtedly has a negative impact on their access to education and results." Concerns about how these problems affected their capacity to instruct and maintain student interest were also voiced by the faculty.

The emotional difficulties of teaching and learning in times of war were discussed. Both students and faculty expressed anxiety, sadness, and a persistent sense of insecurity. Instructors worried about the safety of their students, their own families, and themselves. Students struggled to focus, feeling overwhelmed by the pressure of living and studying through constant uncertainty. One student shared: "There is a feeling that it will never end. I am forced to work in a profession I did not dream of. I have to work while studying… I feel emotionally drained, like a garbage bag in a dumpster. Exhaustion is increasing, and I can't sleep properly even when I rest all day. I wake up 4-5 times, not even from sirens, just by myself... Like I'm 70 years old…" Another student shared, "you know, they say that a person is born with nothing and dies with nothing, but I am afraid only of living with nothing now. That my best years will be wasted (…). Some people are very ideological now, some are enthusiastic about a new goal, some are covered in cruelty, and all of this scares me so much. I don't want any of it. I'm just like a robot. What's even scarier, I smile, I try to be warm, and it's very easy to switch me (…) I'm afraid that I might already be losing my mind - that's my biggest problem."

Faculty, too, voiced the emotional toll of teaching under such conditions. Some spoke of the difficulty of maintaining a positive atmosphere during lessons or of reworking course material to accommodate the fragile emotional state of their students. One faculty member observed, "Students' psychological state required the course to be reformatted." Another wrote, "To maintain a positive and motivational atmosphere during lectures, both for oneself and for the students" had become one of the most significant challenges.

The shift to distance education brought its own set of complications. Faculty noted that many students were not prepared for independent learning. They expressed frustration with the performative aspects of online learning. As one teacher said bluntly, "The higher education system was not prepared to work in the new reality."

Students echoed these sentiments. While some saw potential in distance learning after an adjustment period, others lamented the loss of the student experience they had envisioned. "Offline learning has taken away all the charm of romanticized student life," one student reflected. "Motivational problems and disillusionment… are key (for me)."

Communication, too, was a recurring challenge. Faculty described the difficulties of teaching students scattered across different cities and regions, with minimal visual or real-time interaction. "There were difficulties in communicating with students due to changes in their location," one instructor noted. "Minimal or no visual contact… the need to quickly change the method of presenting the material."

Compounding all of this were the disruptions of daily life. Air raids, safety concerns, and displacement were part of the background noise to every lecture and every study session. Faculty described the strain of teaching while knowing students were in danger—or were themselves at risk. "Some are in a state of almost constant danger from military threats," one instructor explained. Students echoed the same fears, speaking of family members in occupied territories and the stress of worrying about loved ones while trying to study.

Yet despite everything—despite exhaustion, fear, grief, and a profound sense of dislocation—students and faculty continued. Their words combine to create a tapestry of perseverance, adversity, commitment, and fatigue. They paint a realistic picture of what it's like to teach and learn during a time of war.

Instead of evoking pity, these tales inspire action. They highlight the need for support systems that go beyond technology and content delivery. They call on educational institutions, governments, and the international community to provide mental health support, teacher training, digital infrastructure, and policies that prioritize student and faculty well-being. Most of all, they remind us that education is not just about knowledge transmission—it is also a lifeline, a routine that restores meaning and purpose in the midst of chaos.

This chapter is not just a reflection of what has changed in Ukrainian higher education—it is a testament to the unyielding spirit of those who refuse to let learning be another casualty of war.

Understanding the Multifaceted Disruptions in Higher Education

This chapter creates a detailed picture of the challenges that faculty and students of Ukrainian higher education faced during the war. This chapter also reveals the intricate connections between systemic, emotional, and technical disruptions that have affected the educational process and highlights the pressing need for intervention. These disruptions are not unique to Ukraine; analogous higher education crises have occurred in other war-ravaged regions, such as Syria and Afghanistan, where institutions witnessed closures, teacher displacement, and curriculum interruption (Dryden-Peterson, 2011). This chapter's content highlights conflict implications in Ukraine and broader implications in other war-affected nations, as well as potential interventions that can be utilized to minimize school disruptions in future conflicts.

The Psychological Toll on Faculty and Students

In addition to infrastructural and financial challenges, the psychological toll on staff and students has also been enormous. Mental health difficulties such as anxiety, stress, and trauma have been shared among university staff and students, a fact that has also been witnessed in war-torn regions such as Iraq and Yemen (Kayyali, 2024). A 2023 study has highlighted the emotional toll on educators, with increased stress, anxiety, and trauma rates due to the war and accompanying disruption (Al Gharaibeh et al., 2023). Instructors and students in Ukraine also experienced heightened psychological trauma as a result of the war-induced instability. These experiences resonate with those documented in other contexts. Without appropriate psychological support systems, such an effect can last a lifetime on the country's workforce and educational system. Therefore, policymakers and teachers must prioritize student and teacher resilience programs (Nikolaiev, 2023).

The Pandemic as a Precursor to Educational Disturbance

The COVID-19 pandemic had already caused unprecedented disruptions in Ukraine's higher education system long before the war. Universities were compelled to make the quick switch to DE, which exposed significant deficiencies in infrastructure resilience, pedagogical responsiveness, and digital readiness. The digital divide also emerged in rural towns, where students lacked broadband connectivity and technology (Stukalo & Simakhova, 2020). The teachers were also unprepared to teach online and struggled to make space on online platforms without proper training and support (Falfushynska et al., 2021).

The pandemic also developed psychological distress in students and educators, which was reflected in digital fatigue, burnout, and reduced learning achievement (Kostikova et al., 2023). All these intensified during the war because most learners had to be displaced or learn under high-stress environments. Investments in teacher capacity, digital infrastructure, and mental health are urgently needed to address these enduring vulnerabilities.

Digital Transformation as a Resilience Catalyst

The pandemic and war emergency caused a sudden digital revolution in Ukrainian higher education. The ongoing digital learning changes, which were initially a reaction, have the potential to transform education in a sustainable way.

A systematic study by Nieto-Taborda and Luppicini (2024) found that post-pandemic higher education digital transformation involves strategic-administrative, teaching-learning, technical, and socio-cultural dimensions. The corresponding multifaceted strategies in Ukraine have been critical to solving the exacerbated challenges of war and COVID-19 pandemic. Nonetheless, several significant barriers include the unavailability of consistent Internet in combat zones, limited access to learning equipment, and cybersecurity hazards presented by online learning (Tavrovetska & Veldbrekht, 2023). It follows that robust policy frameworks, pedagogy preparation programs, and infrastructure investment are necessary for Ukraine's digital transformation.

International Solidarity and Support

In areas affected by conflict, international partnerships have been essential to maintaining educational continuity. Institutions such as Harvard University (Huang & Yu, 2024) and the University of Toronto (University of Toronto, 2022) have started relief programs for displaced Ukrainian students through scholarships and online studies. Similarly, the University of Cambridge's Trinity College created a £250,000 relief fund to aid Ukrainian students, thus demonstrating the College's dedication to global academic solidarity (The Times, 2024).

These initiatives offer both immediate assistance and long-term stability, reiterating the value of international academic networks in preserving education during crises. However, the Ukrainian higher education system must not rely on international aid. Instead, home policies must encourage cooperation among universities, the private sector, and international agencies to create long-term sustainability.

Challenges in Rebuilding Higher Education

Post-conflict restoration of higher education involves reinstating infrastructural damage, reversing brain drain, and curriculum reform to stay abreast of evolving social requirements. Post-war studies of recovery in higher education in Rwanda and Bosnia underscore the imperative of investing in the autonomy of universities, developing human resources, and equitable access to education (Heleta, 2015). Ukraine must adopt similar measures to revive the educational system under the guidance of European norms (OECD, 2025).

Moreover, intellectual capital has been heavily lost due to migration as several scholars have found their way into Poland, Germany, and Canada in search of stability (Jaroszewicz et al., 2025). Ukraine must work towards enticing the displaced intellectuals with lucrative compensation, research allowances, and possibilities for cooperation elsewhere in the world. The nation faces prolonged academic stagnation and further deterioration of its higher education institutions if this challenge is not avoided.

Implications for the Workforce and Economic Recovery

There are wider societal repercussions when HEIs, students, and staff are displaced. Workforce preparation, economic stability, and demographic transformation have all been impacted as millions of Ukrainians have displaced themselves to find refuge and pursue higher education abroad. The loss of educated human capital could significantly hinder post-war reconstruction, making special reintegration programs and workforce development policies critical to national recovery.

The future of students' careers is most concerning. The majority will face a war economy with few career opportunities and no concept of professional stability. However, those who have learned to adapt to digital education and working from home environments can have a future in the global labour market. Policymakers must explore workforce development programs that suit the economic recovery and employment creation in strategic sectors in Ukraine.

A Call to Action

This book chapter highlights the need for the support of the global community, policymakers, and educators to enable assistance in rebuilding Ukraine's higher education system. As challenging and unprecedented as the war-induced problems have been, however, the latter has also created a window of opportunity for future education in Ukraine to be rebalanced. Digitalization, mental health action, international cooperation, and workforce reintroduction policies must be at the top of policy agendas. To guarantee that its HEIs endure and grow stronger and more resilient in the future, Ukraine must prioritise efforts to strengthen resilience, improve the online educational system, and foster scholarly collaboration.

CONCLUSION

The war in Ukraine has altered higher education, necessitating a prompt response from the university, faculty, and students. The disruption has revealed challenges in access, equity, mental health, DE, institutional readiness, and human resilience. However, it has also encouraged innovation since universities have been forced to adopt new student engagement and retention approaches.

The crisis also provides the opportunity to reshape the education sector to become more responsive and proactive. Recovery must include reinforcing scholarly networks, creating innovative learning solutions, and prioritizing the health and well-being of learners and faculty. The ability to turn adversity into reform presents an opportunity to influence Ukraine's higher education going forward and offer valuable guidance for education systems worldwide facing similar challenges.

The job prospects of conflict-affected students are uncertain. Most will graduate into a labour market devastated by conflict with fewer career choices and unclear career opportunities. The psychosocial impact of war interrupted education, and restricted social contacts may detract students' motivation and career ambitions.

However, these experiences may also render them more adaptable to the changes and enable them to help revive the economy and institutions of Ukraine.

This chapter illustrates the vulnerabilities that the conflict imposed and suggests recovery patterns. Ukraine can create a more resilient and stronger higher education system by embracing an integrated approach that is intermediate between short-term adaptation and long-term planning. Beyond Ukraine, these results provide a model for other conflict-affected countries to protect educational institutions and build staff and student resilience.

REFERENCES

Akgün, B., Alpaydın, Y., Akgün, B., & Alpaydın, Y. (2024). *Global Agendas and Education Reforms: A Comparative Study* (Akgün, B., & Alpaydın, Y., Eds.; 1st ed.). Palgrave Macmillan., DOI: 10.1007/978-981-97-3068-1

Al Gharaibeh, F., Ahmad, I., & Malkawi, R. (2023). Impact of the Russia-Ukraine War on Education and International Students. *Journal of International Women's Studies*, *25*(6), 1–17. https://vc.bridgew.edu/jiws/vol25/iss6/12%0AThis

Berezhna, S., & Prokopenko, I. (2020). Higher Education Institutions in Ukraine during the Coronavirus, or COVID-19, Outbreak: New Challenges vs New Opportunities. *Revista Românească Pentru Educaţie Multidimensională*, *12*(2), 130–135. DOI: 10.18662/rrem/12.1sup2/256

Bernardo, M. A. C., & Baranovich, D.-L. (2014). Higher education in the heart of armed conflict: The pivotal role of student affairs. *International Journal of Educational Development*, *35*, 78–85. https://doi.org/https://doi.org/10.1016/j.ijedudev.2012.08.007. DOI: 10.1016/j.ijedudev.2012.08.007

Cabinet of Ministers of Ukraine. (2022a). *Approval of the Strategy for the Development of Higher Education in Ukraine for 2022-2032*. Legislation of Ukraine.

Cabinet of Ministers of Ukraine. (2022b). *On Amendments to Certain Laws of Ukraine Regarding State Guarantees in the Conditions of Martial Law, State of Emergency, or State of Emergency. Legislation of Ukraine.* https://zakon.rada.gov.ua/laws/show/2126-20#Text

Carter, V. (2013). Book Review - Disruptive Innovation in Technology and Engineering Education: A Review of the Three Works by Clayton Christensen and Colleagues. *Journal of Technology Education*, *24*(2). https://doi.org/https://doi.org/10.21061/jte.v24i2.a.7

Christensen, C. M. (2006). The Ongoing Process of Building a Theory of Disruption. *Journal of Product Innovation Management*, *23*(1), 39–55. DOI: 10.1111/j.1540-5885.2005.00180.x

Christensen, C. M., Horn, M. B., Caldera, L., & Soares, L. (2011). *Disrupting College: How Disruptive Innovation Can Deliver Quality and Affordability to Postsecondary Education.*

Davies, L. (2005). Schools and war: Urgent agendas for comparative and international education. *Compare: A Journal of Comparative Education*, *35*(4), 357–371. DOI: 10.1080/03057920500331561

Diwakar, V. (2015). The Effect of Armed Conflict on Education: Evidence from Iraq. *The Journal of Development Studies*, *51*(12), 1702–1718. DOI: 10.1080/00220388.2015.1056786

Dryden-Peterson, S. (2011). *Refugee Education: A Global Review*. https://www.unhcr.org/media/refugee-education-global-review-sarah-dryden-peterson-november-2011

Elhag, H. E. E. A., & Baleela, R. M. H. (2025). Preliminary investigation and analysis of the impact of the Sudan War on higher education and scientific research sectors. *International Journal of Educational Development*, *112*, •••. https://doi.org/https://doi.org/10.1016/j.ijedudev.2024.103190

European University Association. (2022). *Support to Ukrainian higher education*. https://eua.eu/resources/publications/1034:support-to-ukrainian-higher-education.html

Falfushynska, H. I., Buyak, B. B., Tereshchuk, H. V., Torbin, G. M., & Kasianchuk, M. M. (2021). Strengthening of e-learning at the leading Ukrainian pedagogical universities in the time of COVID-19pandemic. *CTE Workshop Proceedings*, 261–273.

Friedman, E., & Trines, S. (2019). *Education in Ukraine*. World Education News + Reviews. https://wenr.wes.org/2019/06/education-in-ukraine

Galynska, O., & Bilous, S. (2022). Remote learning during the war: Challenges for higher education in Ukraine. *International Science Journal of Education & Linguistics*, *1*(5), 1–6. DOI: 10.46299/j.isjel.20220105.01

Harmash, O. (2025). *Ukraine aid groups cut services, scramble for cash after US funding shock*. Reuters. https://www.reuters.com/world/europe/ukraine-aid-groups-cut-services-scramble-cash-after-us-funding-shock-2025-01-30/?utm_source=chatgpt.com

Heleta, S. (2015). *How to rebuild higher education in countries torn apart by war.* The Conversation. https://theconversation.com/how-to-rebuild-higher-education-in-countries-torn-apart-by-war-48791?utm_source=chatgpt.com

Herrera, L. J. P., & Byndas, O. (2023). "You sway on the waves like a boat in the ocean": The effects of interrupted education on Ukrainian higher education refugee students in Poland. *Cogent Education*, *10*(2), 2264009. Advance online publication. DOI: 10.1080/2331186X.2023.2264009

Hodges, C., Moore, S., Lockee, B., Trust, T., & Bond, A. (2020). T*he Difference Between Emergency Remote Teaching and Online Learning*. Educause Review. https://er.educause.edu/articles/2020/3/the-difference-between-emergency-remote-teaching-and-online-learning

Horn, M. B., & Staker, H. (2014). *Blended: Using Disruptive Innovation to Improve Schools* (1st ed.). Jossey-Bass.

Huang, E. R., & Yu, C. J. (2024). *Harvard Scholars Without Borders Supports Displaced Ukrainian, Russian Scholars Amid War*. The Harvard Crimson. https://www.thecrimson.com/article/2024/2/28/program-displaced-scholars/

Institute of Educational Analytics of the Ministry of Education and Science of Ukraine. (2022). *Higher and Professional Pre-higher Education under Martial Law*.

International Science Council. (2022). *Conference on the Ukraine Crisis: Responses from the European Higher Education and Research Sectors*. DOI: 10.24948/2022.04

Ivanenko, N. (2024). Challenges of Ukrainian Higher Education in Times of War. *International Higher Education*. https://ihe.bc.edu/pub/i3oxjya7/release/2

Jaroszewicz, M., Shkoda, T., & Ovchynnikova, O. (2025). Migration Trajectories of Ukrainian Scholars Abroad: Forced Academic Mobility. *Population Space and Place*, *31*(2), e70011. Advance online publication. DOI: 10.1002/psp.70011

Kahanec, M., Leu-Severynenko, S., Novosad, A., & Stadnyi, Y. (2022). Education reforms during and after the war. In Sologoub, I., di Mauro, B. W., & Gorodnichenko, Y. (Eds.), *Rebuilding Ukraine: Principles and policies* (1st ed.). CEPR Press., https://cepr.org/publications/books-and-reports/rebuilding-ukraine-principles-and-policies

Kayyali, M. (2024). The Impact of War on Higher Education: The Context of Wars in Syria, Iraq, Ukraine, and Afghanistan. In *Rebuilding Higher Education Systems Impacted by Crises: Navigating Traumatic Events, Disasters, and More* (1st ed., pp. 275–284). IGI Global. DOI: 10.4018/979-8-3693-1926-0.ch016

Kichuk, Y., Biliuk, O., Kapinus, O., Fedenko, S., & Livak, P. (2024). Innovation in Higher Education in Wartime: Using Modern Technologies. *Academia (Caracas)*, *35–36*, 178–199. DOI: 10.26220/aca.5008

Kostikova, I., Holubnycha, L., Sytnykova, Y., Shchokina, T., Soroka, N., & Tarasova, S. (2023). University Teachers' Psychological State During the Stress of Online Education After Pandemic in Wartime in Ukraine. *International Journal of Emerging Technologies in Learning*, *18*(13), 152–162. DOI: 10.3991/ijet.v18i13.39825

Kreso, A. P. (2008). The war and post-war impact on the educational system of Bosnia and Herzegovina. *International Review of Education*, *54*(3–4), 353–374. DOI: 10.1007/s11159-008-9087-y

Kurapov, A., Pavlenko, V., Drozdov, A., Bezliudna, V., Reznik, A., & Isralowitz, R. (2023). Toward an Understanding of the Russian-Ukrainian War Impact on University Students and Personnel. *Journal of Loss and Trauma*, *28*(2), 167–174. DOI: 10.1080/15325024.2022.2084838

Kutsyuruba, B. (2011). Education in the period of post-Soviet transition in Ukraine. *Demokratizatsiya (Washington, D.C.)*, *19*(3), 287–310.

Kyiv School of Economics. (2022). *Report on the total amount of damage caused to the infrastructure of Ukraine*. https://kse.ua/about-the-school/news/as-of-november-2022-the-total-amount-of-losses-caused-to-the-infrastructure-of-ukraine-increased-to-almost-136-billion/

Lavrysh, Y., Lytovchenko, I., Lukianenko, V., & Golub, T. (2022). Teaching during the wartime: Experience from Ukraine. *Educational Philosophy and Theory*, •••, 1–8. DOI: 10.1080/00131857.2022.2098714

Liu, S. X. (2022). How war-related deprivation affects political participation: Evidence from education loss in Liberia. *Journal of Peace Research*, *59*(3), 353–366. DOI: 10.1177/00223433211019460

Maiev, A. (2023). Education during the war: challenges for Ukraine. *Teoria Şi Practica Administrării Publice*, 415–420.

Marston, J., & Tsolakis, M. (2022). *Ukraine Points Up the Threat to Education During War*. ReliefWeb. https://reliefweb.int/report/world/ukraine-points-threat-education-during-war

Milton, S. A. (2013). *The Neglected Pillar of Recovery: A study of Higher Education in Post-War Iraq and Libya*. University of York.

Ministry of Education and Science of Ukraine. (2022a). *Letter No. 1/3417-22: On the Conclusion of the Academic Year*. Government Portal. https://osvita.ua/legislation/Vishya_osvita/85933/

Ministry of Education and Science of Ukraine. (2022b). *Serhiy Shkarlet: All educational institutions are recommended to halt the educational process and declare a vacation for two weeks.* Government Portal. https://www.kmu.gov.ua/news/rozyasnennya-mon-shchodo-roboti-zakladiv-osviti-u-mezhah-pravovogo-rezhimu-voyennogo-stanu

Morrice, L. (2022). Will the war in Ukraine be a pivotal moment for refugee education in Europe? *International Journal of Lifelong Education*, *41*(3), 251–256. DOI: 10.1080/02601370.2022.2079260

Nieto-Taborda, M. L., & Luppicini, R. (2024). Accelerated Digital Transformation of Higher Education in the Wake of COVID-19: A Systematic Literature Review. *International Journal of Changes in Education.* https://doi.org/https://doi.org/10.47852/bonviewIJCE42023125

Nikolaiev, Y., Riy, G., & Shemelynets, I. (2023). *War in Ukraine: Reshaping the Higher Education Sector - Analytical Report.* file:///C:/Users/pfidalgo/OneDrive - Emirates College for Advanced Education/Desktop/Readings/War in Ukraine_Report.pdf

OECD. (2025). *Transforming General Upper Secondary Education in Ukraine.* https://doi.org/DOI: 978-92-64-96427-3 (PDF)

Poirier, T. (2012). The effects of armed conflict on schooling in Sub-Saharan Africa. *International Journal of Educational Development, 32*(2), 341–351. https://doi.org/https://doi.org/10.1016/j.ijedudev.2011.06.001. DOI: 10.1016/j.ijedudev.2011.06.001

Reimers, F. M., & Schleicher, A. (2020). *A framework to guide an education response to the COVID-19 Pandemic of 2020.* https://www.oecd.org/en/publications/2020/03/a-framework-to-guide-an-education-response-to-the-covid-19-pandemic-of-2020_a5bcbd8a.html

Rosenberg, M., & Miquelino, L. P. (2002). *e-Learning - estratégias para a transmissão do conhecimento na era digital.* Milton Mira de Assunpção Filho .

Sheather, J. (2022). As Russian troops cross into Ukraine, we need to remind ourselves of the impact of war on health. *BMJ (Clinical Research Ed.), 376*(February), o499. Advance online publication. DOI: 10.1136/bmj.o499 PMID: 35217578

Shemyakina, O. (2011). The Effect of Armed Conflict on Accumulation of Schooling: Results from Tajikistan. *Journal of Development Economics, 95*(2), 186–200. DOI: 10.1016/j.jdeveco.2010.05.002

Shevchenko, V. V. (2019). The reform of the higher education of Ukraine in the conditions of the military-political crisis. *International Journal of Educational Development, 65*, 237–253. DOI: 10.1016/j.ijedudev.2018.08.009

Skokova, L. (2022). War-affected landscapes as a social problem. In Головахи, Є., & Макеєва, С. (Eds.), *Ukrainian Society in Wartime* (1st ed., pp. 66–76). Institute of Sociology of the National Academy of Sciences of Ukraine.

Stukalo, N., & Simakhova, A. (2020). COVID-19 Impact on Ukrainian Higher Education. *Universal Journal of Educational Research, 8*(8), 3673–3678. DOI: 10.13189/ujer.2020.080846

Sultanova, L. (2020). Impact of the COVID-19 pandemic on the Development of Education Systems at Global, European and National Dimensions. *AdultEducation: Theory, Experience. Prospects, 18*(2). https://doi.org/ttps://doi.org/10.35387/od.2(18).2020.171-183

Svyrydenko, D. (2017). Divided Universities: The Postcolonial Experience of Contemporary Ukrainian Higher Education. *Future Human Image, 1*(7), 128–135.

Tavrovetska, N., & Veldbrekht, O. (2023). Psychological aspects of online learning implementation at Ukrainian universities. *Scientific Bulletin of Mukachevo State University Series. Pedagogy and Psychology, 9*(2). Advance online publication. DOI: 10.52534/msu-pp2.2023.38

The Times. (2024, November 15). Trinity College master 'regrets' £250k fund for Ukrainian students. *The Times.*

Torlone, F., & Ishchenko, A. (2024). Learning in higher education while being at war. *Journal of Educational Sciences, XXV*(1), 56–71. Advance online publication. DOI: 10.35923/JES.2024.1.03

UNESCO. (2020). *Education in a post-COVID world: Nine ideas for public action.* News. https://www.unesco.org/en/articles/education-post-covid-world-nine-ideas-public-action

UNESCO. (2023). *Timeline of UNESCO's commitment to protect Ukraine's education and heritage.* UNESCO Actions for Ukraine. https://www.unesco.org/en/ukraine-war/actions-timeline

UNESCO. (2023b). *Ensuring teaching and learning continues in Ukraine. Emergencies.* https://www.unesco.org/en/emergencies/education/ukraine

United Nations. (2023a). Early Recovery Efforts in Ukraine. In *UN Annual Results Report 2022.* chrome-extension://efaidnbmnnnibpcajpcglclefindmkaj/https://ukraine.un.org/sites/default/files/2023-05/UN_report_2023%28English%29.pdf

University of Sheffield. (2022). *University of Sheffield twins with university in Kyiv to help staff and students affected by war.* News. https://www.sheffield.ac.uk/news/university-sheffield-twins-university-kyiv-help-staff-and-students-affected-war

University of Toronto. (2022). *U of T to accept displaced students from Ukraine with significant boost to its Scholars-at-Risk fund.* News and Stories. https://defygravitycampaign.utoronto.ca/news-and-stories/u-of-t-to-accept-displaced-students-from-ukraine-with-significant-boost-to-its-scholars-at-risk-fund/

World Bank. (2022). *Education: Impact of the War in Ukraine.*

Yelinska, A. M. (2020). Some Issues on Organization of Educational Process in Higher Educational Institutions During Pandemic. *Aktual'ni Problemy Suchasnoï Medytsyny : Visnyk Ukraïns'koï Medychnoï Stomatolohichnoï Akademiï, 20*(3), 235–238. DOI: 10.31718/2077-1096.20.3.235

Yeter, Ö., Rabagliati, H., & Özge, D. (2022). Threat of War on Cognitive Development of Refugee Children. *East European Journal of Psycholinguistics, 9*(2), 144–159. DOI: 10.29038/eejpl.2022.9.2.yet

Yuzyk, O., Yuzyk, M., Bilanych, L., Vitalii, H., Bilanych, H., & Myroslava, F. (2022). Distance Learning in Higher Education Institutions in Conditions of Quarantine and Military Conflicts. *International Journal of Computer Science and Network Security, 22*(4), 741–749. https://doi.org/https://doi.org/10.22937/IJCSNS.2022.22.4.87 different

Zakharchuk, N. (2020). Contextualizing the Development of Ukrainian Higher Education: Between Soviet Legacies and European Regionalization. *East/West. Journal of Ukrainian Studies, 7*(2), 173–199. DOI: 10.21226/ewjus616

Zaplatynskyi, V. (2022). Changes in education in Ukraine in connection with the Russian-Ukrainian war. *Reviewed Conference Proceedings of the 15th International Scientific Conference*, 278–293.

KEY TERMS AND DEFINITIONS

Blended Learning: An educational approach that combines face-to-face classroom methods with online activities and digital content, offering flexibility in instructional delivery.

Conflict-Affected Education: Education systems and learning experiences disrupted or transformed by armed conflict, including safety, infrastructure, and emotional wellbeing challenges.

Distance Education (DE): A method of delivering education to students who are not physically present in a traditional classroom setting, often relying on digital platforms and asynchronous communication.

Educational Resilience: The capacity of education systems, teachers, and students to adapt and continue functioning effectively during and after crises, such as armed conflict or pandemics.

Instructional Adaptation: The process of modifying teaching strategies, materials, or delivery modes in response to evolving challenges such as war, displacement, or infrastructural disruption.

Learning Disruption: Interruptions to the educational process caused by external factors such as conflict, disaster, or health emergencies, often resulting in gaps in learning outcomes or access.

Psychosocial Impact: The emotional and mental effects experienced by individuals due to traumatic events or prolonged stress, such as war, including anxiety, exhaustion, and loss of motivation.

Compilation of References

Agalianos, A., Whitty, G., & Noss, R. (2006). The Social Shaping of Logo. *Social Studies of Science*, *36*(2), 241–267. DOI: 10.1177/0306312706053809

Agosto, V., & Roland, E. (2018). Intersectionality and educational leadership: A critical review. *Review of Research in Education*, *42*(1), 255–285. DOI: 10.3102/0091732X18762433

Aguilar, E. (2018). *Onward: Cultivating emotional resilience in educators*. John Wiley & Sons. DOI: 10.1002/9781119441731

Ahammad, M. F., Glaister, K. W., Sarala, R. M., & Glaister, A. J. (2018). Strategic talent management in emerging markets. *Thunderbird International Business Review*, *60*(1), 5–8. DOI: 10.1002/tie.21927

Ahmad, S. F., Rahmat, M. K., Mubarik, M. S., Alam, M. M., & Hyder, S. I. (2021). Artificial intelligence and its role in education. *Sustainability (Basel)*, *13*(22), 12902. DOI: 10.3390/su132212902

Akgün, B., Alpaydın, Y., Akgün, B., & Alpaydın, Y. (2024). *Global Agendas and Education Reforms: A Comparative Study* (Akgün, B., & Alpaydın, Y., Eds.; 1st ed.). Palgrave Macmillan., DOI: 10.1007/978-981-97-3068-1

Al Gharaibeh, F., Ahmad, I., & Malkawi, R. (2023). Impact of the Russia-Ukraine War on Education and International Students. *Journal of International Women's Studies*, *25*(6), 1–17. https://vc.bridgew.edu/jiws/vol25/iss6/12%0AThis

Al Khalili, T. (2025). Parental strategies and educational inequalities during severe economic, political and refugee crises in conflict-affected Lebanon. *Asia Pacific Education Review*, *26*(2), 553–566. DOI: 10.1007/s12564-025-10040-7

Alathari, L., Drysdale, D., Blair, A., McGarry, J., Canilletti, C., Snook, A., & Driscoll, S. (2018). *Enhancing school safety using a threat assessment model: An operational guide to preventing target school violence*. U.S. Department of Homeland Security. https://www.secretservice.gov/sites/default/files/reports/2020-10/USSS_NTAC_Enhancing_School_Safety_Guide.pdf

Aljbour, A., French, E., & Ali, M. (2022). An evidence-based multilevel framework of talent management: A systematic review. *International Journal of Productivity and Performance Management*, *71*(8), 3348–3376. DOI: 10.1108/IJPPM-02-2020-0065

Alkaher, I., & Gan, D. (2020). The role of school partnerships in promoting education for sustainability and social capital. *The Journal of Environmental Education*, *51*(6), 416–433. DOI: 10.1080/00958964.2020.1711499

AlKnawy, B. (2018). Leadership in times of crisis. BMJ Leader, leader-2018.

Allenby, B., & Sarewitz, D. (2011). *The Techno-Human Condition*. MIT Press. DOI: 10.7551/mitpress/8714.001.0001

American Federation of Teachers. (2022). *AFT report: Pandemic intensified job insecurity for adjuncts.* https://www.aft.org/news/aft-report-pandemic-intensified-job-insecurity-adjuncts

Anderson, C., & Krajcik, J. (2021). Co-creation in education: Building institutional resilience through collaborative solutions. *Journal of Educational Change*, *22*(3), 341–360.

Anderson, L., Smith, J., & Johnson, K. (2019). Teacher leadership and innovation: Building professional learning communities in schools. *Journal of Educational Change*, *20*(4), 475–492.

Anlesinya, A., Dartey-Baah, K., & Amponsah-Tawiah, K. (2019). Strategic talent management scholarship: A review of current foci and future directions. *Industrial and Commercial Training*, *51*(5), 299–314. DOI: 10.1108/ICT-11-2018-0095

An, T., & Li, Y. (2014). On Theoretical Prospect of Educational Technology From the Perspective of Philosophy of Technology. *Educational Research*, *4*, 37–42.

Arapi, P., Moumoutzis, N., Mylonakis, M., & Christodoulakis, S. (2007). A pedagogy-driven personalization framework to support adaptive learning experiences. In *Proceedings of the Seventh IEEE International Conference on Advanced Learning Technologies (ICALT 2007)* (pp. 96–97). DOI: 10.1109/ICALT.2007.27

Arthur, W. B. (2009). *The Nature of Technology: What it is and How it Evolves*. Penguin.

Aspelund, K., & Reed, P. (2019). Career and technical education and workforce readiness. *Journal of Career and Technical Education, 34*(1), 45–60.

Astor, R., Meyer, H., Benbenishty, R., Marachi, R., & Rosemond, M. (2005). School safety interventions: Best practices and programs. *Children & Schools, 27*(1), 17–32. DOI: 10.1093/cs/27.1.17

Atiyeh, B., Emsieh, S., Hakim, C., & Chalhoub, R. (2023). A narrative review of artificial intelligence (AI) for objective assessment of aesthetic endpoints in plastic surgery. *Aesthetic Plastic Surgery, 47*(6), 2862–2873. DOI: 10.1007/s00266-023-03328-9

Austin, L., Fisher Liu, B., & Jin, Y. (2012). How audiences seek out crisis information: Exploring the social-mediated crisis communication model. *Journal of Applied Communication Research, 40*(2), 188–207. DOI: 10.1080/00909882.2012.654498

Australia Gonski Institute for Education. (2020). *Scenario planning in education leadership: Responding to future challenges*. University of New South Wales.

Ayas, K., & Zeniuk, N. (2001). Project-based learning: Building communities of reflective practitioners. *Management Learning, 32*(1), 61–76. DOI: 10.1177/1350507601321005

Bachari, E. E., Abelwahed, E. H., & Adnani, M. E. (2011). E-learning personalization based on dynamic learners' preference. *International Journal of Computer Science and Information Technologies, 3*(3), 200–216. DOI: 10.5121/ijcsit.2011.3314

Bagchi, M. J., & Pattnaik, A. (2022). Talent Analytics and its Affect on Employee Retention-Employee Perspective. *Journal of Positive School Psychology, 6*(3), 7311–7321.

Bagga, S. K., Gera, S., & Haque, S. N. (2023). The mediating role of organizational culture: Transformational leadership and change management in virtual teams. *Asia Pacific Management Review, 28*(2), 120–131. DOI: 10.1016/j.apmrv.2022.07.003

Ball, D. L., & Forzani, F. M. (2020). Teacher professional learning and development: What works? *Educational Researcher, 49*(7), 456–464.

Bandyopadhyay, S., Bardhan, A., Dey, P., Bhattacharyya, S., Bandyopadhyay, S., Bardhan, A., ... & Bhattacharyya, S. (2021). Education Ecosystem: A Chronological Perspective. Bridging the Education Divide Using Social Technologies: Explorations in Rural India, 43-75.

Barab, S., & Luehmann, A. (2021). Curriculum integration in environmental education: Promoting ecological literacy through interdisciplinary teaching. *The Journal of Environmental Education, 52*(2), 123–135.

Barad, K. (2019). After the end of the world: Entangled nuclear colonialisms, matters of force, and the material force of justice. *Theory & Event, 22*(3), 524–550.

Barge, J. K., & Loges, W. E. (2018). Community-based school programs and family engagement: Developing a model of educational success. *Journal of Family and Consumer Sciences, 110*(3), 18–28. DOI: 10.14307/JFCS.110.3.18

Barron, B., Schwartz, D., Vye, N., Moore, A., Petrosino, A., Zech, L., & Bransford, J. (2021). Doing with understanding: Lessons from research on problem- and project-based learning. *Journal of the Learning Sciences, 7*(3-4), 271–311. DOI: 10.1080/10508406.1998.9672056

Barzman, D. (2018, May 1). *Pilot study validates AI to help predict school violence.* Cincinnati Children's. https://scienceblog.cincinnatichildrens.org/pilot-study-validates-ai-to-help-predict-school-violence/

Barzman, D., Ni, Y., Griffey, M., Patel, B., Warren, A., Latessa, E., & Sorter, M. (2017). A pilot study on developing a standardized and sensitive school violence risk assessment with manual annotation. *The Psychiatric Quarterly, 88*(3), 447–457. DOI: 10.1007/s11126-016-9458-7 PMID: 27528455

Bass, B. M., & Riggio, R. E. (2006). *Transformational leadership* (2nd ed.). Psychology Press. DOI: 10.4324/9781410617095

Bataille, G. M., & Cordova, D. I. (Eds.). (2023). *Managing the unthinkable: Crisis preparation and response for campus leaders.* Taylor & Francis.

Baturay, M. H. (2015). An overview of the world of MOOCs. *Procedia: Social and Behavioral Sciences, 174*, 427–433. DOI: 10.1016/j.sbspro.2015.01.685

Baumgartner, F. R., Daniely, T., Huang, K., Johnson, S., Love, A., May, L., Mcgloin, P., Swagert, A., Vattikonda, N., & Washington, K. (2021). Throwing away the key: The unintended consequences of "tough-on-crime" laws. *Perspectives on Politics, 19*(4), 1233–1246. DOI: 10.1017/S153759272100164X

Ba, Y., Ming, W., & Zhang, H. (2025). Unlocking academic success: How growth mindset interventions enhance student performance through self-belief and effort regulation. *Acta Psychologica, 256*, 104977. DOI: 10.1016/j.actpsy.2025.104977 PMID: 40209651

Bayne, S. (2015). What's the Matter with 'Technology-Enhanced Learning'? *Learning, Media and Technology*, *40*(1), 5–20. DOI: 10.1080/17439884.2014.915851

Beane, J. A. (1997). *Curriculum integration: Designing the core of democratic education*. Teachers College Press, PO Box 20, Williston, VT 05495-0020.

Beckera, R., & Birdib, A. (2018). Flipping the Classroom: Old Ideas, New Technologies. *International Review of Economics Education*, *29*, 1–5. DOI: 10.1016/j.iree.2018.06.001

Becker, S. A., Cummins, M., Davis, A., Freeman, A., Hall, C., Ananthanarayanan, V., & Koenig, J. (2020). *NMC Horizon Report: 2020 Higher Education Edition*. EDUCAUSE.

Beerkens, M., & van der Hoek, M. (2022). Academic leaders and leadership in the changing higher education landscape. In *Research Handbook on Academic Careers and Managing Academics* (pp. 121–136). Edward Elgar Publishing. DOI: 10.4337/9781839102639.00017

Berezhna, S., & Prokopenko, I. (2020). Higher Education Institutions in Ukraine during the Coronavirus, or COVID-19, Outbreak: New Challenges vs New Opportunities. *Revista Românească Pentru Educaţie Multidimensională*, *12*(2), 130–135. DOI: 10.18662/rrem/12.1sup2/256

Berkowitz, M. W., & Bier, M. C. (2019). What works in character education: A research-driven guide for educators. *Journal of Moral Education*, *48*(4), 378–392.

Bernardo, M. A. C., & Baranovich, D.-L. (2014). Higher education in the heart of armed conflict: The pivotal role of student affairs. *International Journal of Educational Development*, *35*, 78–85. https://doi.org/https://doi.org/10.1016/j.ijedudev.2012.08.007. DOI: 10.1016/j.ijedudev.2012.08.007

Beyer, B., Levy, G., Boulandier, K., & Senior, T. (2025). Student arrested for making online threats to Coral Springs Charter School faces judge. *WSVN 7News*. https://wsvn.com/news/local/broward/student-arrested-for-making-online-threats-to-coral-springs-charter-school-faces-judge/

Biesta, G., Heugh, K., Cervinkova, H., Rasinski, L., Osborne, S., Forde, D., Wrench, A., Carter, J., Anders Safstrom, C., Soong, H., O'Keeffe, S., Paige, K., Rigney, I., O'Toole, L., Hattam, R., Peters, M. A., & Tesar, M. (2021). Philosophy of education in a new key: Publicness, Social Justice, and Education; A South-North Conversation. *Educational Philosophy and Theory*. Advance online publication. DOI: 10.1080/00131857.2021.1929172

Bishop, R., Berryman, M., Cavanagh, T., & Teddy, L. (2009). Te Kotahitanga: Addressing educational disparities facing Māori students in New Zealand. *Teaching and Teacher Education, 25*(5), 734–742. DOI: 10.1016/j.tate.2009.01.009

Blodgett, C., & Dorado, J. (2016). A trauma-informed approach to systems change: A case study of Spokane public schools. *Children & Schools, 38*(4), 207–215.

Blumenfeld, P., Soloway, E., Marx, R., Krajcik, J. S., Guzdial, M., & Palincsar, A. S. (1991). Motivating project-based learning: Sustaining the doing, supporting the learning. *Educational Psychologist, 26*(3-4), 369–398. DOI: 10.1080/00461520.1991.9653139

Bonds, M. D. S. (2022). *Navigating Memorialization and Commemoration on US Campuses: Approaches to Crisis Recovery*. Routledge. DOI: 10.4324/9781003165491

Borko, H. (2018). Professional development and teacher learning: Mapping the terrain. *Educational Researcher, 47*(5), 252–258.

Bower, M., Howe, C., McCredie, N., Robinson, A., & Grover, D. (2014). Augmented reality in education: Cases, places and potentials. *Educational Media International, 51*(1), 1–15. DOI: 10.1080/09523987.2014.889400

Bozkurt, A., Jung, I., Xiao, J., Vladimirschi, V., Schuwer, R., Egorov, G., & Paskevicius, M. (2020). A global outlook to the interruption of education due to COVID-19 pandemic: Navigating in a time of uncertainty and crisis. *Asian Journal of Distance Education, 15*(1), 1–126. DOI: 10.5281/zenodo.3878572

Bozkurt, A., & Sharma, R. C. (2020). Emergency remote teaching in a time of global crisis due to CoronaVirus pandemic. *Asian Journal of Distance Education, 15*(1), i–vi. https://www.asianjde.com/ojs/index.php/AsianJDE/article/view/447

Braden, A. (1995). Review of Instructional Technology: The Definition and Domains of the Field. *Educational Technology Research and Development, 43*(1), 81–83. DOI: 10.1007/BF02300483

Bradshaw, C. P., Cohen, J., Espelage, D. L., & Nation, M. (2021). Addressing school safety through comprehensive school climate approaches. *School Psychology Review, 50*(2-3), 221–236. DOI: 10.1080/2372966X.2021.1926321

Braidotti, R. (2019). *Posthuman knowledge*. Polity Press.

Brallier, S., & Gaudreau, P. (2020). Micro-credentials in teacher professional development: A new paradigm for lifelong learning. *Professional Development in Education, 46*(4), 555–570.

Brennan Center for Justice & Center for Democracy and Technology. (n.d.). *Social media monitoring in K–12 schools: Civil and human rights concerns*. Brennan Center for Justice. https://www.brennancenter.org/our-work/research-reports/social-media-monitoring-k-12-schools-civil-and-human-rights-concerns

Brillantes, A. B., & Fernandez, M. T. (2011). Restoring trust and building integrity in government: Issues and concerns in the Philippines and areas for reform. *International Public Management Review, 12*(2), 55–80.

Bringle, R. G., Clayton, P. H., & Price, M. F. (2009). Partnerships in service learning and civic engagement. Partnerships: *A Journal of Service-Learning and Civic Engagement, 1*(1), 1–20.

Bringle, R. G., & Hatcher, J. A. (2009). Innovative practices in service-learning and civic engagement. *New Directions for Higher Education, 2009*(147), 37–46. DOI: 10.1002/he.356

Brinko, K. T. (2019). Reflective journaling as a tool for teacher development. *Teacher Education Quarterly, 46*(4), 75–90.

Brookfield, S. D. (2017). *Becoming a critically reflective teacher* (2nd ed.). Jossey-Bass.

Brown, S. (2024). Some area schools are closed for the day in response to school threats. *Cape Girardeau Fox23 Now*. https://www.kbsi23.com/news/some-area-schools-are-closed-for-the-day-in-response-to-school-threats/

Brown, P., & Souto-Otero, M. (2020). The misalignment between education and employment. *Oxford Review of Education, 46*(4), 425–444. DOI: 10.1080/03054985.2019.1697397

Bryk, A. S., Gomez, L. M., Grunow, A., & LeMahieu, P. G. (2015). *Learning to improve: How America's schools can get better at getting better*. Harvard Education Press.

Buchanan, R. A., Forster, D. J., Douglas, S., Nakar, S., Boon, H. J., Heath, T., Heyward, P., D'Olimpio, L., Ailwood, J., Eacott, S., Smith, S., Peters, M., & Tesar, M. (2021). Philosophy of education in a new key: Exploring new ways of teaching and doing ethics in education in the 21st century. *Educational Philosophy and Theory*. Advance online publication. DOI: 10.1080/00131857.2021.1880387

Buheji, M. (2018). *Understanding the power of resilience economy: An interdisciplinary perspective to change the world attitude to socio-economic crisis*. Mohamed Buheji.

Burke, C., & Bloss, C. (2020). Social media surveillance in schools: Rethinking public health interventions in the digital age. *Journal of Medical Internet Research, 22*(11), e22612. DOI: 10.2196/22612 PMID: 33179599

Burke, R. (2021). Anticipatory action learning, leadership, strategy and foresight: Creating a successful future while enhancing results today. *Journal of Futures Studies, 25*(3), 85–92.

Burley, B. (2025). Social media as an investigative tool: OSINT strategies for law enforcement. *Police1.* https://www.police1.com/investigations/social-media-as-an-investigative-tool-osint-strategies-for-law-enforcement

Burt, G. (2023). *Evolution of scenario planning: Theory and practice from disorder to order.* De Gruyter., DOI: 10.1515/9783110792065

Bush, T. (2020). Theories of educational leadership and management.

Byun, S., & Joung, S. (2020). The effects of gamification in education: A systematic review. *Educational Technology Research and Development, 68*(4), 1915–1935.

Cabinet of Ministers of Ukraine. (2022a). *Approval of the Strategy for the Development of Higher Education in Ukraine for 2022-2032.* Legislation of Ukraine.

Cabinet of Ministers of Ukraine. (2022b). *On Amendments to Certain Laws of Ukraine Regarding State Guarantees in the Conditions of Martial Law, State of Emergency, or State of Emergency. Legislation of Ukraine.* https://zakon.rada.gov.ua/laws/show/2126-20#Text

Cable, D. (2013). Traditional on-boarding of staff requires a shake-up. *Strategic HR Review, 12*(6). Advance online publication. DOI: 10.1108/SHR-06-2013-0068

Cahill, H., & Leccardi, C. (2020). Reframing resilience. In *Youth and the new adulthood: Generations of change* (pp. 67–81). Springer Singapore.

Campbell, C., Malkus, N., & Miller, L. (2016). *Evaluation of the Teacher Learning and Leadership Program (TLLP).* Ontario Ministry of Education.

Campbell, K. (1999). The promise of computer-based learning: Designing for inclusivity. *IEEE Technology and Society Magazine, 18*(4), 28–34. DOI: 10.1109/44.808846

Candrasari, R., Yorman, Y., Mayasari, N., Yulia, R., & Lake, F. (2023). Visionary leadership in education management: Leading toward optimal achievement in the era of independent learning. [INJOE]. *Indonesian Journal of Education, 3*(3), 451–467.

Capano, G. (2020). Policy design and state capacity in the COVID-19 emergency in Italy: If you are not prepared for the (un)expected, you can be only what you already are. *Policy and Society*, *39*(3), 326–344. DOI: 10.1080/14494035.2020.1783790 PMID: 35039724

Cardona, M. A., Rodríguez, R. J., & Ishmael, K. (2023). *Artificial intelligence and the future of teaching and learning*. Lumina Foundation.

Carter, V. (2013). Book Review - Disruptive Innovation in Technology and Engineering Education: A Review of the Three Works by Clayton Christensen and Colleagues. *Journal of Technology Education*, *24*(2). https://doi.org/https://doi.org/10.21061/jte.v24i2.a.7

CAST. (2018). *Universal design for learning guidelines (Version 2.2).* http://udlguidelines.cast.org

CAST. (2018). *Universal Design for Learning Guidelines version 2.2.*

CAST. (2020). *Universal Design for Learning: Theory and practice*. CAST Professional Publishing.

CAST. (2021). *Universal Design for Learning Guidelines version 2.2.*

Cattaneo, L. (2021). Community engaged learning and student civic development. *Frontiers in Education*, *6*, 626984. Advance online publication. DOI: 10.3389/feduc.2021.626984

Cerna-Turoff, I., Fischer, H. T., Mansourian, H., & Mayhew, S. (2021). The pathways between natural disasters and violence against children: A systematic review. *BMC Public Health*, *21*(1), 1249. DOI: 10.1186/s12889-021-11252-3 PMID: 34247619

Chan, C. K. Y., & Tsi, L. H. Y. (2023). The AI revolution in education: Will AI replace or assist teachers in higher education? arXiv:2305.01185 [cs.CY]. https://doi.org//arXiv.2305.01185DOI: 10.48550

Chang, C., & Lin, S. (2023). Enhancing teacher self-awareness through reflective writing. *International Journal on Teaching and Learning in Higher Education*, *35*(1), 45–59.

Chaplinsky v. New Hampshire, 315 U.S. 568 (1942).

Charlton, C. T., Moulton, S., Sabey, C. V., & West, R. (2021). A systematic review of the effects of schoolwide intervention programs on student and teacher perceptions of school climate. *Journal of Positive Behavior Interventions*, *23*(3), 185–200. DOI: 10.1177/1098300720940168

Chaudhry, M. Y., & Malik, A. M. (2014). Intersection of pedagogy and emerging technologies to enhance student-centred learning in higher education.

Chen, J. J., & Perez, C. (2023). Enhancing assessment and personalized learning through artificial intelligence. *Childhood Education, 99*(6), 72–79. DOI: 10.1080/00094056.2023.2282903

Chen, X., & Li, L. (2020). Family involvement in school-based decision making: The role of school leadership and community engagement. *International Journal of Leadership in Education, 23*(3), 343–360. DOI: 10.1080/13603124.2019.1585597

Chen, X., Xie, H., & Hwang, G.-J. (2021). Effects of adaptive learning systems on students' learning outcomes: A meta-analysis. *Computers & Education, 168*, 104212.

Cherubini, L. (2009). The New Teacher Induction Program in Ontario: Supporting early career educators. *Canadian Journal of Education, 32*(3), 531–554.

Chiang, F. K., Zhu, G. X., Wang, Q., Cui, Z. F., Cai, S., & Yu, S. Q. (2016). Research and Trends in Mobile Learning from 1976 to 2013: A Content Analysis of Patents in Selected Databases. *British Journal of Educational Technology, 47*(6), 1006–1019. DOI: 10.1111/bjet.12311

Christensen, C. M., Horn, M. B., Caldera, L., & Soares, L. (2011). *Disrupting College: How Disruptive Innovation Can Deliver Quality and Affordability to Postsecondary Education.*

Christensen, C. M. (2006). The Ongoing Process of Building a Theory of Disruption. *Journal of Product Innovation Management, 23*(1), 39–55. DOI: 10.1111/j.1540-5885.2005.00180.x

Christensen, C. M., Horn, M. B., & Johnson, C. W. (2011). *Disrupting class: How disruptive innovation will change the way the world learns.* McGraw-Hill.

Christensen, C., Horn, M., & Johnson, C. (2008). *Disrupting Class: How disruptive innovation will change the way the world learns.* McGraw-Hill.

Chua Reyes, V., Hamid, O., & Hardy, I. (2022). When reforms make things worse: School leadership responses to poverty, disasters, and cultures of crises in the Philippine education system. *International Journal of Leadership in Education, 25*(2), 331–344. DOI: 10.1080/13603124.2021.2009038

Cipriano, C., Barnes, T., Jones, S. M., & Osher, D. (2020). Educating the whole child: Improving school climate to support student success. *Phi Delta Kappan, 101*(8), 8–14. DOI: 10.1177/0031721720926230

Clegg, S., & Steel, J. (2002). "Flexibility as Myth? New Technologies and Post-Fordism in Higher Education." *Proceedings of the Networked Learning Conference*.

Cohen, L. E., & Felson, M. (1979). Social change and crime rate trends: A routine activity approach. *American Sociological Review*, *44*(4), 588–608. DOI: 10.2307/2094589

Collie, R. J., Shapka, J. D., & Perry, N. E. (2020). School climate and social–emotional learning: Predicting teacher stress, job satisfaction, and teaching efficacy. *Journal of Educational Psychology*, *112*(4), 664–678.

Coombs, W. T. (2007). Protecting organization reputations during a crisis: The development and application of situational crisis communication theory. *Corporate Reputation Review*, *10*(3), 163–176. DOI: 10.1057/palgrave.crr.1550049

Cornell, D. (2020). Threat assessment as a school violence prevention strategy. *Criminology & Public Policy*, *19*(1), 235–252. DOI: 10.1111/1745-9133.12471

Cornish, L. (Ed.). (2024). *(Re)learning as Reflective Practitioners: Insights for Other Professions from Reflective Practice in Teacher Education*. Springer., DOI: 10.1007/978-3-031-60211-5

Counterman v. Colorado, 600 U.S. 66 (2023).

Counts, J., Randall, K. N., Ryan, J. B., & Katsiyannis, A. (2018). School resource officers in public schools: A national review. *Education & Treatment of Children*, *41*(4), 405–430. DOI: 10.1353/etc.2018.0023

Crane, A., & Matten, D. (2022). *Business ethics: Managing corporate citizenship and sustainability in the age of globalization* (6th ed.). Oxford University Press.

Crawford, J., Cowling, M., & Allen, K.-A. (2023). Leadership is needed for ethical ChatGPT: Character, assessment, and learning using artificial intelligence (AI). *Journal of University Teaching & Learning Practice*, *20*(3), 457–476. DOI: 10.53761/1.20.3.02

Creasman, B. K. (2022). *Prioritizing health and well-being: Self-care as a leadership strategy for school leaders*. Rowman & Littlefield. https://rowman.com/ISBN/9781475867367/Prioritizing-Health-and-Well-Being-Self-Care-as-a-Leadership-Strategy-for-School-Leaders

Crow, M. M., & Dabars, W. B. (2020). *The fifth wave: The evolution of American higher education*. Johns Hopkins University Press. DOI: 10.1353/book.73164

Cruz, R. D. D., & Ormilla, R. C. G. (2022). Disaster risk reduction management implementation in the public elementary schools of the department of education, Philippines. *International Journal of Disaster Risk Management, 4*(2), 1–15. DOI: 10.18485/ijdrm.2022.4.2.1

Cuban, L. (2001). *Oversold and Underused: Computers in the Classroom*. Harvard University Press. DOI: 10.4159/9780674030107

Cueller, M. J., & Theriot, M. T. (2017). Exploring the effects of parental and community involvement on the implementation of school safety strategies in United States high schools. *Journal of School Public Relations, 36*(3), 320–345. DOI: 10.3138/jspr.36.3.320

Cybersecurity and Infrastructure Security Agency. (2023). *Social media threats guidance for school staff and authorities (infographic)*. U.S. Department of Homeland Security. https://www.cisa.gov/sites/default/files/2023-12/Social%20Media%20Threat%20Guidance%20for%20School%20Staff%20and%20Authorities%20Infographic_508.pdf

Czerniewicz, L. (2010). Educational Technology–Mapping the Terrain with Bernstein as Cartographer. *Journal of Computer Assisted Learning, 26*(6), 523–534. DOI: 10.1111/j.1365-2729.2010.00359.x

Czerniewicz, L., Agherdien, N., Badenhorst, J., Belluigi, D., Chambers, T., Chili, M., de Villiers, M., Felix, A., Gachago, D., Gokhale, C., Ivala, E., Kramm, N., Madiba, M., Mistri, G., Mgqwashu, E., Pallitt, N., Prinsloo, P., Solomon, K., Strydom, S., & Wissing, G. (2020). A wake-up call: Equity, inequality and COVID-19 emergency remote teaching and learning. *Postdigital Science and Education, 2*(3), 946–967. DOI: 10.1007/s42438-020-00187-4 PMID: 40477091

D'Auria, G., & De Smet, A. (2020). Leadership in a crisis: Responding to the coronavirus outbreak and future challenges. *Psychology (Irvine, Calif.), 22*(2), 273–287.

D'Errico, F. (2007). The Origin of Humanity and Modern Cultures: Archaeology's View. *Diogenes, 54*(2), 122–133. DOI: 10.1177/0392192107077652

Dalcher, D. (2020). Leadership in times of crisis: What's different now. *PM World Journal, 9*(5), 1–18.

Daniel, S. J. (2020). Education and the COVID-19 pandemic. *Prospects, 49*(1), 91–96. DOI: 10.1007/s11125-020-09464-3 PMID: 32313309

Darling-Hammond, L., Hyler, M. E., & Gardner, M. (2020). *Effective teacher professional development*. Learning Policy Institute. https://learningpolicyinstitute.org/product/effective-teacher-professional-development-report

Darling-Hammond, L., & DePaoli, J. (2020). Why school climate matters and what can be done to improve it. *State Education Standard, 20*(2), 7.

Darling-Hammond, L., Flook, L., Cook-Harvey, C., Barron, B., & Osher, D. (2020). Implications for educational practice of the science of learning and development. *Applied Developmental Science, 24*(2), 97–140. DOI: 10.1080/10888691.2018.1537791

Darling-Hammond, L., Hyler, M. E., & Gardner, M. (2017). *Effective teacher professional development*. Learning Policy Institute. DOI: 10.54300/122.311

Darling-Hammond, L., Schachner, A., & Edgerton, A. K. (2020). *Restarting and Reinventing School: Learning in the Time of COVID and Beyond*. Learning Policy Institute.

Davies, L. (2005). Schools and war: Urgent agendas for comparative and international education. *Compare: A Journal of Comparative Education, 35*(4), 357–371. DOI: 10.1080/03057920500331561

Dayson, C., Ayo, J., Equipado, B., Polon, J., Buban, R., & Digo, G. (2024). A Systematic Review and Meta-synthesis of Indigenous Filipino Values in Educational Leadership. [Humanities And Social Sciences]. *Journal Of Community Development Research, 17*(4), 1–22. DOI: 10.14456/jcdr-hs.2024.24

De Anda, A. (2024). Experiences with Crisis and Resilience in Education in Emergencies: Voices and Lenses of Teachers in Nicaragua (Doctoral dissertation, Concordia University, St. Paul).

de La Hoz-Ruiz, J., Khalil, M., Domingo Segovia, J., & Liu, Q. (2024). Learning analytics for enhanced professional capital development: A systematic review. *Frontiers in Psychology, 15*, 1302658. DOI: 10.3389/fpsyg.2024.1302658 PMID: 38318080

de Souza Zanirato Maia, J., Bueno, A. P., & Sato, J. R. (2023). Applications of artificial intelligence models in educational analytics and decision making: A systematic review. SSRN *Electronic Journal*. DOI: 10.3390/world4020019

Dede, C., Edwards, L., & Huang, Y. (2019). Micro-credentials for teacher learning: Trends and challenges. *Journal of Digital Learning in Teacher Education, 35*(3), 151–160.

del Valle, J. L. (2022). Tough teachers actually care: An ethnographic look into the'problematic'role of teachers as figures of authority under learner-centered education. *The Australian Journal of Teacher Education, 47*(6), 19–36. DOI: 10.14221/ajte.2022v47n6.2

Delaliarte, M., Linaugo, J. D., & Madrigal, D. V. (2024). Fire Safety Awareness and Practices of Science, Technology, Engineering, and Mathematics Students in a Philippine Public Secondary School. *Formatif: Jurnal Ilmiah Pendidikan MIPA, 14*(1).

Deleuze, G. (1995). *Negotiations*. Columbia University Press.

Delos Reyes, Z. M. A. (2019). Assessment of School Critical Safety Concerns. *Delos Reyes, ZMA, 1*, 22–32.

DeMatthews, D., Carrola, P., Reyes, P., & Knight, D. (2021). School leadership burnout and job-related stress: Recommendations for district administrators and principals. *The Clearing House: A Journal of Educational Strategies, Issues and Ideas, 94*(4), 159–167. DOI: 10.1080/00098655.2021.1894083

Deng, L. (2018). Artificial intelligence in the rising wave of deep learning: The historical path and future outlook. *IEEE Signal Processing Magazine, 35*(1), 180–177. DOI: 10.1109/MSP.2017.2762725

Department of Education. (2009). *DepEd Order No. 72, s. 2009 – Inclusive Education as Strategy for Increasing Participation Rate of Children*. https://www.deped.gov.ph/2009/07/06/do-72-s-2009-inclusive-education-as-strategy-for-increasing-participation-rate-of-children/

Department of Education. (2012). *DepEd Order No. 40, s. 2012 – DepEd Child Protection Policy*. https://www.deped.gov.ph/2012/05/14/do-40-s-2012-deped-child-protection-policy/

Department of Education. (2022, December 7). *DepEd highlights Karapatan, Kalidad, Kalinga at 4th National Summit on Child's Rights in Education*. https://www.deped.gov.ph/2022/12/07/deped-highlights-karapatan-kalidad-kalinga-at-4th-national-summit-on-childs-rights-in-education/

Dépelteau, F. (2008). Relational Thinking: A Critique of co-Deterministic Theories of Structure and Agency. *Sociological Theory, 26*(1), 51–73. DOI: 10.1111/j.1467-9558.2008.00318.x

Derreth, R. T., Nelson, K. E., Nguy n, C. H., Adams, A. W., & Levin, M. B. (2024). Community partner experiences in a service-learning development program. *Journal of Higher Education Outreach & Engagement, 28*(2), 45–63.

Devlin, D. N., & Fisher, B. W. (2021). An examination of school resource officers as an approach to reduce social disturbances in schools: Evidence from a national longitudinal study. *Journal of School Violence, 20*(2), 228–240. DOI: 10.1080/15388220.2021.1875843

Dewey, J. (1916). Democracy and Education: An Introduction to the Philosophy of Education. New York, *NY: the Free Press.*

DiMaggio, P., & Bonikowski, B. (2020). Digital divide: The impact of technology on inequalities in education. *Annual Review of Sociology*, *46*, 441–463.

Diwakar, V. (2015). The Effect of Armed Conflict on Education: Evidence from Iraq. *The Journal of Development Studies*, *51*(12), 1702–1718. DOI: 10.1080/00220388.2015.1056786

Dizikes, P. (2018, March 8). Study: On Twitter, false news travels faster than true stories. *MIT News*. https://news.mit.edu

Dogan, M. E., Goru Dogan, T., & Bozkurt, A. (2023). The use of artificial intelligence (AI) in online learning and distance education processes: A systematic review of empirical studies. *Applied Sciences (Basel, Switzerland)*, *13*(5), 56–88. DOI: 10.3390/app13053056

Domínguez, A., Saenz-de-Navarrete, J., Marcos, L. D., Sanz, L. F., Pagés-Arévalo, C., & Martínez, J. (2013). Gamifying learning experiences: Practical implications and outcomes. *Computers & Education*, *63*, 380–392. DOI: 10.1016/j.compedu.2012.12.020

Donelan, H., & Kear, K. (2024). *Online group projects in higher education: Persistent challenges and implications for practice*. Technology, Knowledge and Learning.

Donnelly, K. (2007). Australia's Adoption of Outcomes Based Education: A Critique. *Issues in Educational Research*, *17*(2), 183–206.

Dorado, J., Martinez, M., McArthur, L., & Leibovitz, T. (2019). Secondary traumatic stress and trauma-informed schools. *Psychology in the Schools*, *56*(6), 1005–1019.

Doubeni, C. A., Adams, S. A., Alcaraz, K. I., Artiga, S., Bailey, S., Carney, J., Daniels, R., DeVoe, J., Dulin, M., Fuchs, S., George, S., Gennarelli, R., Geller, S., Goodman, M., Hamm, M., Harvey, H., Jatlaoui, T., Jones, S., Khan, M., & Zgierska, A. (2022). Community engagement education in academic health centers, colleges, and universities. *Journal of Clinical and Translational Science*, *6*(1), e192. DOI: 10.1017/cts.2022.424

Douglas, E. S., & Slate, J. R. (2025). A post-pandemic analysis of career and technical education performance by student gender. *American Journal of STEM Education*, *18*, 15–50. DOI: 10.32674/jk1c1x22

Drigas, A., Papoutsi, C., & Skianis, C. (2023). Being an emotionally intelligent leader through the nine-layer model of emotional intelligence—The supporting role of new technologies. *Sustainability (Basel), 15*(10), 8103. DOI: 10.3390/su15108103

Dryden-Peterson, S. (2011). *Refugee Education: A Global Review*. https://www.unhcr.org/media/refugee-education-global-review-sarah-dryden-peterson-november-2011

Dufour, R., Dufour, R., & Eaker, R. (2008). *Revisiting professional learning communities at work*. Solution Tree, Llc.

Dunn, R. (2020). Adaptive leadership: Leading through complexity. *International Studies in Educational Administration, 48*(1), 31–38.

Dusek, V. (2006). *Philosophy of Technology: An Introduction*. Blackwell Publishing.

Dweck, C. S. (2006). *Mindset: The new psychology of success*. Random House.

Edwards, A., Martin, A., & Knight, P. (2019). Reflective practice and teacher resilience: Expanding adaptive capacity in schools. *Teaching and Teacher Education, 79*, 96–104.

El-Bawab, N. (2022). New evidence alleges Ethan Crumbley exhibited more warning signs ahead of school shooting. *ABC News*. https://abcnews.go.com/US/evidence-alleges-ethan-crumbley-exhibited-warning-signs-ahead/story?id=90321148

Elhag, H. E. E. A., & Baleela, R. M. H. (2025). Preliminary investigation and analysis of the impact of the Sudan War on higher education and scientific research sectors. *International Journal of Educational Development, 112*, •••. https://doi.org/https://doi.org/10.1016/j.ijedudev.2024.103190

Ellen, M. (2000). *The school shooter: A threat assessment perspective*. Federal Bureau of Investigation. https://www.ojp.gov/ncjrs/virtual-library/abstracts/school-shooter-threat-assessment-perspective

Ellington, B. L. (2019). *Preventing targeted school violence using a threat assessment model: An exploration of school counselors' prevention and intervention practices* (Order No. 27731899). Available from ProQuest Central. (2380641460). Retrieved from https://ezproxy.semo.edu:2443/login?url=https://www.proquest.com/dissertations-theses/preventing-targeted-school-violence-using-threat/docview/2380641460/se-2

Enriquez, V. G. (1992). *From colonial to liberation psychology: The Philippine experience*. University of the Philippines Press.

Epstein, J. L. (2018). *School, family, and community partnerships: Preparing educators and improving schools* (3rd ed.). Routledge.

Epstein, J. L., & Sheldon, S. B. (2019). Present and accounted for: Improving student attendance with family and community involvement. *The Journal of Educational Research, 112*(1), 40–48. DOI: 10.1080/00220671.2018.1466029

Eschenbacher, S., & Fleming, T. (2021). Toward a critical pedagogy of crisis. European journal for Research on the Education and Learning of Adults, 12(3), 295-309.

Esposito, A., Sangrà, A., & Maina, M. (2013). "Chronotopes in Learner-Generated Contexts. A Reflection About the Interconnectedness of Temporal and Spatial Dimensions to Provide a Framework for the Exploration of Hybrid Learning Ecologies of Doctoral e-Researchers." *LC Research Paper Series,* 6: 15–28.

Estes, J. S., Dailey-Hebert, A., & Choi, D. H. (2017). Integrating technological innovations to enhance the teaching-learning process.

European University Association. (2022). *Support to Ukrainian higher education.* https://eua.eu/resources/publications/1034:support-to-ukrainian-higher-education.html

Eva, N., Wolfram Cox, J., Tse, H., & Lowe, K. B. (2021). *The concept of collective leadership and how it could lead us out of a pandemic.* Monash Impact. https://www2.monash.edu/impact/articles/covid-19/the-concept-of-collective-leadership-and-how-it-could-lead-us-out-of-a-pandemic/

Eynon, R. (2015). The Quantified Self for Learning: Critical Questions for Education. *Learning, Media and Technology, 40*(4), 407–411. DOI: 10.1080/17439884.2015.1100797

Falfushynska, H. I., Buyak, B. B., Tereshchuk, H. V., Torbin, G. M., & Kasianchuk, M. M. (2021). Strengthening of e-learning at the leading Ukrainian pedagogical universities in the time of COVID-19pandemic. *CTE Workshop Proceedings*, 261–273.

Falgoust, G., Winterlind, E., Moon, P., Parker, A., Zinzow, H., & Madathil, K. (2022). Applying the uses and gratifications theory to identify motivational factors behind young adults' participation in viral social media challenges on TikTok. *Human Factors in Healthcare, 2*, 100014. DOI: 10.1016/j.hfh.2022.100014

Farlow, J. (2022, May 19). SRO success story: School resource officer's fast suspect identification avoids disruption from social media threat. *National Association of School Resource Officers*. https://www.nasro.org/news/2022/05/19/news-releases/sro-success-story-school-resource-officer-s-fast-suspect-identification-avoids-disruption-from-social-media-threat/

Farrell, S., & Ives, P. (2021). Professional dialogue and peer coaching: Enhancing teacher reflective practices. *Journal of Teacher Education, 72*(1), 36–50.

Federal Bureau of Investigation. (n.d.). *Threat intimidation guide*. https://www.fbi.gov/file-repository/threat-intimidation-guide-english-022322.pdf/view

Feenberg, A. (2002). *Transforming Technology: A Critical Theory Revisited*. Oxford University Press. DOI: 10.1093/oso/9780195146158.001.0001

Felten, P., & Lambert, L. M. (2020). *Relationship-rich education: How human connections drive success in college*. Jhu Press. DOI: 10.1353/book.78561

Fernandez, M. G. H., Lopez, G. D., Mesa, M. L. R. A., Fernandez, K. T. G., Madrid, B. J., Dans, L. F., & Lina, S. G. A. (2022). An outcomes-based evaluation of the mindfulness for safe schools program. *Acta Medica Philippina*, *56*(15). Advance online publication. DOI: 10.47895/amp.v56i15.6157

Fisher, B., & Hennessy, E. (2016). School resource officers and exclusionary discipline in U.S. high schools: A systematic review and meta-analysis. *Adolescent Research Review*, *1*(3), 217–233. DOI: 10.1007/s40894-015-0006-8

Fisher, B., Petrosino, A., Sutherland, H., Guckenburg, S., Fronius, T., Benitez, I., & Earl, K. (2023). School-based law enforcement strategies to reduce crime, increase perceptions of safety, and improve learning outcomes in primary and secondary schools: A systematic review. *Campbell Systematic Reviews*, *19*(4), e1360. DOI: 10.1002/cl2.1360 PMID: 38024781

Florian, L., & Black-Hawkins, K. (2018). Exploring inclusive pedagogy. *British Educational Research Journal*, *44*(3), 1–16.

Fowler, C. (2019). Virtual reality and learning: Where is the pedagogy? *British Journal of Educational Technology*, *50*(2), 437–450.

Fragouli, E., & Tourlaki, T. (2020). Leading change and crisis: How transformational changes can be successful nowadays? *Global Research in Higher Education*, *3*(3), 22–41. DOI: 10.22158/grhe.v3n3p22

Frei, F. X., & Morriss, A. (2020). Begin with trust. *Harvard Business Review*, *98*(3), 112–121.

Freire, P. (2005). *Pedagogy of the Oppressed (30th Anniversary Edition)*. Continuum.

Friedman, E., & Trines, S. (2019). *Education in Ukraine*. World Education News + Reviews. https://wenr.wes.org/2019/06/education-in-ukraine

Friedman, T. (2014). *The world is flat: A brief history of the twenty-first century* (Updated and expanded edition). Picador.

Friesen, N. (2009). *Re-thinking e-Learning Research: Foundations, Methods, and Practices*. Peter Lang.

Fullan, M. (2016). *The new meaning of educational change* (5th ed.). Teachers College Press.

Fullan, M., & Quinn, J. (2018). *Coherence: The right drivers in action for schools, districts, and systems*. Corwin.

Fullan, M., Quinn, J., & McEachen, J. (2021). *Deep learning: Engage the world change the world*. Corwin.

Fuller, A., & Unwin, L. (2017). Apprenticeship and the future of vocational education. *Journal of Vocational Education and Training*, *69*(1), 1–5. DOI: 10.1080/13636820.2016.1274362

Furco, A. (2001). Service-learning: A balanced approach to experiential education. *New Directions for Higher Education*, *2001*(114), 25–39.

Galynska, O., & Bilous, S. (2022). Remote learning during the war: Challenges for higher education in Ukraine. *International Science Journal of Education & Linguistics*, *1*(5), 1–6. DOI: 10.46299/j.isjel.20220105.01

Gamage, K. A. A., Dehideniya, S. C. P., Xu, Z., & Tang, X. (2023). ChatGPT and higher education assessments: More opportunities than concerns? *Journal of Applied Learning and Teaching*, *6*(2), 654–679. DOI: 10.37074/jalt.2023.6.2.32

Garcia, A., & Lee, C. H. (2020). Equity-centered approaches to educational technology.

Garg, S., Tsipras, D., Liang, P. S., & Valiant, G. (2022). What can transformers learn in-context? a case study of simple function classes. *Advances in Neural Information Processing Systems*, *35*, 30583–30598. DOI: 10.52202/068431-2217

Garroutte, E., & Westcott, K. (2013). The story is a living being: Companionship with stories in Anishinaabe Studies. In Doerfler, J., Sinclair, N. J., & Stark, H. K. (Eds.), *Centering Anishinaabeg studies: Understanding the world through stories* (pp. 61–80).

Gaston, P. L. (2023). *Higher education accreditation: How it's changing, why it must*. Taylor & Francis. DOI: 10.4324/9781003445074

Gay, G. (2010). *Culturally responsive teaching: Theory, research, and practice* (2nd ed.). Teachers College Press.

Gehlen, A. (1988). *Man: His Nature and Place in the World*. Columbia University Press.

George, A. S. (2024). The Symbiotic Relationship Between Visionary and Pragmatic Leadership in Propelling Organizational Success. *Partners Universal International Innovation Journal*, *2*(3), 57–79.

Ghaemi, F., & Mirsaeed, S. J. (2017). The impact of inquiry-based learning approach on critical thinking skill of EFL students.

Ghavami, N., Thornton, B. E., & Graham, S. (2021). School police officers' roles: The influence of social, developmental and historical contexts. *Journal of Criminal Justice*, *72*, 101724. DOI: 10.1016/j.jcrimjus.2020.101724 PMID: 32958967

Ghosh, O & Kumar, B. (2025). The Impact of Remote Experimentation on Student Engagement. *Revolutionizing Education With Remote Experimentation and Learning Analytics,* 523-544, DOI: DOI: 10.4018/979-8-3693-8593-7.ch031

Glusac, D., Karuović, D., & Milanov, D. (2015). Massive open online courses: Pedagogical overview. In *Proceedings of the 16th International Carpathian Control Conference (ICCC)* (pp. 142–146).

Goleman, D. (1995). *Emotional intelligence: Why it can matter more than IQ*. Bantam Books.

Goleman, D., Boyatzis, R., & McKee, A. (2002). *Primal leadership: Realizing the power of emotional intelligence*. Harvard Business Press.

Golob, B. (2022). Finding me in social me-dia: Teaching students to use social networking platforms legally and mindfully. *Communication Teacher*, *37*(2), 151–158. DOI: 10.1080/17404622.2022.2118340

González-DeHass, A. R., Willems, P. P., & Holbein, M. F. (2017). Examining the role of parental involvement in the academic success of Latino children. *Educational Psychology Review*, *29*(2), 235–257. DOI: 10.1007/s10648-016-9364-0

Gonzalez, T., Vélez, V., & Núñez, C. (2019). Restorative justice in Oakland Unified School District: Reducing suspensions and fostering equity. *Journal of School Violence*, *18*(1), 23–39.

Gottfried, M. A., & Bozick, R. (2019). Supporting the transition to college and careers through career technical education. *Educational Evaluation and Policy Analysis*, *41*(2), 191–213. DOI: 10.3102/0162373719832237

Graham, C. R., Burgoyne, N., Cantrell, S., Smith, L., & Harris, R. (2019). Using technology-enhanced interdisciplinary collaboration to support STEM education. *Journal of Educational Computing Research*, *57*(5), 1176–1198.

Grant, M. M. (2011). Learning, beliefs, and products: Students' perspectives with project-based learning. *The Interdisciplinary Journal of Problem-Based Learning*, *5*(2), 37–69. DOI: 10.7771/1541-5015.1254

Grazia, V., & Molinari, L. (2021). School climate multidimensionality and measurement: A systematic literature review. *Research Papers in Education*, *36*(5), 561–587. DOI: 10.1080/02671522.2019.1697735

Greco, S., & Cinganotto, L. (2023). Re-thinking education in the age of AI. *Journal of E-Learning and Knowledge Society,* I-IV Pages. DOI: 10.20368/1971-8829/1135873

Greenberg, M. T., Brown, J. L., & Abenavoli, R. M. (2021). Teacher stress and health: Effects on teachers, students, and schools. *EdArXiv*.

Gregory, A., Clawson, K., Davis, A., & Gerewitz, J. (2018). The promise of restorative practices to transform teacher-student relationships and achieve equity in school discipline. *Journal of Educational & Psychological Consultation*, *28*(4), 344–370.

Gupta, M., & Sharma, A. (2021). Fear of missing out: A brief overview of origin, theoretical underpinnings and relationship with mental health. *World Journal of Clinical Cases*, *9*(19), 4881–4889. DOI: 10.12998/wjcc.v9.i19.4881 PMID: 34307542

Gupta, N., & Rohil, M. K. (2017). Exploring possible applications of augmented reality in education. In *Proceedings of the 4th International Conference on Signal Processing and Integrated Networks (SPIN)* (pp. 437–441). DOI: 10.1109/SPIN.2017.8049989

Hahn, J. (2021, December 16). Schools in the U.S. taking precautions over disturbing TikTok threat for Dec. 17: 'I'm scared'. *People Magazine*. https://people.com/human-interest/schools-taking-precautions-over-tiktok-violence-warning-despite-no-credible-threats/

Halagatti, M., Gadag, S., Mahantshetti, S., Hiremath, C. V., Tharkude, D., & Banakar, V. (2023). Artificial intelligence: The new tool of disruption in educational performance assessment. In Tyagi, P., Grima, S., Sood, K., Balamurugan, B., Özen, E., & Eleftherios, T. (Eds.), *Smart analytics, artificial intelligence and sustainable performance management in a global digitalised economy (110A)* (pp. 261–287). Emerald Publishing Limited., DOI: 10.1108/S1569-37592023000110A014

Hammar Chiriac, E., Forsberg, C., & Thornberg, R. (2024). Teacher teams: A safe place to work on creating and maintaining a positive school climate. *Social Psychology of Education*, *27*(4), 1775–1795. DOI: 10.1007/s11218-023-09880-1

Hammond, B. (2015, April 28). Teacher 'terrified' by surprise active shooter drill at Oregon school. *The Oregonian*. https://www.oregonlive.com/pacific-northwest-news/2015/04/teacher_terrified_by_surprise.html

Harari, Y. N. (2014). *Sapiens: A brief history of humankind*. Random House.

Haraway, D. (1988). Situated Knowledges: The Science Question in Feminism and the Privilege of Partial Perspective. *Feminist Studies, 14*(3), 575–599. DOI: 10.2307/3178066

Hargreaves, A., & Fullan, M. (2012). *Professional capital: Transforming teaching in every school*. Teachers College Press.

Hargreaves, A., & O'Connor, M. T. (2018). *Collaborative professionalism: When teaching together means learning for all*. Corwin. DOI: 10.4135/9781506328133

Harmash, O. (2025). *Ukraine aid groups cut services, scramble for cash after US funding shock*. Reuters. https://www.reuters.com/world/europe/ukraine-aid-groups-cut-services-scramble-cash-after-us-funding-shock-2025-01-30/?utm_source=chatgpt.com

Haron, H., Yusof, A. R., Samad, H., Ismail, N., Juanita, A., & Yusof, H. (2019). The platform of MOOC (massive open online course) on open learning: Issues and challenges.

Harrer, S. (2023). Attention is not all you need: The complicated case of ethically using large language models in healthcare and medicine. *EBioMedicine, 90*, 104512. DOI: 10.1016/j.ebiom.2023.104512

Harris, A., & Goodall, J. (2018). Parental involvement and academic achievement: A systematic review of research. *Journal of Educational Psychology, 110*(1), 17–32. DOI: 10.1037/edu0000209

Harry, A. (2023). *Role of AI in education. Interdisciplinary Journal and Hummanity*. INJURITY.

Hartley, M. (2011). Reclaiming democratic purposes in higher education. *Educational Theory, 61*(4), 391–406.

Hazlett, B. (2025). Resiliency: An Exploration of Non-Cognitive Characteristics of First-Generation College Students as It Relates to Academic Success (Doctoral dissertation, Northeastern University).

Headden, S., & Cowan, J. (2019). Formative assessment and differentiated instruction. *Educational Leadership, 76*(3), 48–53.

Hegel, G. (1967). *The Phenomenology of the Mind* (Baillie, J. B., Trans.). Harper Torchbooks.

Heidegger, M. (1959). *An Introduction to Metaphysics* (Manheim, R., Trans.). Yale University Press.

Heidegger, M. (1971). *Poetry, Language, Thought* (Hofstader, A., Trans.). Harper and Row.

Heidegger, M. (1977). *The Question Concerning Technology, and Other Essays*. Harper Torchbooks.

Heidegger, M. (1996). *Being and Time* (Stambaugh, J., Trans.). State University of New York Press.

Heidegger, M. (2000). *Elucidations of Holderlin's Poetry* (Hoeller, K., Trans.). Humanity Books.

Heifetz, R., Grashow, A., & Linsky, M. (2009). *The practice of adaptive leadership: Tools and tactics for changing your organization and the world*. Harvard Business Press.

Heleta, S. (2015). *How to rebuild higher education in countries torn apart by war.* The Conversation. https://theconversation.com/how-to-rebuild-higher-education-in-countries-torn-apart-by-war-48791?utm_source=chatgpt.com

Helle, L., Tynjälä, P., & Olkinuora, E. (2006). Project-based learning in post-secondary education: Theory, practice and rubber sling shots. *Higher Education*, *51*(2), 287–314. DOI: 10.1007/s10734-004-6386-5

Heller, M. (2024, September 13). Jackson R-2 School District closed Friday due to threat investigation. *KFVS12*. https://www.kfvs12.com/2024/09/13/jackson-r-2-school-district-closed-friday-threat-investigation/

Henderson, A. T., & Mapp, K. L. (2020). *A new wave of evidence: The impact of school, family, and community connections on student achievement revisited*. Educational Policy Institute.

Henderson, M., Alluvia, A., & Glazer, N. (2022). Sustaining teacher professional development through micro-credentials and communities of practice. *Teaching and Teacher Education*, *112*, 103615.

Hendrickx, M. M. H. G., Thurlings, M. C. G., & den Brok, P. (2025). Teachers' collaborative knowledge building in professional learning communities: Connecting interaction patterns to learning gains. *European Journal of Psychology of Education*, *40*(1), 39. Advance online publication. DOI: 10.1007/s10212-024-00938-y

Herrera, L. J. P., & Byndas, O. (2023). "You sway on the waves like a boat in the ocean": The effects of interrupted education on Ukrainian higher education refugee students in Poland. *Cogent Education, 10*(2), 2264009. Advance online publication. DOI: 10.1080/2331186X.2023.2264009

Hew, K., & Cheung, W. (2014). Students' and Instructors' Use of Massive Open Online Courses (MOOCs): Motivations and Challenges. *Educational Research Review, 12*, 45–58. DOI: 10.1016/j.edurev.2014.05.001

Hoachlander, G., & Hamilton, S. (2020). Career pathways and workforce development in modern education. *Education Policy Analysis Archives, 28*(75).

Hodges, C., Moore, S., Lockee, B., Trust, T., & Bond, A. (2020). T*he Difference Between Emergency Remote Teaching and Online Learning*. Educause Review. https://er.educause.edu/articles/2020/3/the-difference-between-emergency-remote-teaching-and-online-learning

Hodges, C., Moore, S., Lockee, B., Trust, T., & Bond, A. (2020). The difference between emergency remote teaching and online learning. *EDUCAUSE Review, 27*, •••. https://er.educause.edu/articles/2020/3/the-difference-between-emergency-remote-teaching-and-online-learning

Hoidn, S., & Reusser, K. (2020). Foundations of student-centered learning and teaching. In *The Routledge international handbook of student-centered learning and teaching in higher education* (pp. 17–46). Routledge. DOI: 10.4324/9780429259371-3

Holmes, W., Bialik, M., & Fadel, C. (2019). *Artificial intelligence in education: Promises and implications for teaching and learning*. Center for Curriculum Redesign.

Holmes, W., & Miao, F. (2023). *Guidance for generative AI in education and research.* UNESCO Publishing.

Home Office. (2015). *Prevent duty guidance for England and Wales: Guidance for specified authorities in England and Wales on the duty of schools and other bodies to have due regard to the need to prevent people from being drawn into terrorism.* UK Government. https://www.gov.uk/government/publications/prevent-duty-guidance

Hooda, M., Rana, C., Dahiya, O., Rizwan, A., & Hossain, M. S. (2022). Artificial intelligence for assessment and feedback to enhance student success in higher education. *Mathematical Problems in Engineering, 2022*, 5215722. Advance online publication. DOI: 10.1155/2022/5215722

Horn, M. B., & Staker, H. (2014). *Blended: Using Disruptive Innovation to Improve Schools* (1st ed.). Jossey-Bass.

Hosseini, Z., & Kinnunen, J. (2021). Integration of pedagogy into technology: A practical paradigm. In *Education and New Developments 2021*. DOI: 10.36315/2021end086

Howard, S., & Johnson, B. (2018). Resilient teachers: Resisting stress and burnout. *Social Psychology of Education*, *21*(5), 1253–1273.

Hsieh, C. (2016). The impact of performance-based assessments on student engagement and academic achievement. *Educational Assessment Journal*, *21*(3), 223–240.

Hsieh, H. F., Lee, D. B., Zimmerman, M. A., Pomerantz, N., Cunningham, M. C., Messman, E., Stoddard, S. A., Grodzinski, A. R., & Heinze, J. E. (2022). The effectiveness of the Say Something Anonymous Reporting System in preventing school violence: A cluster randomized control trial in 19 middle schools. *Journal of School Violence*, *21*(4), 413–428. DOI: 10.1080/15388220.2022.2105858

Huang, E. R., & Yu, C. J. (2024). *Harvard Scholars Without Borders Supports Displaced Ukrainian, Russian Scholars Amid War*. The Harvard Crimson. https://www.thecrimson.com/article/2024/2/28/program-displaced-scholars/

Huang, K., Ball, C., Francis, J., Ratan, R. A., Boumis, J. K., & Fordham, J. A. (2019). Augmented versus virtual reality in education: An exploratory study examining science knowledge retention when using augmented reality/virtual reality mobile applications. *Cyberpsychology, Behavior and Social Networking*, *22*(2), 105–110. DOI: 10.1089/cyber.2018.0150 PMID: 30657334

Huang, M., & Liu, T. F. (2016). Education Facing to Thinking of Heidegger and Its Theoretical Predicaments. *Educational Research*, *12*, 12–21.

Hung, R., Zhengmei, P., Kato, M., Nishihira, T., Okabe, M., Di, X., Kwak, D.-J., Hwang, K., Tschong, Y., Chien, C.-H., Peters, M. A., & Tesar, M. (2020). Philosophy of education in a new key: East Asia. *Educational Philosophy and Theory*. Advance online publication. DOI: 10.1080/00131857.2020.1772028

Ibrahim, H., Asim, R., Zaffar, F., Rahwan, T., & Zaki, Y. (2023). Rethinking homework in the age of artificial intelligence. *IEEE Intelligent Systems*, *38*(2), 24–27. DOI: 10.1109/MIS.2023.3255599

Ihde, D. (1990). *Technology and the Lifeworld*. Indiana University Press. DOI: 10.2979/3108.0

Ihde, D. (2003). A Phenomenology of Technics. In Kaplan, D. (Ed.), *Readings in the Philosophy of Technology* (pp. 137–160). Rowman & Littlefield Publisher.

Institute of Educational Analytics of the Ministry of Education and Science of Ukraine. (2022). *Higher and Professional Pre-higher Education under Martial Law*.

International Science Council. (2022). *Conference on the Ukraine Crisis: Responses from the European Higher Education and Research Sectors*. DOI: 10.24948/2022.04

Ismail, N. (2006). Inquiry-based learning: A new approach to classroom learning.

Ivanenko, N. (2024). Challenges of Ukrainian Higher Education in Times of War. *International Higher Education*. https://ihe.bc.edu/pub/i3oxjya7/release/2

Jabar, M., Kasilag, R., Collado, Z., & Jamoral, R. (2023). Family capital and parental involvement among parents in public elementary and secondary schools in the Philippines: Perspectives of parents and children. *Asia Pacific Journal of Education*, *43*(2), 555–571. DOI: 10.1080/02188791.2021.1944841

Jackson, B., Diliberti, M., Moore, P., & Schwartz, H. (2023). *Teachers' views on school safety: Consensus on many security measures, but stark division about arming teachers*. RAND Corporation. https://www.rand.org/pubs/research_reports/RRA2641-1.html

Jackson, A. Y., & Mazzei, L. A. (2012). *Thinking with theory in qualitative research*. Routledge Press.

Jackson, L., Alston, K., Bialystok, L., Blum, L., Burbules, N. C., Chinnery, A., Hansen, D. T., Hytten, K., Mayo, C., Norris, T., Stitzlein, S. M., Thompson, W. C., Waks, L., Peters, M. A., & Tesar, M. (2020a). Philosophy of education in a new key: Snapshot 2020 from the United States and Canada. *Educational Philosophy and Theory*. Advance online publication. DOI: 10.1080/00131857.2020.1821189

Jackson, Z. I. (2020b). *Becoming Human: Matter and meaning in an anti-black world*. Duke University Press.

Jandric, P., Jaldemark, J., Hurley, Z., Bartram, B., Matthews, A., Jopling, M., Manero, J., MacKenzie, M., Irwin, J., Rothmüller, N., Green, B., Ralston, S. J., Pyyhtinen, O., Hayes, S., Wright, J., Peters, M. A., & Tesar, M. (2021, December 06). Philosophy of education in a new key: Who remembers Greta Thunberg? Education and environment after the coronavirus. *Educational Philosophy and Theory*, *53*(14), 1421–1441. Advance online publication. DOI: 10.1080/00131857.2020.1811678

Januszewski, A., & Persichitte, K. (2008). A History of the AECT's Definitions of Educational Technology. In Januszewski, A., & Molenda, M. (Eds.), *Educational Technology: A Definition with Commentary* (pp. 259–282). Routledge.

Jaroszewicz, M., Shkoda, T., & Ovchynnikova, O. (2025). Migration Trajectories of Ukrainian Scholars Abroad: Forced Academic Mobility. *Population Space and Place*, *31*(2), e70011. Advance online publication. DOI: 10.1002/psp.70011

Jennings, P. A., Frank, J. L., Snowberg, K. E., Coccia, M. A., & Greenberg, M. T. (2020). Improving classroom learning environments by cultivating awareness and resilience in education (CARE): Results of a randomized controlled trial. *School Psychology Quarterly, 35*(4), 471–484. PMID: 24015983

Jennings, P. A., & Greenberg, M. T. (2009). The prosocial classroom: Teacher social and emotional competence in relation to student and classroom outcomes. *Review of Educational Research, 79*(1), 491–525. DOI: 10.3102/0034654308325693

Jerab, D., & Mabrouk, T. (2023). The role of leadership in changing organizational culture. Available at *SSRN* 4574324. DOI: 10.2139/ssrn.4574324

Jocano, F. L. (2000). *Filipino value system: A cultural definition*. Punlad Research House.

Johnson, M. (2024, December 4). Jackson, MO school district sued over year-long student suspension. *KFVS12*. https://www.kfvs12.com/2024/12/04/jackson-mo-school-district-sued-over-year-long-student-suspension/

Johnson, K., & Mims, C. (2018). Growth mindset in teacher professional development. *Educational Review, 70*(5), 624–638.

Johnson, L., & Wells, J. (2021). Global digital citizenship education: Developing skills for a connected world. *Journal of Global Education and Research, 5*(1), 22–38.

Johnson, L., & Yates, K. (2020). Blended learning simulations: Preparing educators for crisis management. *Educational Technology Research and Development, 68*(4), 2123–2141.

Johnston, K., & Berger, J. G. (2021). Guide to adaptive challenges and action learning.

Kahanec, M., Leu-Severynenko, S., Novosad, A., & Stadnyi, Y. (2022). Education reforms during and after the war. In Sologoub, I., di Mauro, B. W., & Gorodnichenko, Y. (Eds.), *Rebuilding Ukraine: Principles and policies* (1st ed.). CEPR Press., https://cepr.org/publications/books-and-reports/rebuilding-ukraine-principles-and-policies

Kai, S., Almeda, M. V., Baker, R. S., Heffernan, C., & Heffernan, N. (2018). Decision tree modeling of wheel spinning and productive persistence in skill builders. *Journal of Educational Data Mining, 10*(1), 789–822. DOI: 10.5281/zenodo.3344810

Kalonga, R., & Mwila, P. M. (2025). Effect of Class Size on Quality of Education in Public Secondary Schools in Temeke Municipality, Tanzania. [ARJI]. *Action Research Journal Indonesia, 7*(2), 251–277.

Kant, I. (1990). *Kant on Education*. Heath and Co.

Kaput, K. (2018). Evidence for Student-Centered Learning. Education evolving.

Karaman, M. A., Eşici, H., Tomar, İ. H., & Aliyev, R. (2021). COVID-19: Are school counseling services ready? Students' psychological symptoms, school counselors' views, and solutions. *Frontiers in Psychology*, *12*, 647740. DOI: 10.3389/fpsyg.2021.647740 PMID: 33868121

Kariippanon, K. E., Cliff, D. P., Lancaster, S. L., Okely, A. D., & Parrish, A. M. (2018). Perceived interplay between flexible learning spaces and teaching, learning and student wellbeing. *Learning Environments Research*, *21*(3), 301–320. DOI: 10.1007/s10984-017-9254-9

Kato, M., Saito, N., Matsushita, R., Ueno, M., Izawa, S., Maruyama, Y., Sugita, H., Ono, F., Muroi, R., Miyazaki, Y., Yamana, J., Peters, M. A., & Tesar, M. (2020). Philosophy of education in a new key: Voices from Japan. *Educational Philosophy and Theory*. Advance online publication. DOI: 10.1080/00131857.2020.1802819

Kayyali, M. (2022). Post-COVID-19 Pandemic Main Challenges with e-Learning.

Kayyali, M. (2024). The Impact of War on Higher Education: The Context of Wars in Syria, Iraq, Ukraine, and Afghanistan. In *Rebuilding Higher Education Systems Impacted by Crises: Navigating Traumatic Events, Disasters, and More* (1st ed., pp. 275–284). IGI Global. DOI: 10.4018/979-8-3693-1926-0.ch016

Kayyali, M. (2024a). Crisis Response and Higher Education: Social Innovations in Times of Global Challenges. In Social Innovations in Education, Environment, and Healthcare (pp. 1-17). IGI Global.

Kayyali, M. (Ed.). (2024b). Building Resiliency in Higher Education: Globalization, Digital Skills, and Student Wellness: Globalization, Digital Skills, and Student Wellness. IGI Global.

Kayyali, M. (Ed.). (2024b). Rebuilding Higher Education Systems Impacted by Crises: Navigating Traumatic Events, Disasters, and More: Navigating Traumatic Events, Disasters, and More. IGI Global.

Kayyali, M. (Ed.). (2024c). Building Resiliency in Higher Education: Globalization, Digital Skills, and Student Wellness: Globalization, Digital Skills, and Student Wellness. IGI Global.

Kayyali, M. (2020). Post COVID-19: New era for higher education systems. *International Journal of Applied Science and Engineering*, *8*(2), 131–145. DOI: 10.30954/2322-0465.2.2020.6

Kayyali, M. (2023). *Virtual Universities: An Overview & Trends*. Online Submission.

Kayyali, M. (2025). A). The Impact of Personalized Learning on Student Engagement and Achievement in STEAM. In *Integrating Personalized Learning Methods Into STEAM Education* (pp. 51–78). IGI Global Scientific Publishing. DOI: 10.4018/979-8-3693-7718-5.ch003

Kayyali, M. (2025). B). Leadership's Role in Fostering a Resilient Organizational Culture. In *Enhancing Resilience in Business Continuity Management* (pp. 267–294). IGI Global Scientific Publishing. DOI: 10.4018/979-8-3693-8809-9.ch009

Kellogg, D., & Van Maanen, K. (2020). Enhancing family-school collaboration: The role of community-based interventions in rural schools. *Rural Education Journal, 39*(1), 98–115. DOI: 10.1080/09120201.2020.174654

Kelly, J. M. (2025). The Glitch Is the Machine: Technological Mediation and Identity in The Collected Works of Rowan Black (Doctoral dissertation, Drew University).

Kem, D. (2022). Personalised and adaptive learning: Emerging learning platforms in the era of digital and smart learning. *International Journal of Social Science and Human Research, 5*(2). Advance online publication. DOI: 10.47191/ijsshr/v5-i2-02

Kichuk, Y., Biliuk, O., Kapinus, O., Fedenko, S., & Livak, P. (2024). Innovation in Higher Education in Wartime: Using Modern Technologies. *Academia (Caracas), 35–36*, 178–199. DOI: 10.26220/aca.5008

Kimmons, R., & Hall, C. (2020). Technology integration in education: A critical examination of TPACK in teacher preparation. *Journal of Digital Learning in Teacher Education, 36*(2), 78–90.

Kingsbury, I. (2021). Online learning: How do brick and mortar schools stack up to virtual schools? *Education and Information Technologies, 26*(6), 6567–6588. DOI: 10.1007/s10639-021-10450-1 PMID: 33551662

Kirkwood, A., & Price, L. (2014). Technology-Enhanced Learning and Teaching in Higher Education: What is 'Enhanced' and How Do We Know? A Critical Literature Review. *Learning, Media and Technology, 39*(1), 6–36. DOI: 10.1080/17439884.2013.770404

Kirkwood, A., & Price, L. (2020). Technology-enhanced learning and teaching in higher education: What is 'enhanced' and how do we know? A critical literature review. *Learning, Media and Technology, 45*(2), 145–163.

Kluch, Y., Taylor, E. A., Wright-Mair, R., & Boring, D. (2022). "I felt like... they left me to fend for myself": A mixed-method examination of sense of belonging among minoritized groups in sport academic programs. *Sport Management Education Journal, 17*(1), 1–16. DOI: 10.1123/smej.2021-0014

Kode, S., & Nori, K. V. (2016). Enhancing IT education: Education technology for teacher training. In *Proceedings of the International Conference on Learning and Teaching in Computing and Engineering (LaTICE)* (pp. 94–98). DOI: 10.1109/LaTiCE.2016.4

Koh, E., & Doroudi, S. (2023). Learning, teaching, and assessment with generative artificial intelligence: Towards a plateau of productivity. *Learning: Research and Practice*, *9*(2), 109–116. DOI: 10.1080/23735082.2023.2264086

Koh, J. H. L., Chai, C. S., & Wong, B. (2019). Teachers' 21st century competencies and readiness for educational technologies: A Singapore study. *Educational Technology Research and Development*, *67*, 293–310.

Kohler, K. (2021). *Strategic foresight: Knowledge, tools, and methods for the future*. CSS Risk and Resilience Reports.

Konishi, C., Wong, T. K., Persram, R. J., Vargas-Madriz, L. F., & Liu, X. (2022). Reconstructing the concept of school climate. *Educational Research*, *64*(2), 159–175. DOI: 10.1080/00131881.2022.2056495

Konopka, C. L., Adaime, M. B., & Mosele, P. H. (2015). Active teaching and learning methodologies: Some considerations. *Creative Education*, *6*(14), 1536–1545. DOI: 10.4236/ce.2015.614154

Kori, K. (2021). Inquiry-based learning in higher education. In *Technology Supported Active Learning*. DOI: 10.1007/978-981-16-2082-9_4

Kostikova, I., Holubnycha, L., Sytnykova, Y., Shchokina, T., Soroka, N., & Tarasova, S. (2023). University Teachers' Psychological State During the Stress of Online Education After Pandemic in Wartime in Ukraine. *International Journal of Emerging Technologies in Learning*, *18*(13), 152–162. DOI: 10.3991/ijet.v18i13.39825

Kouzes, J. M., & Posner, B. Z. (2024). *The student leadership challenge: Five practices for becoming an exemplary leader*. John Wiley & Sons.

Kremser, J. (n.d.). *Do metal detectors improve school security?* Forensic Security & Protection. https://forensicsp.com/2023/07/27/do-metal-detectors-improve-school-security/

Kreso, A. P. (2008). The war and post-war impact on the educational system of Bosnia and Herzegovina. *International Review of Education*, *54*(3–4), 353–374. DOI: 10.1007/s11159-008-9087-y

KRON4. (n.d.). Teachers shot execution-style with rubber bullets during active shooter drill. *KRON4*. https://www.kron4.com/news/teachers-shot-execution-style-with-rubber-bullets-during-active-shooter-drill/

Kryshtanovych, M., Akimova, L., Akimov, O., Parkhomenko-Kutsevil, O., & Omarov, A. (2022). Features of creative burnout among educational workers in public administration system. *Creativity Studies*, *15*(1), 116–129. DOI: 10.3846/cs.2022.15145

Kuntz, J. C. (2021). Resilience in times of global pandemic: Steering recovery and thriving trajectories. *Applied Psychology = Psychologie Appliquée*, *70*(1), 188–215. DOI: 10.1111/apps.12296 PMID: 33362330

Kurapov, A., Pavlenko, V., Drozdov, A., Bezliudna, V., Reznik, A., & Isralowitz, R. (2023). Toward an Understanding of the Russian-Ukrainian War Impact on University Students and Personnel. *Journal of Loss and Trauma*, *28*(2), 167–174. DOI: 10.1080/15325024.2022.2084838

Kusumawati, R., & Saputra, A. D. (2023, August). FASCA Leadership: Alternative Leadership Styles in an Era of Change. In International Conference on Intelligent Networking and Collaborative Systems (pp. 295-305). Cham: Springer Nature Switzerland. DOI: 10.1007/978-3-031-40971-4_28

Kutsyuruba, B. (2011). Education in the period of post-Soviet transition in Ukraine. *Demokratizatsiya (Washington, D.C.)*, *19*(3), 287–310.

Kyiv School of Economics. (2022). *Report on the total amount of damage caused to the infrastructure of Ukraine*. https://kse.ua/about-the-school/news/as-of-november-2022-the-total-amount-of-losses-caused-to-the-infrastructure-of-ukraine-increased-to-almost-136-billion/

Lachlan, K., Spence, P., Lin, X., Najarian, K., & Del Greco, M. (2016). Social media and crisis management: CERC, search strategies, and Twitter content. *Computers in Human Behavior*, *54*, 647–652. DOI: 10.1016/j.chb.2015.05.027

Ladson-Billings, G. (1995). Toward a theory of culturally relevant pedagogy. *American Educational Research Journal*, *32*(3), 465–491. DOI: 10.3102/00028312032003465

Ladson-Billings, G. (2021). Culturally relevant pedagogy: Asking a different question. *Teachers College Record*, *123*(1), 101–120.

Ladson-Billings, G. (2021). I'm here for the hard re-set: Post pandemic pedagogy to preserve our culture. *Equity & Excellence in Education*, *54*(1), 68–78. DOI: 10.1080/10665684.2020.1863883

Ladyman, J. (2007). Ontological, Epistemological, and Methodological Positions. In Kuipers, T. A. F. (Ed.), *General Philosophy of Science* (pp. 303–376). North Holland. DOI: 10.1016/B978-044451548-3/50008-2

Lahti, H., Kokkonen, M., Hietajärvi, L., Lyyra, N., & Paakkari, L. (2024). Social media threats and health among adolescents: Evidence from the Health Behaviour in School-aged Children study. *Child and Adolescent Psychiatry and Mental Health*, *18*(1), 62. DOI: 10.1186/s13034-024-00754-8 PMID: 38812043

Langendahl, P., Cook, M., & Mark-Herbert, C. (2016). Gamification in higher education.

Larmer, J., & Mergendoller, J. R. (2019). *Gold standard PBL: Essential project design elements*. Buck Institute for Education.

Larrivee, B. (2000). Transforming teaching practice: Becoming the critically reflective teacher. *Reflective Practice*, *1*(3), 293–307. DOI: 10.1080/713693162

Latchem, C. (2014). BJET Editorial: Opening up the Educational Technology Research Agenda. *British Journal of Educational Technology*, *41*(1), 3–11. DOI: 10.1111/bjet.12122

Latour, B. (2004). Why Has Critique Run Out of Steam? From Matters of Fact to Matters of Concern. *Critical Inquiry*, *30*(2), 225–248. DOI: 10.1086/421123

Latour, B. (2005). *Reassembling the Social*. Oxford University Press. DOI: 10.1093/oso/9780199256044.001.0001

Lavrysh, Y., Lytovchenko, I., Lukianenko, V., & Golub, T. (2022). Teaching during the wartime: Experience from Ukraine. *Educational Philosophy and Theory*, •••, 1–8. DOI: 10.1080/00131857.2022.2098714

Lee, J., & Soylu, M. Y. (2023). *ChatGPT and assessment in higher education*. Center for 21st Century Universities. https://www.c21u.gatech.edu/sites/default/files/publication/2023/03/C21U%20ChatGPT%20White%20Paper_ Final.pdf

Lee, A., Chen, Y., & Wang, H. (2021). Growth mindset and adaptive teaching in the 21st century classroom. *Teaching and Teacher Education*, *97*, 103217.

Lee, C., & Green, M. (2023). Culturally responsive teaching in diverse classrooms: New directions. *Journal of Education*, *21*(2), 78–94.

Leithwood, K., Harris, A., & Hopkins, D. (2020). Seven strong claims about successful school leadership revisited. *School Leadership & Management*, *40*(1), 5–22. DOI: 10.1080/13632434.2019.1596077

Leithwood, K., Harris, A., & Strauss, T. (2020). *Leading school turnaround: How successful leaders transform low-performing schools*. Jossey-Bass.

Leithwood, K., & Louis, K. S. (Eds.). (2012). *Linking leadership to student learning*. Jossey-Bass.

Lemov, D. (2020). *Teaching in the online classroom: Surviving and thriving in the new normal*. John Wiley & Sons.

Lewin, D., & Lundie, D. (2016). Philosophies of digital pedagogy. *Studies in Philosophy and Education*, *3*(35), 235–240. DOI: 10.1007/s11217-016-9514-7

Lewis, M., Thompson, G., & Rodriguez, A. (2024). Innovation and resilience in uncertain educational landscapes: The role of teacher collaboration. *International Journal of Educational Research*, *115*, 102097.

Liboon, C. A., Gutierrez, R. A. E., & Dimagiba, A. G. (2024). Reciprocity in Research and Evaluation: Conceptualizing Utang Na Loob, Pakikipagkapwa, and Alalay as Filipina American Educational Researchers. In *Theories Bridging Ethnography and Evaluation: Making Transformative, Intersectional, and Comparative Connections* (pp. 105-132). Emerald Group Publishing Limited.

Lieberman, A., & Miller, L. (2019). Teacher leadership as a lever for change: Promoting professional collaboration and student success. *Educational Leadership*, *76*(8), 36–42.

Lim, C., Tan, C., & Chee, Y. (2021). Smart classrooms and student learning: A Singapore case study. *International Journal of Educational Technology in Higher Education*, *18*(1), 50. PMID: 34778537

Lim, R. P. S. (2024). Filipino values of secondary school heads on school-based management: Basis for values enhancement program. *International Journal of Management* [IJMTS]. *Technology and Social Sciences*, *9*(1), 115–132.

Lindgren, R., Tscholl, M., Wang, S., & Johnson, E. (2016). Enhancing Learning and Engagement Through Embodied Interaction Within a Mixed Reality Simulation. *Computers & Education*, *95*, 174–187. DOI: 10.1016/j.compedu.2016.01.001

Lingard, B., Wyatt-Smith, C., & Heck, E. (2021). Transforming schooling through digital disruption: Big data, policy, teaching, and assessment 1. In *Digital disruption in teaching and testing* (pp. 1–44). Routledge.

Liu, S. X. (2022). How war-related deprivation affects political participation: Evidence from education loss in Liberia. *Journal of Peace Research*, *59*(3), 353–366. DOI: 10.1177/00223433211019460

London, M. (2023). Causes and consequences of adaptive leadership: A model of leaders' rapid responses to unexpected events. Psychology of leaders and leadership, 26(1), 22.

Lower-Hoppe, L. M., Brgoch, S. M., Chen, Y., & Sutherland, S. (2021). Inquiry-based learning in action.

Macias, O. (n.d.). *Combating social media threat hoaxes: A unified approach*. Association of California School Administrators. https://content.acsa.org/combating-social-media-threat-hoaxes-a-unified-approach/

Madrid, B. J., Lopez, G. D., Dans, L. F., Fry, D. A., Duka-Pante, F. G. H., & Muyot, A. T. (2020). Safe schools for teens: Preventing sexual abuse of urban poor teens, proof-of-concept study-Improving teachers' and students' knowledge, skills and attitudes. *Heliyon*, *6*(6), e04080. DOI: 10.1016/j.heliyon.2020.e04080 PMID: 32566775

Maiev, A. (2023). Education during the war: challenges for Ukraine. *Teoria Şi Practica Administrării Publice*, 415–420.

Malbas, M., Kilag, O. K., Diano, F., Jr., Tiongzon, B., Catacutan, A., & Abendan, C. F. (2023). In Retrospect and Prospect: An Analysis of the Philippine Educational System and the Impact of K-12 Implementation. *Excellencia: International Multi-disciplinary Journal of Education (2994-9521)*, *1*(4), 283-294.

Malik, G., Tayal, D. K., & Vij, S. (2018). An analysis of the role of artificial intelligence in education and teaching. In *Advances in Intelligent Systems and Computing*.

Malkus, N. (2020). *The evolution of career and technical education in U.S. high schools*. American Enterprise Institute.

Manla, V. H. (2021). School climate: Its impact on teachers' commitment and school performance. *Journal of World Englishes and Educational Practices*, *3*(2), 21–35. DOI: 10.32996/jweep.2021.3.2.3

Mann, J., & Bowen, G. (2021). Emphasizing the impact of service-learning and community engagement partnerships amid challenges facing higher education. *The International Journal of Research on Service-Learning and Community Engagement*, *9*(1). Advance online publication. DOI: 10.37333/001c.31379

Marker, M. (2018). There is no place of nature; there is only the nature of place: Animate landscapes as methodology for inquiry in the Coast Salish territory. *International Journal of Qualitative Studies in Education : QSE*, *31*(6), 453–464. DOI: 10.1080/09518398.2018.1430391

Marrero, A. (2024). In the Crosshairs: Navigating Political Pressures as a New Leader. In *Pursuing Equity and Success for Marginalized Educational Leaders* (pp. 101-117). IGI Global.

Marston, J., & Tsolakis, M. (2022). *Ukraine Points Up the Threat to Education During War*. ReliefWeb. https://reliefweb.int/report/world/ukraine-points-threat-education-during-war

Martínez-Comesaña, M., Rigueira-Díaz, X., Larrañaga-Janeiro, A., Martínez-Torres, J., Ocarranza-Prado, I., & Kreibel, D. (2023). Impact of artificial intelligence on assessment methods in primary and secondary education: Systematic literature review. *Revista de Psicodidáctica* (English Ed.), 28(2), 93–103. DOI: 10.1016/j.psicoe.2023.06.002

Martinez, L., Brown, T., & Kim, S. (2023). Trauma-informed approaches in teacher training: Simulation as a preparation tool. *Journal of School Psychology*, *95*, 1–15.

Martin, M. A., & Kemple, K. M. (2020). Building capacity for school-family partnerships: A longitudinal study of an urban district's implementation of a parent involvement program. *Journal of Educational Administration*, *58*(4), 457–475. DOI: 10.1108/JEA-10-2018-0197

Marx, K. (1964). *Economic and Philosophic Manuscripts of 1844*. International Publishers.

Marx, K. (1975). *Economic and Philosophical Manuscripts of 1844* (Livingstone, R., & Benton, G., Trans.). Penguin.

Maslach, C., & Leiter, M. P. (2016). Burnout. In Fink, G. (Ed.), *Stress: Concepts, cognition, emotion, and behavior* (pp. 351–357). Academic Press.

Mason, M. (2008). Complexity Theory and the Philosophy of Education. *Educational Philosophy and Theory*, *40*(1), 4–18. DOI: 10.1111/j.1469-5812.2007.00412.x

Mattessich, P. W., & Johnson, K. M. (2018). Collaboration: What makes it work.

Mayes, T. (1995). Learning Technology and Groundhog day. In Strang, W., Simpson, V. B., & Slater, D. (Eds.), *Hypermedia at Work: Practice and Theory in Higher Education* (pp. 21–37). University of Kent Press.

Mazer, J., Thompson, B., Cherry, J., Russell, M., Payne, H., Kirby, E., & Pfohl, W. (2015). Communication in the face of a school crisis: Examining the volume and content of social media mentions during active shooter incidents. *Computers in Human Behavior*, *53*, 238–248. DOI: 10.1016/j.chb.2015.06.040

McMillian, R. (2023). Using critical discourse analysis to operationalize discursive violence in school closure education reform policy. *International Journal of Qualitative Studies in Education : QSE*, •••, 1–24. DOI: 10.1080/09518398.2023.2258099

Means, B., & Neisler, J. with Langer Research Associates. (2020). *Suddenly Online: A National Survey of Undergraduates During the COVID-19 Pandemic*. San Mateo, CA: Digital Promise.

Medina-Garrido, J. A., Biedma-Ferrer, J. M., & Rodríguez-Cornejo, M. V. (2021). I quit! Effects of work-family policies on the turnover intention. *International Journal of Environmental Research and Public Health, 18*(4), 1893. DOI: 10.3390/ijerph18041893

Messiou, K. (2012). *Confronting marginalisation in education: A framework for promoting inclusion.*

Metzler, L., Lazaros, E., Davison, C., & Truell, A. (2023). Social media challenges are causing disruptions in schools. *Issues in Information Systems, 24*(1), 84–92. DOI: 10.48009/1_iis_2023_107

Meyers, E. M., & Lieberman, A. (2021). Teacher voice and agency in education reform: Strategies for empowerment. *International Journal of Leadership in Education, 24*(5), 605–622.

Miller, J. (2024). Perspective: Bridging the digital divide: The crucial partnership between law enforcement and cybersecurity professionals. *Homeland Security Today.US*. https://www.hstoday.us/featured/perspective-we-need-a-new-information-sharing-model-for-cyber-defense/

Miller, G. E., & Ives, K. S. (Eds.). (2023). *Leading the eLearning transformation of higher education: Leadership strategies for the next generation*. Taylor & Francis.

Miller, M. A., & Johnson, R. M. (2021). Fostering family-school partnerships: Effects of teacher communication practices on student achievement in early elementary grades. *School Community Journal, 31*(2), 75–95. DOI: 10.21747/0903-1068/31.2.75

Milton, S. A. (2013). *The Neglected Pillar of Recovery: A study of Higher Education in Post-War Iraq and Libya*. University of York.

Ministry of Education and Science of Ukraine. (2022a). *Letter No. 1/3417-22: On the Conclusion of the Academic Year*. Government Portal. https://osvita.ua/legislation/Vishya_osvita/85933/

Ministry of Education and Science of Ukraine. (2022b). *Serhiy Shkarlet: All educational institutions are recommended to halt the educational process and declare a vacation for two weeks.* Government Portal. https://www.kmu.gov.ua/news/rozyasnennya-mon-shchodo-roboti-zakladiv-osviti-u-mezhah-pravovogo-rezhimu-voyennogo-stanu

Minocha, S., & Reeves, A. (2010). Design of Learning Spaces in 3D Virtual Worlds: An Empirical Investigation of Second Life. *Learning, Media and Technology*, *35*(2), 111–137. DOI: 10.1080/17439884.2010.494419

Mitcham, C. (2006). From Phenomenology to Pragmatism: Using Technology as an Instrument. In Selinger, E. (Ed.), *Postphenomenology: A Critical Companion to Ihde* (pp. 21–33). State University of New York Press.

Mitchell, M. (2022). Building resilient organizations: Toward joy and durable power in a time of crisis. *Convergence.*

Monsillion, J., Zebdi, R., & Romo-Desprez, L. (2023). School mindfulness-based interventions for youth, and considerations for anxiety, depression, and a positive school climate—A systematic literature review. *Children (Basel, Switzerland)*, *10*(5), 861. DOI: 10.3390/children10050861 PMID: 37238409

Montag, C., Duke, É., & Markowetz, A. (2021). Toward psycho-informatics: The study of online behavior as a window to psychological phenomena. *Current Opinion in Behavioral Sciences*, *18*, 1–8.

Moore, P., Jackson, B., Leschitz, J., Wolters, N., Goode, T., Diliberti, M., & Pham, P. (2024). *Developing practical responses to social media threats against K–12 schools: An overview of trends, challenges, and current approaches.* RAND Corporation. https://www.rand.org/pubs/research_reports/RRA1077-5.html

Moqbel, M. S. S., & Al-Kadi, A. M. T. (2023). Foreign language learning assessment in the age of ChatGPT: A theoretical account. *Journal of English Studies in Arabia Felix*, *2*(1), 76–88. DOI: 10.56540/jesaf.v2i1.62

Morrice, L. (2022). Will the war in Ukraine be a pivotal moment for refugee education in Europe? *International Journal of Lifelong Education*, *41*(3), 251–256. DOI: 10.1080/02601370.2022.2079260

Mpungose, C. B. (2020). Emergent transition from face-to-face to online learning in a South African University in the context of the Coronavirus pandemic. *Humanities & Social Sciences Communications*, *7*(1), 113. Advance online publication. DOI: 10.1057/s41599-020-00603-x

Murgatroyd, S. (2023). Rethinking teaching in the age of artificial intelligence. *Revista Paraguaya De Educación A Distancia* (REPED), 4(2), 4–10. DOI: 10.56152/reped2023-vol4num2-art1

Murray, M. D. (2023). Generative AI art: Copyright infringement and fair use. *SMU Science and Technology Law Review, 26*(2), 259. DOI: 10.25172/smustlr.26.2.4

Mwamwenda, T., & Mokoena, A. (2019). Environmental education and sustainability: Impact of student projects on water conservation in South Africa. *International Journal of Environmental and Science Education, 14*(9), 567–579.

National Child Traumatic Stress Network. (2019). *Creating school active shooter/intruder drills: A trauma-informed approach.* https://www.nctsn.org/resources/creating-school-active-shooter-intruder-drills

New York State School Boards Association. (2019, December 3). *Armed security in schools: A report on current trends and policy recommendations.* https://www.nyssba.org/clientuploads/nyssba_pdf/Reports/armed-security-report-12032019.pdf

Ng, C. F., & Goh, H. K. (2023). Artificial intelligence in education: Smart classrooms and pedagogical integration. *Computers & Education, 181*, 104519.

Nicholson, J., & Kurucz, E. (2019). Relational leadership for sustainability: Building an ethical framework from the moral theory of 'ethics of care'. *Journal of Business Ethics, 156*(1), 25–43. DOI: 10.1007/s10551-017-3593-4

Niemi, A. M. (2007). What are effective strategies to support student engagement and learning.

Nieto-Taborda, M. L., & Luppicini, R. (2024). Accelerated Digital Transformation of Higher Education in the Wake of COVID-19: A Systematic Literature Review. *International Journal of Changes in Education.* https://doi.org/https://doi.org/10.47852/bonviewIJCE42023125

Nikolaiev, Y., Riy, G., & Shemelynets, I. (2023). *War in Ukraine: Reshaping the Higher Education Sector - Analytical Report.* file:///C:/Users/pfidalgo/OneDrive - Emirates College for Advanced Education/Desktop/Readings/War in Ukraine_Report.pdf

Nikolopoulou, K. (2024a). Generative artificial intelligence in higher education: Exploring ways of harnessing pedagogical practices with the assistance of ChatGPT. *International Journal of Changes in Education, 1*(2), 103–111. DOI: 10.47852/bonviewIJCE42022489

Nilson, L. B., & Zimmerman, B. J. (2023). *Creating self-regulated learners: Strategies to strengthen students' self-awareness and learning skills*. Routledge. DOI: 10.4324/9781003443803

Nolan, K., & O'Higgins, N. (2020). Educating for global citizenship: Reflections on theory and practice. *Journal of Global Ethics*, *16*(1), 1–17.

Norrish, J., & Brunzell, T. (2023). How is trauma-informed education implemented within classrooms? A synthesis of trauma-informed education programs. *The Australian Journal of Teacher Education*, *48*(3). Advance online publication. DOI: 10.14221/1835-517X.6159

Northouse, P. G. (2022). *Leadership: Theory and practice* (9th ed.). Sage Publications.

Ntuli, E. (2019). *Active learning strategies in technology integrated K-12 classrooms*. Early Childhood Development. DOI: 10.4018/978-1-5225-7507-8.ch035

O'Malley, P., Rainford, J., & Thompson, A. (2009). Transparency during public health emergencies: From rhetoric to reality. *Bulletin of the World Health Organization*, *87*(8), 614–618. DOI: 10.2471/BLT.08.056689 PMID: 19705012

OECD. (2019). *Trends shaping education 2019*. OECD Publishing.

OECD. (2020). *The impact of COVID-19 on education: Insights from Education at a Glance 2020.* Paris: OECD Publishing. https://web-archive.oecd.org/2020-09-08/562941-the-impact-of-covid-19-on-education-insights-education-at-a-glance-2020.pdf

OECD. (2021). *The state of school education: One year into the COVID pandemic*. OECD Publishing.

OECD. (2025). *Transforming General Upper Secondary Education in Ukraine*. https://doi.org/DOI: 978-92-64-96427-3 (PDF)

Ogor, E. N. (2007). Student academic performance monitoring and evaluation using data mining techniques. In *Electronics, Robotics and Automotive Mechanics Conference (CERMA 2007)* (pp. 354–359). DOI: 10.1109/CERMA.2007.4367712

Oigara, J. N. (2013). Integrating technology in teacher education programs.

Olawoyin, R. (2018). Emotional intelligence: Assessing its importance in safety leadership. *Professional Safety*, *63*(08), 41–47.

Oliver, M. (2011). Technological Determinism in Educational Technology Research: Some Alternative Ways of Thinking about the Relationship Between Learning and Technology. *Journal of Computer Assisted Learning, 27*(5), 373–384. DOI: 10.1111/j.1365-2729.2011.00406.x

Oliver, M. (2013). Learning Technology: Theorising the Tools We Study. *British Journal of Educational Technology, 44*(1), 31–43. DOI: 10.1111/j.1467-8535.2011.01283.x

Oliver, M. (2015). From Openness to Permeability: Reframing Open Education in Terms of Positive Liberty in the Enactment of Academic Practices. *Learning, Media and Technology, 40*(3), 365–384. DOI: 10.1080/17439884.2015.1029940

Oliver, M., & Conole, G. (2003). Evidence-Based Practice and e-Learning in Higher Education: Can We and Should We? *Research Papers in Education, 18*(4), 385–397. DOI: 10.1080/0267152032000176873

Ontario Ministry of Education. (2009). *Equity and Inclusive Education Strategy*. Government of Ontario.

Ontario Ministry of Education. (2018). *Safe and accepting schools: Policy and program requirements.* Government of Ontario.http://www.edu.gov.on.ca/eng/safeschools/

Orchard, J., Gaydon, P., Williams, K., Bennett, P., D'Olimpio, L., C¸elik, R., Shah, Q., Neusiedl, C., Suissa, J., Peters, M. A., & Tesar, M. (2020). Philosophy of education in a new key: A 'covid collective' of the philosophy of education society of Great Britain (PESGB). *Educational Philosophy and Theory*. Advance online publication. DOI: 10.1080/00131857.2020.1838274

Ouyang, F., Zheng, L., & Jiao, P. (2022). Artificial intelligence in online higher education: A systematic review of empirical research from 2011 to 2020. *Education and Information Technologies, 27*(6), 7893–7925. DOI: 10.1007/s10639-022-10925-9

Overstreet, S., & Chafouleas, S. M. (2016). Trauma-informed schools: Introduction to the special issue. *School Mental Health, 8*(1), 1–6. DOI: 10.1007/s12310-016-9184-1

Oxley, L., & Morris, P. (2020). Global citizenship: A typology for distinguishing its multiple conceptions. *British Journal of Educational Studies, 68*(2), 159–182.

Papastephanou, M., Zembylas, M., Bostad, I., Benhur Oral, S., Drousioti, K., Kouppanou, A., Strand, T., Wain, K., & Peters, M. A. (2022, July 03). (Open Reviewer) & Tesar, M. (Open Reviewer) (2020). Philosophy of education in a new key: Education for justice now. *Educational Philosophy and Theory, 54*(8), 1083–1098. Advance online publication. DOI: 10.1080/00131857.2020.1793539

Paredes, C. S., & Medina, S. A. (2016). The role of family and community in supporting student success in schools with high levels of cultural diversity. *Education Policy Analysis Archives*, *24*(79), 1–21. DOI: 10.14507/epaa.24.2659

Patil, S., & Dharwadkar, D. N. (2020). Improving students' engagement through active learning strategies: Case study based active review sessions and skillathon. *Journal of Engineering Education Transformations*.

Pearce, J., & Chiavaroli, N. (2023). Rethinking assessment in response to generative artificial intelligence. *Medical Education*, *57*(10), 889–891. DOI: 10.1111/medu.15092

Pedaste, M., Mäeots, M., Siiman, L., Jong, T., Riesen, S., Kamp, E., Manoli, C., Zacharia, Z., & Tsourlidaki, E. (2015). Phases of Inquiry-Based Learning: Definitions and the Inquiry Cycle. *Educational Research Review*, *14*(2), 47–61. DOI: 10.1016/j.edurev.2015.02.003

Pera, A. (2019). Towards effective workforce management: Hiring algorithms, big data-driven accountability systems, and organizational performance. *Psychosociological Issues in Human Resource Management*, *7*(2), 19–24. DOI: 10.22381/PIHRM7220193

Perkins, M. (2023). Academic integrity considerations of AI large language models in the post-pandemic era: ChatGPT and beyond. *Journal of University Teaching & Learning Practice*, *20*(2), 345–367. DOI: 10.53761/1.20.02.07

Perrotta, C., & Evans, M. A. (2013). Orchestration, Power, and Educational Technology: A Response to Dillenbourg. *Computers & Education*, *69*, 520–522. DOI: 10.1016/j.compedu.2013.04.007

Peters, A. (2010). Elements of successful mentoring of a female school leader. *Leadership and Policy in Schools*, *9*(1), 108–129. DOI: 10.1080/15700760903026755

Peters, M. A., Arndt, S., Tesar, M., Jackson, L., Hung, R., Mika, C., Ozolins, J. T., Teschers, C., Orchard, J., Buchanan, R., Madjar, A., Novak, R., Besley, T., & Sturm, S. (2020). Philosophy of education in a new key. A collective project of the PESA executive. *Educational Philosophy and Theory*. Advance online publication. DOI: 10.1080/00131857.2020.1759194

Peterson, J., Densley, J., & Erickson, G. (2021). Presence of armed school officials and fatal and nonfatal gunshot injuries during mass school shootings, United States, 1980–2019. *JAMA Network Open*, *4*(2), e2037394. DOI: 10.1001/jamanetworkopen.2020.37394 PMID: 33591364

Peterson, J., Densley, J., Reidman, D., Spaulding, J., & Malicky, H. (2024). An exploration of K-12 school shooting threats in the United States. *Journal of Threat Assessment and Management, 11*(2), 106–120. DOI: 10.1037/tam0000215

Peterson, J., Densley, J., Spaulding, J., & Higgins, S. (2023). How mass public shooters use social media: Exploring themes and future directions. *Social Media + Society, 9*(1), 20563051231155101. Advance online publication. DOI: 10.1177/20563051231155101

Pezze, B. D. (2006). Heidegger on Gelassenheit. *Minerva, 10*, •••. Retrieved March 15, 2020, from https://www.minerva.mic.ul.ie/vol10/Heidegger.html

Philippine Information Agency. (2024, February 26). *Sorsogon schools train peer counselors to aid students facing mental stress*. https://mirror.pia.gov.ph/news/2024/02/26/sorsogon-schools-train-peer-counselors-to-aid-students-facing-mental-stress

Philippines, U. N. I. C. E. F. (2021). *The state of Filipino children's mental health*. https://www.unicef.org/philippines/reports/state-filipino-childrens-mental-health

Poirier, T. (2012). The effects of armed conflict on schooling in Sub-Saharan Africa. *International Journal of Educational Development, 32*(2), 341–351. https://doi.org/https://doi.org/10.1016/j.ijedudev.2011.06.001. DOI: 10.1016/j.ijedudev.2011.06.001

Popescu, E. (2010). Adaptation provisioning with respect to learning styles in a web-based educational system: An experimental study. *Journal of Computer Assisted Learning, 26*(4), 243–257. DOI: 10.1111/j.1365-2729.2010.00364.x

Pranjić, S. S. (2021). Development of a caring teacher-student realationship in higher education. *The Journal of Education, Culture, and Society, 12*(1), 151–163. DOI: 10.15503/jecs2021.1.151.163

Price, H., & Khubchandani, J. (2019). School firearm violence prevention practices and policies: Functional or folly? *Violence and Gender, 6*(3), 154–167. DOI: 10.1089/vio.2018.0044

Priestley, M., Biesta, G., & Robinson, S. (2020). *Teacher agency: An ecological approach*. Bloomsbury Academic.

Priestley, M., & Minty, S. (2013). Curriculum for excellence and teachers' work: A Scottish perspective. *Journal of Curriculum Studies, 45*(6), 779–801.

Puentedura, R. R. (2009). Transformation, technology, and education.

Quereshi, A. (n.d.). *Locked out of the classroom: How implicit bias contributes to disparities in school discipline*. LDF Thurgood Marshall Institute. https://dignityinschools.org/wp-content/uploads/2017/12/Bias_Reportv2017_30_11_FINAL.pdf

Radfar, A., Ferreira, M. M., Sosa, J. P., & Filip, I. (2021). Emergent crisis of COVID-19 pandemic: Mental health challenges and opportunities. *Frontiers in Psychiatry*, *12*, 631008. DOI: 10.3389/fpsyt.2021.631008 PMID: 34349675

Radianti, J., Majchrzak, T. A., Fromm, J., & Wohlgenannt, I. (2020). A systematic review of immersive virtual reality applications for higher education: Design elements, lessons learned, and research agenda. *Computers & Education*, *147*, 103778. DOI: 10.1016/j.compedu.2019.103778

Rao, K., Okolo, C., & Thurlow, M. L. (2020). Impact of Universal Design for Learning on students with disabilities: A meta-analysis. *Journal of Special Education Technology*, *35*(1), 3–15. PMID: 38911489

Reeves, D. (2019). *Leading change in your school: How to conquer myths, build commitment, and get results* (3rd ed.). ASCD.

Regan, P. M., & Jesse, J. (2019). Ethical challenges of ed-tech, big data and personalized learning: Twenty-first century student sorting and tracking. *Ethics and Information Technology*, *21*(2), 85–99.

Regan, P. M., & Jesse, J. (2019). Ethical challenges of edtech, big data and personalized learning: Twenty-first century student sorting and tracking. *Ethics and Information Technology*, *21*(3), 167–179. DOI: 10.1007/s10676-018-9492-2

Reich, K., Garrison, J., & Neubert, S. (2016). Complexity and Reductionism in Educational Philosophy—John Dewey's Critical Approach in 'Democracy and Education' Reconsidered. *Educational Philosophy and Theory*, *48*(10), 997–1012. DOI: 10.1080/00131857.2016.1150802

Reimers, F. M., & Schleicher, A. (2020). *A framework to guide an education response to the COVID-19 Pandemic of 2020.* https://www.oecd.org/en/publications/2020/03/a-framework-to-guide-an-education-response-to-the-covid-19-pandemic-of-2020_a5bcbd8a.html

Reimers, F. M., Amaechi, U., Banerji, A., & Wang, M. (Eds.). (2022). *Education to build back better: What can we learn from education reform for a post-pandemic world?* Springer., DOI: 10.1007/978-3-030-93951-9

Reimers, F. M., & Chung, C. K. (2018). *Education for shared prosperity: Principles and practices for educating children and youth to engage with the world.* Harvard Education Press.

Reimers, F. M., & Schleicher, A. (2020). *A framework to guide an education response to the COVID-19 pandemic of 2020.* OECD Publishing., DOI: 10.1787/6ae21003-

Reveley, J. (2013). Understanding Social Media Use as Alienation: A Review and Critique. *E-Learning and Digital Media*, *1*(10), 83–94. DOI: 10.2304/elea.2013.10.1.83

Reyes, C. M. (2015). Improving the quality of basic education in the Philippines: Challenges and solutions. *Philippine Institute for Development Studies Discussion Paper Series*, No. 2015-55.

Reynolds, H., & Astor, R. (2024). School policing, climate, and safety: The challenges for local policy makers in urban districts. *Education and Urban Society*, *56*(3), 286–308. DOI: 10.1177/00131245221121798 PMID: 39991466

Reyteran, R. (2021). Teaching in IP Schools: An Analysis of Teachers' Experiences as Input to Pre-Service Teacher Education. *International Journal of Language and Literary Studies*, *3*(4), 131–142. DOI: 10.36892/ijlls.v3i4.758

Ribeiro, N., & Subrahmanyam, S. (2021). Emotional challenges of teaching online during the pandemic. In Uddin, M., Sony, K. K., & Verma, N. (Eds.), *Psychosocial issues in COVID-19 pandemic* (pp. 244–255). SHINE Publication. RED.

Rice, C. (2023). *From Pandemic to Renewal: Practices for a World Shaken by Crisis*. InterVarsity Press.

Riddle, T., & Sinclair, S. (2019). Racial disparities in school-based disciplinary actions are associated with county-level rates of racial bias. *Proceedings of the National Academy of Sciences of the United States of America*, *116*(17), 8255–8260. https://www.jstor.org/stable/26703499. DOI: 10.1073/pnas.1808307116 PMID: 30940747

Rigano, C. (2018). Using Artificial Intelligence to Address Criminal Justice Needs. *National Institute of Justice*. https://nij.ojp.gov/topics/articles/using-artificial-intelligence-address-criminal-justice-needs

Rios, M., & Neas, S. (2024). Community-engaged learning in graduate education: A review and call to action. *The International Journal of Research on Service-Learning and Community Engagement*, *12*(1), 7. Advance online publication. DOI: 10.37333/001c.127560

Rizzotto, J. S., & França, M. T. A. (2022). Indiscipline: The school climate of Brazilian schools and the impact on student performance. *International Journal of Educational Development, 94*, 102657. Roldán, S. M., Marauri, J., Aubert, A., & Flecha, R. How Inclusive Interactive Learning Environments Benefit Students Without Special Needs. *Frontiers in Psychology, 12*, 661427. DOI: 10.3389/fpsyg.2021.661427

Robinson, K., Harris, A., & Rainer, D. (2019). Mindsets in education reform: Changing perspectives. *Educational Leadership*, *76*(3), 44–51.

Rock, A. (2024). School shooting threats continue to skyrocket on social media after Apalachee shooting. *Campus Safety Magazine*. https://www.campussafetymagazine.com/news/school-shooting-threats-skyrocket-on-social-media-after-apalachee-shooting/

Rodriguez, A. (2020). Inclusive education reforms in British Columbia: Teacher collaboration and student success. *Canadian Journal of Education*, *43*(1), 124–142.

Roffey, S. (2020). *Positive relationships: Evidence-based practice across the world*. Springer.

Roll, I., Aleven, V., McLaren, B. M., & Koedinger, K. R. (2011). Improving students' help-seeking skills using metacognitive feedback in an intelligent tutoring system. *Learning and Instruction*, *21*(2), 267–280. DOI: 10.1016/j.learninstruc.2010.07.004

Rosa, M., & Gifford, E. (2023). Parental involvement in urban schools: A cross-cultural examination of engagement strategies. *Journal of Urban Education*, *62*(4), 361–378. DOI: 10.1007/s11524-022-00442-2

Rosén Rasmussen, L. (2021). Building Pedagogies. A historical study of teachers' spatial work in new school architecture. *Education Inquiry*, *12*(3), 225–248. DOI: 10.1080/20004508.2020.1857495

Rosenberg, M., & Miquelino, L. P. (2002). *e-Learning - estratégias para a transmissão do conhecimento na era digital*. Milton Mira de Assunpção Filho .

Rosiek, J. L., Snyder, J., & Pratt, S. L. (2020). The new materialisms and Indigenous theories of non-human agency: Making the case for respectful anti-colonial engagement. *Qualitative Inquiry*, *26*(3–4), 331–346. DOI: 10.1177/1077800419830135

Roth, K., Mollvik, L., Alshoufani, R., Adami, R., Dineen, K., Majlesi, F., Peters, M. A., & Tesar, M. (2020). Philosophy of education in a new key: Constraints and possibilities in present times with regard to dignity. *Educational Philosophy and Theory*. Advance online publication. DOI: 10.1080/00131857.2020.1851189

Roy, A. (2020). The pandemic is a portal. Financial Times. Available at: https://www.ft.com/content/10d8f5e8-74eb-11ea-95fe-fcd274e920ca

Rudolph, J., Ismail, M. F. B. M., & Popenici, S. (2024). Higher education's generative artificial intelligence paradox: The meaning of chatbot mania. *Journal of University Teaching & Learning Practice*, *21*(6), 1–35. DOI: 10.53761/54fs5e77

Saborío-Taylor, S., & Rojas-Ramírez, F. (2024). Universal design for learning and artificial intelligence in the digital era: Fostering inclusion and autonomous learning. *International Journal of Professional Development. Learners and Learning, 6*(2), ep2408. Advance online publication. DOI: 10.30935/ijpdll/14694

Sahlberg, P. (2015). *Finnish lessons 2.0: What can the world learn from educational change in Finland?* Teachers College Press.

Salinas-Navarro, D. E., Vilalta-Perdomo, E., Michel-Villarreal, R., & Montesinos, L. (2024). Using generative artificial intelligence tools to explain and enhance experiential learning for authentic assessment. *Education Sciences, 14*(1), 887–900. DOI: 10.3390/educsci14010083

Saltmarsh, J., & Hartley, M. (2011). *To serve a larger purpose: Engagement for democracy and the transformation of higher education*. Temple University Press.

Salza, P., Musmarra, P., & Ferrucci, F. (2019). Agile methodologies in education: A review. In Parsons, D., & MacCallum, K. (Eds.), *Agile and lean concepts for teaching and learning: Bringing methodologies from industry to the classroom* (pp. 25–45). Springer Singapore., DOI: 10.1007/978-981-13-2751-3_2

Sanders, M. G., & Harvey, A. (2019). Community and family partnerships: Enhancing student achievement in high-poverty schools. *Educational Policy, 50*(6), 785–804. DOI: 10.1111/edpo.12297

Sandoval, W., Vigotsky, A., & Callahan, J. (2020). Interdisciplinary collaboration in education: Fostering innovation through shared inquiry. *Educational Researcher, 49*(7), 486–495.

Sardoc, M., Coady, C. A. J., Bufacchi, V., Moghaddam, F. M., Cassam, Q., Silva, D., Miscevic, N., Andrejc, G., Kodelja, Z., Vezjak, B., Peters, M. A., & Tesar, M. (2020). Philosophy of education in a new key: On radicalization and violent extremism. *Educational Philosophy and Theory*. Advance online publication. DOI: 10.1080/00131857.2020.1861937

Sathishkumar, A. S. (2012). Emerging technology of smart class teaching in school education: A literature review. *International Journal of Scientific Research, 3*, 446–449. DOI: 10.15373/22778179/August2014/137

Sayed, Y., Cooper, A., & John, V. M. (2021). Crises and disruptions: Educational reflections,(re) imaginings, and (re) vitalization. Journal of Education (University of KwaZulu-Natal), (84), 9-30.

Schildkamp, K., & Visscher, A. J. (2016). The role of family and community partnerships in school improvement: A framework for monitoring student progress. *Educational Management Administration & Leadership, 44*(2), 179–194. DOI: 10.1177/1741143214566754

Schleicher, A. (2020). *The impact of COVID-19 on education: Insights from Education at a Glance 2020*. OECD. https://www.improvingthestudentexperience.com/library/covid19/the-impact-of-covid-19-on-education-insights-education-at-a-glance-2020.pdf

Schön, D. A. (1983). *The reflective practitioner: How professionals think in action*. Basic Books.

Schulz, W., Ainley, J., Fraillon, J., Kerr, D., & Losito, B. (2022). *International civic and citizenship education study 2022: Assessment framework*. International Association for the Evaluation of Educational Achievement (IEA).

Schussler, D. L., Jennings, P. A., Sharp, J. E., & Frank, J. L. (2022). Factors affecting educational leader wellbeing: Sources of stress and self-care strategies. *Educational Management Administration & Leadership, 50*(3), 456–473. DOI: 10.1177/1741143220931260

Schweit, K. W., & Mancik, A. M. (2017, May 10). School resource officers and violence prevention: Best practices (part two). *FBI Law Enforcement Bulletin*. https://leb.fbi.gov/articles/featured-articles/school-resource-officers-and-violence-prevention-best-practices-part-two

Scribes, S. (2024). A discourse on generative AI. *Medium*. https://medium.com/@sundryscribes/a-discourse-on-generative-ai-4ca4a8911e1a

Second Congressional Commission on Education. (2024). Miseducation: The Failed System of Philippine Education. EDCOMII Year 1 Report. Second Congressional Commission on Education. https://edcom2.gov.ph/media/2024/02/EDCOM-II-Year-One-Report-PDF-022924.pdf

Security, H. Department of. (2024). *Improving school safety through bystander reporting*. U.S. Department of Homeland Security. https://www.cisa.gov/sites/default/files/2023-05/CISA-USSS%20K-12%20Bystander%20Reporting%20Toolkit_508.pdf

Selwyn, N. (2010). Looking Beyond Learning: Notes Towards the Critical Study of Educational Technology. *Journal of Computer Assisted Learning, 26*(1), 65–73. DOI: 10.1111/j.1365-2729.2009.00338.x

Selwyn, N. (2014). Education and 'the Digital'. *British Journal of Sociology of Education, 35*(1), 155–164. DOI: 10.1080/01425692.2013.856668

Selwyn, N. (2016). *Education and technology: Key issues and debates*. Bloomsbury Academic.

Selwyn, N. (2019). *Should robots replace teachers? AI and the future of education*. Polity Press.

Selwyn, N. (2020). What should digital education be? *Learning, Media and Technology, 45*(3), 225–237.

Selwyn, N. (2022). The future of AI and education: Some cautionary notes. *European Journal of Education, 57*(4), 620–631. DOI: 10.1111/ejed.12532

Senge, P. M., Hamilton, H., & Kania, J. (2019). The dawn of system leadership. *Stanford Social Innovation Review, 11*(1), 27–33.

Serafico-Reyes, N. M. A. (2020). Brigada Eskwela: Bayanihan through school-community partnerships (2003–2020). *Philippine Normal University Research Journal, 12*(1), 45–60.

Sergio, M., Gabriel, D., Sancristobal, E., Rosario, G., Manuel, C., & Juan, P. (2011). New Technology Trends in Education: Seven Years of Forecasts and Convergence. *Computers & Education, 57*(3), 1893–1906. DOI: 10.1016/j.compedu.2011.04.003

Shankar, D. A., Tewari, D. V., Rahman, M., Mishra, D. A., & Bajaj, M. K. (2023). Impact of virtual reality (VR) and augmented reality (AR) in education. *Tuijin Jishu/ Journal of Propulsion Technology*.

Sheather, J. (2022). As Russian troops cross into Ukraine, we need to remind ourselves of the impact of war on health. *BMJ (Clinical Research Ed.), 376*(February), o499. Advance online publication. DOI: 10.1136/bmj.o499 PMID: 35217578

Shemyakina, O. (2011). The Effect of Armed Conflict on Accumulation of Schooling: Results from Tajikistan. *Journal of Development Economics, 95*(2), 186–200. DOI: 10.1016/j.jdeveco.2010.05.002

Shevchenko, V. V. (2019). The reform of the higher education of Ukraine in the conditions of the military-political crisis. *International Journal of Educational Development, 65*, 237–253. DOI: 10.1016/j.ijedudev.2018.08.009

Shields, C. M. (2010). Transformative leadership: Working for equity in diverse contexts. *Educational Administration Quarterly, 46*(4), 558–589. DOI: 10.1177/0013161X10375609

Siegel, H. (1983). On the obligations of the professional philosopher of education. *Journal of Thought, 18*(2), 31–37.

Simons, L., & Vanderlinde, R. (2020). Teacher perspectives on parent-school partnerships: A comparative study across Europe. *Teaching and Teacher Education*, *91*, 103049. DOI: 10.1016/j.tate.2020.103049

Simula, B. L., & Scott, T. L. (2021). Disciplining academic identities: Boundaries and identity work among arts and sciences faculty. *Social Currents*, *8*(4), 378–397. DOI: 10.1177/23294965211001401

Skiba, R. J., & Losen, D. J. (2016). From reaction to prevention: Turning the page on school discipline. *American Educator*, *39*(4), 4.

Skokova, L. (2022). War-affected landscapes as a social problem. In Головахи, Є., & Макеєва, С. (Eds.), *Ukrainian Society in Wartime* (1st ed., pp. 66–76). Institute of Sociology of the National Academy of Sciences of Ukraine.

Slade, S., & Prinsloo, P. (2019). Learning analytics ethical issues and dilemmas. *The American Behavioral Scientist*, *63*(10), 1430–1440.

Smith, J. (2022). International education trends: Shaping the future of higher education. *International Journal of Open Publication and Exploration*, *10*(1), 1–5.

Smith, S. (2022). Promoting civic engagement through service learning. *Frontiers of Educational Research*, *4*(1), 27–33. DOI: 10.26549/fer.v4i1.3798

Smylie, M. A., Murphy, J., & Louis, K. S. (2016). Caring school leadership: A multidisciplinary, cross-occupational model. *American Journal of Education*, *123*(1), 1–35. DOI: 10.1086/688166

Solitario, J. (2022). Revisiting kapwa: Filipino ethics, subjectivity, and self-formation. *Philippine Studies: Historical and Ethnographic Viewpoints*, *70*(4), 539–564. DOI: 10.13185/PS2022.70404

Solochek, S. J. (2021, January 11). Pinellas parents take precautions against non-credible school threat. *Tampa Bay Times*. https://www.tampabay.com/news/education/2021/01/11/pinellas-parents-take-precautions-against-non-credible-school-threat/

Sorensen, L. C., Avila-Acosta, M., Engberg, J. B., & Bushway, S. D. (2023). The thin blue line in schools: New evidence on school-based policing across the U.S. *Journal of Policy Analysis and Management*, *42*(4), 941–970. DOI: 10.1002/pam.22498

Sott, M. K., & Bender, M. S. (2025). The Role of Adaptive Leadership in Times of Crisis: A Systematic Review and Conceptual Framework. *Merits*, *5*(1), 2. DOI: 10.3390/merits5010002

Spillane, J. P. (2006). *Distributed leadership*. Jossey-Bass.

Sruthi, S., & Seema Menon, K.P. (2024). Cultural Adaptation in Educational Systems: Strategies for Integrating Diverse Cultural Values. *ISHAL PAITHRKAM*, (39).

Statista. (n.d.). *Security measures taken by public schools in the United States from 2001 to 2022*. https://www.statista.com/statistics/183770/security-measures-at-school/

Steiner, E. D., & Woo, A. (2021). Job-related stress threatens the teacher supply: Key findings from the 2021 state of the U.S. teacher survey. RAND Corporation.

Steiner, E. D., & Woo, A. (2021). *Job-related stress threatens the teacher supply: Key findings from the 2021 State of the U.S. Teacher Survey*. RAND Corporation. https://www.rand.org/pubs/research_reports/RRA1108-1.html

Stohlman, S., Konold, T., & Cornell, D. (2020). Evaluation of threat assessment training for school personnel. *Journal of Threat Assessment and Management*, *7*(2), 105–119. DOI: 10.1037/tam0000142

Stoll, L., & Louis, K. S. (2023). *Professional learning communities: Divergence, depth, and dilemmas*. Routledge.

Stukalo, N., & Simakhova, A. (2020). COVID-19 Impact on Ukrainian Higher Education. *Universal Journal of Educational Research*, *8*(8), 3673–3678. DOI: 10.13189/ujer.2020.080846

Subhash, S., & Cudney, E. A. (2018). Gamified learning in higher education: A systematic review of the literature. *Computers in Human Behavior*, *87*, 192–206. DOI: 10.1016/j.chb.2018.05.028

Subrahmanyam, S. (2021). Attitudes and beliefs of faculty on teaching online during pandemic. In Sony, K. K., Verma, N., & Uddin, M. (Eds.), *Psychosocial issues in Covid-19 pandemic* (pp. 256–268). SHINE Publication. RED.

Subrahmanyam, S. (2024). Implementing agile talent management practices to adapt to changing higher education landscapes. In Kayyali, M. (Ed.), *Cultivating creativity and navigating talent management in academia* (pp. 187–210). IGI Global Scientific Publishing., DOI: 10.4018/979-8-3693-6880-0.ch009

Subrahmanyam, S. (2025). Open Educational Resources and Their Impact on Curriculum Flexibility and Accessibility. In Cruz, M., Queirós, R., & Mascarenhas, D. (Eds.), *Building Teaching Competencies for AI-Driven and Inclusive Learning* (pp. 69–102). IGI Global Scientific Publishing., DOI: 10.4018/979-8-3373-7729-2.ch002

Sullivan, C. (2022, July 19). How superintendents can restore public trust in schools. *Yale Insights*. Edited by T. O'Callahan. https://insights.som.yale.edu/insights/how-superintendents-can-restore-public-trust-in-schools

Sullivan, P. (2022). *Building community support for school safety initiatives*. Education Strategies Press.

Sultanova, L. (2020). Impact of the COVID-19 pandemic on the Development of Education Systems at Global, European and National Dimensions. *AdultEducation: Theory, Experience. Prospects*, *18*(2). https://doi.org/ttps://doi.org/10.35387/od.2(18).2020.171-183

Surry, D., & Baker, F.III. (2016). The Co-Dependent Relationship of Technology and Communities. *British Journal of Educational Technology*, *47*(1), 13–28. DOI: 10.1111/bjet.12349

Sutherland, R., & Howard, S. (2021). Distributed leadership and teacher collaboration: Building adaptive professional cultures. *Journal of Educational Administration*, *59*(2), 141–158.

Su, Z., Zhang, H., McDonnell, D., Ahmad, J., Cheshmehzangi, A., & Yuan, C. (2022). Crisis communication strategies for health officials. *Frontiers in Public Health*, *10*, 796572. DOI: 10.3389/fpubh.2022.796572 PMID: 36483240

Svyrydenko, D. (2017). Divided Universities: The Postcolonial Experience of Contemporary Ukrainian Higher Education. *Future Human Image*, *1*(7), 128–135.

Swaby, A. (2024, October 30). An 11-year-old denied making a threat and was allowed to return to school. Tennessee police arrested him anyway. *ProPublica*. https://www.propublica.org/article/tennessee-school-threat-law-kids-arrested

Sweller, J. (2020). Cognitive load theory and educational technology. *Educational Technology Research and Development*, *68*(1), 1–16. DOI: 10.1007/s11423-019-09701-3

Tan, J., & Lim, M. (2021). Professional learning communities in Singapore schools: Policy and practice. *Asia Pacific Journal of Education*, *41*(4), 492–507.

Tan, S. C., Hu, C., Wong, S. K., & Wettasinghe, C. M. (2003). Teacher training on technology-enhanced instruction: A holistic approach. *Journal of Educational Technology & Society*, •••, 6.

Tavrovetska, N., & Veldbrekht, O. (2023). Psychological aspects of online learning implementation at Ukrainian universities. *Scientific Bulletin of Mukachevo State University Series. Pedagogy and Psychology*, *9*(2). Advance online publication. DOI: 10.52534/msu-pp2.2023.38

Taylor, J., & Dearybury, J. (2021). *The Courageous Classroom: Creating a Culture of Safety for Students to Learn and Thrive*. John Wiley & Sons.

Tennessee Mass Violence Threats Act, TN Pub. Act No. 72-2022.

Tesar, M. (2021). Philosophy as a method: Tracing the histories of intersections of 'philosophy', 'methodology' and 'education. *Qualitative Inquiry*, *27*(5), 544–553. DOI: 10.1177/1077800420934144

Thapa, A., Cohen, J., Guffey, S., & Higgins-D'Alessandro, A. (2013). A review of school climate research. *Review of Educational Research*, *83*(3), 357–385. DOI: 10.3102/0034654313483907

The Times. (2024, November 15). Trinity College master 'regrets' £250k fund for Ukrainian students. *The Times*.

Thomson, I. (2004). Heidegger's Perfectionist Philosophy of Education in Being and Time. *Continental Philosophy Review*, *7*(4), 439–467. DOI: 10.1007/s11007-005-6886-8

Thumlert, K., Castell, S., & Jenson, J. (2015). Short Cuts and Extended Techniques: Rethinking Relations Between Technology and Educational Theory. *Educational Philosophy and Theory*, *47*(8), 786–803. DOI: 10.1080/00131857.2014.901163

Tian, L., & Yu, M. (2021). Investigating school-family partnerships in an urban setting: Parent perceptions and school effectiveness. *Journal of Educational Psychology*, *113*(3), 374–388. DOI: 10.1037/edu0000421

Timperley, H., Wilson, A., Barrar, H., & Fung, I. (2019). *Teacher professional learning and development: Best evidence synthesis iteration*. Ministry of Education.

Tirri, K., & Nokelainen, P. (2021). Teacher well-being and burnout in Finnish schools. *Educational Psychology*, *41*(3), 253–266.

Tocci, C., Stacy, S. T., Siegal, R., Renick, J., LoCurto, J., Lakind, D., Gruber, J., & Fisher, B. W. (2024). Statement on the effects of law enforcement in school settings. *American Journal of Community Psychology*, *73*(3-4), 360–377. DOI: 10.1002/ajcp.12723 PMID: 37994201

Tomlinson, C. A. (2014). *The differentiated classroom: Responding to the needs of all learners* (2nd ed.). ASCD.

Tomlinson, C. A., & Imbeau, M. B. (2019). *Leading and managing a differentiated classroom*. ASCD.

Torlone, F., & Ishchenko, A. (2024). Learning in higher education while being at war. *Journal of Educational Sciences*, *XXV*(1), 56–71. Advance online publication. DOI: 10.35923/JES.2024.1.03

Trajtenberg, M. (2018). *AI as the next GPT: A political-economy perspective*. National Bureau of Economic Research., DOI: 10.3386/w24245

Treviranus, J., & Coombs, N. (2000). Bridging the digital divide in higher education.

Trump, K. S. (1998). *Practical school security: Basic guidelines for safe and secure schools*. Corwin Press.

Trust, T., Carpenter, J. P., & Krutka, D. G. (2020). Leading schools in a digital world: Teacher learning communities and the use of digital tools to support students. *Educational Media International*, *57*(2), 104–120.

Trust, T., Krutka, D. G., & Carpenter, J. P. (2021). "Together we are better": Professional learning networks for teachers. *Computers & Education*, *163*, 104099.

Trust, T., & Whalen, J. (2020). Should teachers be trained in emergency remote teaching? Lessons learned from the COVID-19 pandemic. *Journal of Technology and Teacher Education*, *28*(2), 189–199. https://www.learntechlib.org/primary/p/215995/. DOI: 10.70725/307718pkpjuu

Tsai, F. S., & Wong, K. (2013). The state of massive open online courses (MOOCs) in engineering education: Where do we go from here?

U.S. Secret Service, National Threat Assessment Center. (2021). *Averting targeted school violence: A U.S. Secret Service analysis of plots against schools*. https://www.secretservice.gov/

Ullah, A., & Khan, M. (2023). Political instability and its implications on public policy: A comprehensive analysis of governance transitions and their effect on educational policies in Pakistan. *The Lighthouse Journal of Social Sciences*, *2*(2), 81–96.

UNESCO. (2019). *Global citizenship education: Topics and learning objectives*. UNESCO Publishing.

UNESCO. (2020). *Education in a post-COVID world: Nine ideas for public action*. News. https://www.unesco.org/en/articles/education-post-covid-world-nine-ideas-public-action

UNESCO. (2020). *Education: From disruption to recovery.* UNESCO. https://en.unesco.org/covid19/educationresponse

UNESCO. (2023). *Timeline of UNESCO's commitment to protect Ukraine's education and heritage*. UNESCO Actions for Ukraine. https://www.unesco.org/en/ukraine-war/actions-timeline

UNESCO. (2023b). *Ensuring teaching and learning continues in Ukraine. Emergencies*. https://www.unesco.org/en/emergencies/education/ukraine

UNICEF. (2024). *Education in emergencies*. https://www.unicef.org/education/emergencies

United Nations. (2023a). Early Recovery Efforts in Ukraine. In *UN Annual Results Report 2022*. chrome-extension://efaidnbmnnnibpcajpcglclefindmkaj/https://ukraine.un.org/sites/default/files/2023-05/UN_report_2023%28English%29.pdf

University of Sheffield. (2022). *University of Sheffield twins with university in Kyiv to help staff and students affected by war*. News. https://www.sheffield.ac.uk/news/university-sheffield-twins-university-kyiv-help-staff-and-students-affected-war

University of Toronto. (2022). *U of T to accept displaced students from Ukraine with significant boost to its Scholars-at-Risk fund.* News and Stories. https://defygravitycampaign.utoronto.ca/news-and-stories/u-of-t-to-accept-displaced-students-from-ukraine-with-significant-boost-to-its-scholars-at-risk-fund/

van der Vorst, T., & Jelicic, N. (2019). Artificial intelligence in education: Can AI bring the full potential of personalized learning to education? https://www.econstor.eu/handle/10419/205222

Van Dijk, J. (2020). *The digital divide*. Polity Press.

Varaki, B. S., Sadeqzadeh Qamsari, A., Sefidkhosh, M., Mahdi Sajjadi, S., Mohammadi Chaboki, R., Javidi Kalatehjafarabadi, T., Saffarheidari, H., Mohammadamini, M., Karimzadeh, O., Barkhordari, M., Zarghami-Hamrah, S., Peters, M. A., & Tesar, M. (2021). Philosophy of education in a new key: Reflection on higher education in Iran. *Educational Philosophy and Theory*. Advance online publication. DOI: 10.1080/00131857.2021.1905517

Veldkamp, A., Van De Grint, L., Knippels, M. C. P., & Van Joolingen, W. R. (2020). Escape education: A systematic review on escape rooms in education. *Educational Research Review*, *31*, 100364. DOI: 10.1016/j.edurev.2020.100364

Venegas Marin, S., Schwarz, L., & Sabarwal, S. (2024). Impacts of extreme weather events on education outcomes: A review of evidence. *The World Bank Research Observer, 39*(2), 177–226. DOI: 10.1093/wbro/lkae001

Verbeek, P. (2008). Cyborg Intentionality: Rethinking the Phenomenology of Human–Technology Relations. *Phenomenology and the Cognitive Sciences, 7*(3), 387–395. DOI: 10.1007/s11097-008-9099-x

Vescio, V., Ross, D., & Adams, A. (2018). A review of research on the impact of professional learning communities on teaching practice and student learning. *Teaching and Teacher Education, 24*(1), 80–91. DOI: 10.1016/j.tate.2007.01.004

Vilanova, R., Domínguez, M., Vicario, J. L., Prada, M. A., Barbu, M., Varanda, M. J., Alves, P., Podpora, M., Spagnolini, U., & Paganoni, A. M. (2019). Data-driven tool for monitoring of students' performance. *IFAC-PapersOnLine, 52*(9), 165–170. DOI: 10.1016/j.ifacol.2019.08.188

Visvizi, A. (2018). *The future of innovation and technology in education: Policies and practices for teaching and learning excellence.*

Vogelgesang, L. J., & Astin, A. W. (2000). Comparing the effects of community service and service-learning. *Michigan Journal of Community Service Learning, 7*(1), 25–34. https://quod.lib.umich.edu/m/mjcsl

Voinow, A. V. (2024). Holistic Strategies for a Connected Future: Can we Bridge the Digital Divide by Redefining Smart City Agendas?.

Voogt, J., McKenney, S., & Yang, Y. (2018). Designing international collaborative projects for digital literacy and global competence. *Computers & Education, 123*, 131–143.

Waghid, Y., Davids, N., Mathebula, T., Terblanche, J., Higgs, P., Shawa, L., Herbert Manthalu, C., Waghid, Z., Ngwenya, C., Divala, J., Waghid, F., Peters, M. A., & Tesar, M. (2020). Philosophy of education in a new key: Cultivating a living philosophy of education to overcome coloniality and violence in African Universities. *Educational Philosophy and Theory*. Advance online publication. DOI: 10.1080/00131857.2020.1793714

Walker, J. M. T., & Hoover-Dempsey, K. V. (2017). Parental involvement in education: A longitudinal analysis of involvement, academic achievement, and self-regulation. *Parenting, Science and Practice, 17*(3), 228–242. DOI: 10.1080/15295192.2017.1301152

Warschauer, M., & Matuchniak, T. (2019). New technology and digital worlds: Analyzing evidence of equity in access, use, and outcomes. *Review of Research in Education, 34*(1), 179–225. DOI: 10.3102/0091732X09349791

Wasylyshyn, K. M., & Masterpasqua, F. (2018). Developing self-compassion in leadership development coaching: A practice model and case study analysis. *International Coaching Psychology Review, 13*(1), 21–34. DOI: 10.53841/bpsicpr.2018.13.1.21

Wayman, J. C. (2005). Involving teachers in data-driven decision making: Using computer data systems to support teacher inquiry and reflection. *Journal of Education for Students Placed at Risk, 10*(3), 295–308. DOI: 10.1207/s15327671espr1003_5

Weatherington, I. (2024). Adversity to Action: A Case Study Analysis of Crisis Leadership.

Webster, S. (2009). *Educating for meaningful lives through existential spirituality.* Sense Publishers. DOI: 10.1163/9789460910043

Weiner, M., Wilcox, D., & Lieberman, R. (2022). Teachers as advocates: A framework for teacher leadership. *Education Policy Analysis Archives, 30*(14), 1–24.

Weiss, D. (2008). Human—Technology—World. *Techné (Blacksburg, Va.), 12*(2), 110–119. DOI: 10.5840/techne20081228

Weiss, H. B., & Caspe, M. (2021). Supporting family and community engagement in schools: A study of successful practices in K-12 education. *Journal of Family & Community Engagement, 2*(2), 85–100. DOI: 10.32474/JFCE.2021.02.00003

Welch, K., & Payne, A. (2010). Racial threat and punitive school discipline. *Social Problems, 57*(1), 25–48. DOI: 10.1525/sp.2010.57.1.25

West, A., Chute, D. E., Daniels, J., & Bower, K. M. (2025). Community-engaged research in early home visiting: A scoping review of peer-reviewed literature. *Prevention Science : The Official Journal of the Society for Prevention Research, 26*(5), 703–715. DOI: 10.1007/s11121-025-01812-z

Whitehead, A. N. (1929). *Process and Reality: An Essay in Cosmology*. Macmillan.

Willet, R. (2018). Learning Through Making in Public Libraries: Theories, Practices, and Tensions. *Learning, Media and Technology, 43*(3), 250–262. DOI: 10.1080/17439884.2017.1369107

Williams, E., Armistead, J., & Rude, D. A. (2022). Leading for Innovation: A New Model for 21st–century Leadership. *New Horizons in Adult Education and Human Resource Development, 34*(4), 3–13. DOI: 10.1002/nha3.20366

Williams, J. A.III, Hill-Jackson, V., Caldwell, C., & Craig, C. J. (2022). Teacher recruitment and retention: Local strategies, global inspiration. *Journal of Teacher Education, 73*(4), 333–337. DOI: 10.1177/00224871221118155

Williamson, B., & Eynon, R. (2020). Historical threads, missing links, and future directions in AI in education. *Learning, Media and Technology, 45*(3), 223–235. DOI: 10.1080/17439884.2020.1798995

Wilson, A. D. (2015). The effect of community involvement programs on school violence. *Journal of Multidisciplinary Graduate Research, 3*(1), 1–19. https://www.shsu.edu/academics/education/journal-of-multidisciplinary-graduate-research/documents/2015/Wilson-3.pdf

Wong, E., & Badr, H. (2022). The intersection of family involvement and school culture: Implications for educational success in immigrant communities. *International Journal of Intercultural Relations, 83*, 32–45. DOI: 10.1016/j.ijintrel.2021.11.007

Wong, H. K., & Chia, S. (2023). Building collaborative teacher communities. *International Journal of Educational Research, 115*, 102102.

World Bank. (2022). *Education: Impact of the War in Ukraine.*

Wrathall, A. (2011). We Still Need Education in Thinking: Heidegger's Education in a Technological Age. *Journal of Educational Studies, 7*(1), 3–14.

Wright, S. (2022). Foresight in education leadership: Implementing scenario planning in schools. *Leadership and Policy in Schools, 21*(2), 203–220.

Wright, S., & Lee, M. (2018). Experiential learning and technical education in secondary schools. *The Journal of Educational Research, 111*(5), 561–570. DOI: 10.1080/00220671.2018.1451907

Wu, G. S. (2009). *Lectures on the Philosophy of Technology*. China Renmin University Press.

Xhelilaj, L. K., Petani, R., & Ntalla, M. (2021). Relationship between teacher's burnout, occupational stress, coping, gender and age. *Journal of Educational and Social Research, 11*(4), 275.

Xu, A., Pathak, E., Wallace, E., Gururangan, S., Klein, D., & Sap, M. (2021). Detoxifying language models risks marginalizing minority voices. *arXiv Preprint arXiv:2104.06390*. https://arxiv.org/abs/2104.06390 DOI: 10.18653/v1/2021.naacl-main.190

Yarnall, L., Kozak, S., & Rueda, R. (2016). Adaptive learning and its impact on student success in mathematics courses. *Journal of Learning Analytics, 3*(2), 45–61.

Yazzie-Mintz, E., & Nagaoka, J. (2021). Engaging communities and families in school reform: Lessons from successful programs. *Educational Policy Research Journal*, *49*(2), 140–160. DOI: 10.1080/00357562.2021.1850152

Yelinska, A. M. (2020). Some Issues on Organization of Educational Process in Higher Educational Institutions During Pandemic. *Aktual'ni Problemy Suchasnoï Medytsyny : Visnyk Ukraïns'koï Medychnoï Stomatolohichnoï Akademiï*, *20*(3), 235–238. DOI: 10.31718/2077-1096.20.3.235

Yeter, Ö., Rabagliati, H., & Özge, D. (2022). Threat of War on Cognitive Development of Refugee Children. *East European Journal of Psycholinguistics*, *9*(2), 144–159. DOI: 10.29038/eejpl.2022.9.2.yet

Yu, Z., Gao, M., & Wang, L. (2020). The effect of educational games on learning outcomes, student motivation, engagement and satisfaction. *Journal of Educational Computing Research*, *59*(3), 522–546. DOI: 10.1177/0735633120969214

Yuzyk, O., Yuzyk, M., Bilanych, L., Vitalii, H., Bilanych, H., & Myroslava, F. (2022). Distance Learning in Higher Education Institutions in Conditions of Quarantine and Military Conflicts. *International Journal of Computer Science and Network Security*, *22*(4), 741–749. https://doi.org/https://doi.org/10.22937/IJCSNS.2022.22.4.87 different

Zakharchuk, N. (2020). Contextualizing the Development of Ukrainian Higher Education: Between Soviet Legacies and European Regionalization. *East/West. Journal of Ukrainian Studies*, *7*(2), 173–199. DOI: 10.21226/ewjus616

Zalman, A. (2019). Maximizing the power of strategic foresight. *Joint Force Quarterly*, *95*(4), 14–21.

Zaman, M., Sajjad, S., & Gharsheen, L. U. (2023). Exploring The Dynamics of Teacher-Student Relationships in Overcrowded Secondary School Classrooms. *International Research Journal of Management and Social Sciences*, *4*(4), 245–261.

Zaplatynskyi, V. (2022). Changes in education in Ukraine in connection with the Russian-Ukrainian war. *Reviewed Conference Proceedings of the 15th International Scientific Conference*, 278–293.

Zawacki-Richter, O., Marín, V. I., Bond, M., & Gouverneur, F. (2019). Systematic review of research on artificial intelligence applications in higher education: Where are the educators? *International Journal of Educational Technology in Higher Education*, *16*(1), 16. DOI: 10.1186/s41239-019-0171-0

Zayed University. (2020). *Zayed University switches to hybrid model of teaching, learning for Fall 2020 semester.* ZU News. https://www.zu.ac.ae/main/en/news/2020/august/hybrid20

Zhao, Y., Cao, Y., & Liu, X. (2023). Transformational leadership in education: Supporting professional growth and institutional change. *Educational Management Administration & Leadership, 51*(2), 234–252.

Zhao, Y., & Lei, J. (2022). Professional development for teacher leadership: A meta-analysis. *Review of Educational Research, 92*(1), 54–82.

Zhao, Y., Lei, J., Yan, B., Tan, H. S., & Lai, C. (2021). What makes the difference? A practical analysis of research on the effectiveness of technology integration in education. *Computers & Education, 69*, 403–415.

Zhao, Y., Pugh, K., Sheldon, S., & Byers, J. (2002). Conditions for Classroom Technology Innovations. *Teachers College Record, 104*(3), 482–515. DOI: 10.1111/1467-9620.00170

Zhe, E. J., & Nickerson, A. B. (2007). Effects of an intruder crisis drill on children's knowledge, anxiety, and perceptions of school safety. *School Psychology Review, 36*(3), 501–508. https://www.researchgate.net/publication/340666939. DOI: 10.1080/02796015.2007.12087936

Zimmerman, J., Forlizzi, J., & Evenson, S. (2007). "Research Through Design as a Method for Interaction Design Research in HCI." *Proceedings of the SIGCHI Conference on Human Factors in Computing Systems.* DOI: 10.1145/1240624.1240704

Zootzky, G., & Pfeiffer, A. (2024). Educational transformation through AI: Preparing for a new era of learning. *INTED2024 Proceedings,* 1202–1207. DOI: 10.21125/inted.2024.0382

About the Contributors

Annette Walters is currently an administrator in K-12 education and an adjunct professor in the Northeast, USA. With extensive experience in various educational roles, she has focused on developing and enhancing professional development initiatives centered on teacher training, inclusive practices, behavior management, and student achievement. Utilizing research, she guides instructional practitioners in adopting effective and relevant practices. Dr. Walters has delivered presentations on topics such as Student Access, Data-Driven Practices, Special Education, Staff Development, and Educational Accountability. Her passion lies in classroom management, teacher retention, coaching, and staffing strategies that promote student success. Her diverse experience spans from Pre-K to college, allowing her to share valuable insights to enhance the quality of education for all learners. Additionally, she has served as an adjunct professor for over fifteen years in both traditional and online settings. Dr. Walters holds a doctorate in Education and three graduate degrees in Educational Leadership, Management and Policy; Education; and Special Education with a minor in Developmental Disabilities. She has also worked for several large private organizations.

Vivek Ahuja is a seasoned education administrator with a rich background in teaching and leadership roles at prominent universities in India. Currently, he is serving as the Director of PhD Programs and Accreditation Compliance at FLAME University in Pune, India. Prior to joining FLAME, he worked with Ashoka University, where he led the academics and peer learning aspects of the Young India Fellowship. His experience also includes a stint at Shiv Nadar University, where he led academic program operations at the business school and taught the Global Entrepreneurship Simulation Capstone in the MBA and Executive MBA Programs. Vivek holds a PhD in Business from Aligarh Muslim University, complemented by an MBA from IBS Hyderabad and a Bachelor of Engineering in Information Technology from Rajiv Gandhi Prodyogiki Vishwavidyalaya Bhopal, all India.

He has research interests in the fields of Higher Education and Entrepreneurship. Notably, Vivek has made significant contributions to the academic community by publishing numerous research articles and book chapters in indexed and peer-reviewed journals and editorials.

***

Susan L. Cobarrubias is a highly respected educator with a doctorate in Educational Administration and a master's in Special Education. Her extensive career includes government service, special education teaching, and curriculum development. She's a National Child Protection Specialist and currently serves as a Public Schools District Supervisor in Pasig City, also contributing to the Division's research efforts. Her part-time faculty positions at several universities further demonstrate her commitment to education.

Samuel Duncan is the Graduate Coordinator for the Specialist in Educational Administration Program for Principals & Superintendents, and Assistant Professor in the Department of Leadership, Middle, and Secondary Education. Duncan retired from his role as Superintendent of Schools to join Southeast Missouri State University (SEMO) full-time in August 2024. He served 26 years as a P-12 public school administrator (17 years in central office, and 9 years as a principal). Prior to moving to administration, he taught high school English, journalism, and French, while coaching various high school sports. He began his work at SEMO in 2001 as an adjunct professor upon earning his Ed. D. from the University of Missouri, Columbia in May 2000. Duncan is proud to have followed his mother's footsteps to this place in his career – at the university she loved, where she once taught and was adored by many.

Patricia Fidalgo holds a Ph.D. in Sciences of Education from Nova University of Lisbon, Portugal. Fidalgo is an expert in technology, networks, and multimedia in education and training and has developed most of her research agenda for those subjects. She was responsible for several years for the Emerging Researchers' Group of the European Educational Research Association. Dr. Fidalgo has 27 years of teaching experience in Europe, Africa, and the Middle East. She is currently living in the UAE and is an Associate Professor at the Emirates College for Advanced Education (ECAE), where she teaches educational technology. Dr. Patricia Fidalgo is also head of the Pedagogy, Technology, and Teacher Education Division at the ECAE.

Ethelbert Garcia Galang is a fulltime faculty at the University of the Philippines Diliman - Extension Program in Pampanga and Olongapo (UPDEPPO) where he also earned his degree in Bachelor of Arts in Economics and Master of Management. Prior to his fulltime status, he worked at a Financial Institution for almost 20 years handling Customer Care and Corporate Banking at the same time he has been a Senior Lecturer for over a decade specializing in Economics, Management and Marketing.

Oindrilla Ghosh is a Research Scholar hailing from the School of Law at Christ (Deemed to be University) Bangalore, India, focuses her research on health of tribal communities. Her work includes total 41 publications on Neuroscience on Organizational Behavior at IGI Global and also Environmental Disasters that are indexed in SCOPUS and Web of Sciences conference proceedings. She completed her early education at Loreto Day School, Bowbazar and went on to achieve distinction in her undergraduate (B.A/LL.B) and postgraduate (LL.M) studies at Calcutta University. Having served as an Assistant Professor of Law at Kingston Law College,Barasat, Kolkata, Ghosh has actively participated in numerous National Conferences, International Conferences, UN Global Summits, webinars, and seminars organized by prestigious academic institutions. Her expertise lies in health perspectives, environmental studies, human behavior, and related areas. Additionally, she has conducted various workshops, lectures in educational institutions, and is engaged with NGO's to help the needy.

Marie Grace A. Gomez is Professor 11 at the Counselor Education Area, College of Education, University of the Philippines-Diliman. She teaches counseling and research courses at the undergraduate and graduate levels. She has a bachelor's degree in Political Science, a Master of Arts in Education (Guidance) degree, a Doctor of Philosophy degree in Guidance, and another doctorate in Special Education. She has served as a classroom teacher, guidance counselor, and administrator in the basic and tertiary education settings. Her community extension projects involve counseling and assessment of children, persons with disability, children in conflict with the law, drug surrenderees, development of sexual health literacy programs for teens with disabilities, sexually abused children, medical personnel, women with mental health issues in institutional settings, and persons affected by COVID-19. She has been the lead researcher for the Philippines in a UNESCO research led by Leonard Cheshire UK on Comprehensive Sexuality Education for persons with disabilities. She has received the 2022 Most Outstanding Professional Teacher Award in the field of Academe and Research from the Philippine Federation for Professional Associations.

Joel C. Javiniar is an assistant professor in the Educational Administration Area of UP Diliman College of Education. He is a Licensed Professional Teacher and a Registered Guidance Counselor. He holds a Doctor of Philosophy in Education (Educational Administration) degree from the University of the Philippines Diliman; an MA in Education major in Guidance and Counseling from Ateneo de Manila University; a Bachelor of Arts in Philosophy from San Carlos Major Seminary, Makati City; a Certificate in Executive Education Program at the Lee Kuan Yew School of Public Policy at the National University of Singapore; a Tier 1 Training Certificate from the ASEAN University Network-Quality Assurance (AUNQA) in Bangkok, Thailand; and the UP External Reviewers Training, UP Los Baños, Laguna.

Michelle Kilburn is a full professor in the Department of Criminal Justice, Social Work, and Sociology at Southeast Missouri State University (Cape Girardeau, MO, USA). She currently teaches undergraduate and graduate courses pertaining to the criminal justice system, social justice, law and social control, policy analysis, leadership, research methods, and statistics. Her research is varied focusing primarily on program and policy analysis, victims' perceptions of victim impact panels, the use of social media in law enforcement, inmate leadership development programs, students' perceptions of online learning, and innovative pedagogical strategies in criminal justice. She has also volunteered with the Missouri Department Corrections and assisted in programming for incarcerated persons as well as served as the lead researcher in the state-wide evaluation of the state's probation and parole system.

Oleksandr Kulyk Since 2004, he has taught at Ukrainian universities, and in 2015, he became a full professor in the Department of Philosophy at Oles Honchar Dnipro National University, Ukraine. After Russian troops' full-scale invasion of Ukraine, his city, Dnipro, began to be regularly bombarded by Russian missiles. He and his wife decided to get their children out of harm's way, choosing Ireland as a safe place. In Ireland, Oleksandr Kulyk taught the course Current Ukrainian Thought at Maynooth University during the fall semester of 2022, and since September 2023, he has been a visiting professor at University College Cork.

Binod Kumar is a graduate student known for his academic excellence, with his total 30 scholarly works indexed in Scopus database on Neuroscience in Organizational Behavior, Neuromarketing and AI Applications on Wildlife Conservation. Hailing from Ranchi, Jharkhand, in India, Binod is pursuing a Bachelor of Law at the prestigious Radha Govind University. His passion for technology, innovation, and climate change is evident through his high academic performance and numerous projects. Binod has a keen interest in artificial intelligence, machine learning, neuropsychology, climate change, and disaster management, where he has undertaken several research projects. He is particularly fascinated by the potential of AI to solve real-world problems and is dedicated to learning and applying new techniques in these areas. He volunteers at local NGOs, teaching underprivileged children basic computer skills. With a strong foundation in law and a commitment to making a positive impact, Binod Kumar is poised to become a future leader in the legal industry

R.Bhuvaneswari working as an Assistant Professor of English at Dr.Mahalingam College of Engineering and Technology, Pollachi. To her credit she completed 10 NPTEL / SWAYAM and published few articles in Scopus indexed journals.

R. Velmurugan presently working as an Associate Professor in Commerce, Karpagam Academy of Higher Education, Coimbatore. He started his career as a Project Fellow under UGC Sponsored Major Research Project at NGM College, Pollachi. He has 18 years of teaching experience has 15 years of research experience. To his credit he produced 22 Ph.D. Scholars and 5 M.Phil. Scholars. At present 1 scholar submitter his thesis, 6 Ph.D. Scholars are under his guidance. Moreover, to his credit he published 168 articles in various International Journals, of which 58 articles are published in Scopus Indexed Journals and 15 articles in Web of Science and 76 articles in edited volumes. He also participates and presented 27 papers in various International Seminar / Conference and 54 National Seminar / Conference. He acted as an external examiner for various M.Phil. and Ph.D. Viva voce examinations. He acted as a Vice Chancellor nominee for Doctoral Committee Expert at Bharathiar University, Coimbatore and Periyar University, Salem. He acted as a Chairperson at various National Level Conferences. He also acted as an Expert for Syllabus Scrutiny Committee, Question Paper Scrutiny Committee and Expert members for Board of Studies at various reputed Colleges. Considering his meritorious contribution, he received awards like Best Paper Presenter, Best Research Supervisor and Best Faculty Awards from various research foundations and councils. He too acted as a resource person for 27 Guest Lectures, 30 Workshops and 10 faculty development programmes.

Satya Subrahmanyam is a distinguished academic leader, researcher, and institutional strategist with over 28 years of experience in higher education, business research, leadership and management. He holds two PhDs and a Postdoctoral qualification from the Polytechnic Institute of Leiria, Portugal. Currently serving as Professor and Dean (Research) at Holy Spirit University of Kaslik (USEK), Lebanon, he spearheads research initiatives that enhance academic excellence and institutional reputation. He played a pivotal role in the development of the Catholic University in Erbil, Kurdistan, where he established new departments, led research programs, and collaborated on international educational projects. As a founding member and Managing Partner of Vignan Institute of Technology and Management, India, he contributed to its transformation into a premier institution. With expertise in AI-driven business strategies, HR analytics, and leadership in technology adoption, he has published extensively, guided PhD research, and led training programs in corporate and academic settings.

J.Sudarvel working as an Assistant Professor of Management at Karpagam Academy of Higher Education. He produced five Ph.D. Scholars and four Ph.D. Scholars are under his guidance. Further, to his credit he published more than five articles in Scopus indexed journals and published more than 50 articles in international journals. He is acting as a Subject Experts in Board of studies of various autonomous colleges.

Ravi Thirumalaisamy working as an Assistant Professor, Department of Business and Economics, Modern College of Business and Science. To his credit, he published more than 10 articles in Scopus indexed journals.

Joan Thormann, Emerita Professor in the Division of Educational Technology, taught at Lesley University from 1989 to 2016. She has developed and taught courses online since 1996. Her recent research addresses distance education, online teaching strategies such as synchronous conferencing, student moderating, and incorporating principles of Universal Design for Learning (UDL), information literacy, and student use of technology. She has written numerous articles and has co-authored books about technology and education, including the most recent book, The Complete Step-by-Step Guide to Designing and Teaching Online Courses (2012), published by Teachers College Press at Columbia University.

Emilie Torres is a first-generation college student at Southeast Missouri State University pursuing degrees in Criminal Justice, Psychology, Political Science, Anthropology, and Social Science. She focuses on interdisciplinary research topics such as neurocriminology and the ethical implications of emerging neuroscience.

Index

A

B

C

D

E

F

H

I

K

L

W

www.ingramcontent.com/pod-product-compliance
Lightning Source LLC
LaVergne TN
LVHW062218200926
848484LV00006B/1424